Studies in Advanced Mathematics

Fourier Analysis and Partial Differential Equations

Studies in Advanced Mathematics

JOSÉ GARCÍA-CUERVA EUGENIO HERNÁNDEZ FERNANDO SORIA JOSÉ-LUIS TORREA
Universidad Autónoma de Madrid

Fourier Analysis and Partial Differential Equations

CRC PRESS
Boca Raton   London Tokyo

Library of Congress Cataloging-in-Publication Data

Fourier analysis and partial differential equations : proceedings of the Miraflores de la Sierra
  conference (Madrid), 15–20 June, 1992 / edited by José García-Cuerva . . . [et al.].
      p.   cm. — (Studies in advanced mathematics)
  "Proceedings of the International Conference on Fourier Analysis and Partial Differential
Equations held at Miraflores de la Sierra, June 1992" — Galley.
  Includes bibliographical references.
  ISBN 0-8493-7877-X
  1. Fourier analysis—Congresses.   2. Differential equations, Partial—
Congresses.   I. García-Cuerva, José.   II. International Conference on Fourier Analysis and
Partial Differential Equations (1992: Miraflores de la Sierra, Spain).   III. Series.
QA403.5.F673   1995
515′.2433—dc20   94-34845
                                                                                                       CIP

These Proceedings are dedicated to the memory of
Antoni Zygmund
(1900–1992)
and
Stylianos Pichorides
(1940–1992)

Contents

Preface

These are the Proceedings of the International Conference on Fourier Analysis and Partial Differential Equations held at Miraflores de la Sierra from the 15th to the 20th of June, 1992. This Conference is the fourth of a series, the three previous ones being held in 1979, 1983, and 1987 at El Escorial, near Madrid. The format of these Conferences has always been the same and it has held general approval among the mathematical community. Four outstanding mathematicians in the field of Fourier Analysis and Partial Differential Equations are invited to teach mini-courses. These courses try to present, in three to four 1-hour sessions, a rich and actual piece of mathematics assuming minimal background from the audience and reaching the frontiers of present day research. They are taught primarily for people who are starting graduate work or for mathematicians in other areas who want to become acquainted with our field. In this occasion, the four courses were given by Sagun Chanillo, Michael Christ, Guido Weiss, and Thomas Wolff. The three last authors submitted their lectures for publication and they appear here as the Main Lectures—second, first, and fourth, respectively. Sagun Chanillo declined to submit a manuscript for publication, since his lectures are to appear elsewhere. The third Main Lecture of these Proceedings by Detlef Müller and Fulvio Ricci is the written version of the talks they both delivered at the Conference. The nature of their contribution, which is a general survey with some original material, explains the decision of the editors to make it into a Main Lecture.

Another important part of the Conference consists of some 15 invited one-hour lectures that tend to be of a more specialized nature.

Finally, some sessions of short talks are scheduled to accommodate those wishing to present very recent results.

The Proceedings contain almost all invited lectures and also a few short talks that had a favorable impact on the audience or received good refereeing reports.

We decided to dedicate the Miraflores Conference and its Proceedings to the great mathematician Antoni Zygmund, who passed away in Chicago on May 30, 1992. He is the creator of the famous Chicago School of Fourier Analysis, which has had a tremendous influence in the development of the field since the 1940's, and has also been instrumental in the flourishing of research in Fourier Analysis in Spain. Guido Weiss, a student of Zygmund and a distinguished member of the Chicago School, has written a short paper to honor the memory of Antoni Zygmund. The editors wish to give special thanks for this contribution.

The last of Zygmund's students, Stylianos Pichorides, died while attending the Miraflores Conference. He was an outstanding mathematician, and we have decided to dedicate some space in the Proceedings to honor his memory and review his mathematical work. We want to thank Professor Jean-Pierre Kahane for the article he has written about Pichorides, half in English to deal with Mathematics and half in French to deal with more personal matters.

Finally we want to express our gratitude to the following institutions who provided financial support for the Conference: Universidad Autónoma de Madrid (Vicerrectorado de Investigación), Ministerio de Educación y Ciencia de España (D.G.I.C.Y.T.), Comunidad Autónoma de Madrid, and IBM Corporation.

The Editors
Madrid, July 1993

Part I

Memorial Articles

1

Antoni Zygmund 1900–1992

Guido Weiss

It is only fitting that this conference was originally dedicated to the memory of Antoni Zygmund. This was the fourth of a series of conferences devoted to a branch of analysis that was strongly influenced by him. The series is often referred to as the "El Escorial" conferences, since the first three were, indeed, held in El Escorial in the province of Madrid. For purely logistic reasons this conference was held in Miraflores de la Sierra, also in the province of Madrid. Each of these conferences was organized by a group of analysts that was strongly influenced by the school Zygmund had created. In fact, the first of these conferences (which took place in 1979) was dedicated to him, and the attractive poster that announced it has, as its background, the image of Zygmund. He was proud of this, and he kept this poster in his study at home.

During this past year much has been written about Zygmund, and this is not the only conference that has been held in his honor. I do not want to repeat what has been said in the various articles and dedications involved in this homage. Perhaps it is appropriate to cite the article published in 1989 by the American Mathematical Society (AMS) in Part III of *A Century of Mathematics in America*, pages 343–368. This article is entitled "The School of Antoni Zygmund," and from it we can learn some aspects of his character and the immense influence he had in the world of mathematics. One of the features of this article is that it lists two generations of students: his and those who were trained by his students. This list contains 179 names, and, of course, this number is still growing. The *Selected Papers of Antoni Zygmund*, published by Kluwer Academic Publishers in 1989, presents another good picture of the influence Zygmund had on the development of analysis in the twentieth century. In this article I would like to complement what appears elsewhere and cite some more personal memories I have of this kind, gentle, and most influential man. Hopefully, then, the various articles on Zygmund that have appeared in the last year will, when put together, give a reasonably complete and proper image of the man. The articles mentioned above give a good account of Zygmund's research and his success with students and collaborators. Perhaps not enough has been said about his books and what they mean to mathematics.

I will concentrate most of my remarks around their importance and how they were regarded by Zygmund.

Zygmund was finishing the second edition of his book on trigonometric series when I was working on my Ph.D. thesis as one of his students. Shortly after we obtained our degrees, he asked my wife, Mary Weiss, who was also his student, and me to proofread the galleys of his book. At that time (during the academic year 1958–59) we were young and energetic and we did a very thorough and careful job of proofreading. Influenced by this reading and the legendary Zygmund seminar, we felt we knew reasonably well what the subject called "trigonometric series" was all about. It is interesting to muse upon how much we take for granted in everyday life; we push a button and a television screen lights up and scenes from the other side of the globe are displayed. As we do this we do not think about all the scientific, engineering, and intellectual effort that made this possible. We just make use of the results. Much the same can be said about reading Zygmund's second edition. It was, of course, considerably harder to have proofread the book than to push a button; but having done so, we used the results we learned without giving much thought about how all this material was organized into a meaningful whole. I must confess that I have read very few mathematics books from cover to cover; the first edition of Zygmund's book, *Trigonometrical Series*, is a member of this small set. I do not think that I am unusual among my colleagues. Mathematicians are not scholars in the sense this term is used in other disciplines. They do make use of libraries, books, and journals. Most often they do this to examine a very particular result and will focus only on the material that is closest to this result. I think that they are remarkably good in this endeavor, and this practice does, indeed, advance the development of new mathematics. This activity, however, makes it difficult to obtain a "broad view" of a subject. Zygmund's two books, *Trigonometrical Series* and *Trigonometric Series,* represent a remarkable achievement of bringing together several different areas of mathematics into one coherent whole. The titles of these books are misleading because of their unpretentious simplicity. In his first book, Zygmund gave a structure to many apparently disconnected results and techniques: the considerable work of Hardy and Littlewood on Fourier and trigonometric series, the maximal function, summability theory, the M. Riesz convexity theorem, the study of the properties of Lebesgue spaces and their extensions, the development by F. Riesz of what are known as Hardy spaces, probabilistic methods for studying trigonometric series, convergence results, and differentiability of integrals are just some of the topics that are introduced in *Trigonometrical Series* and then updated and refined in *Trigonometric Series*.

Zygmund was very much aware of the effort he put into the writing of these books. As mentioned in the AMS article cited above, he complained that the writing of *Trigonometric Series* cost him at least thirty research papers. I was present when Littlewood, on a visit to the University of Chicago, told him, "many generations of mathematicians will thank you." He was also aware that Littlewood was right. He told Mary and me after we finished our proofreading, "Now that the second edition is finished I have started working on the third," and showed us extensive notes of expositions of the development of singular integrals and other higher dimensional aspects of Fourier analysis. Each of us who worked with him in the later 1950s and early 1960s was aware of his conviction that the future of Fourier analysis lay in "higher dimensions and other settings associated with higher dimensional Euclidean spaces." He also encouraged his students to carry on his program. This was one of the motivations that Stein and I had for dedicating our book *Introduction to Fourier Analysis on Euclidean Spaces* to him.

As mentioned in some of the other articles that have been written recently in memory of Zygmund, he wrote other books. His book on the theory of functions, written with Saks, and the text on integration, written with Wheeden, are well known and highly regarded. Not at all known is the fact that he wrote a book on differential equations (in Polish), mostly devoted to ordinary differential equations. I could not find a reference to it in any of the lists of his publications; thus, this book may never have been published. I had the good fortune of taking a course he offered on this subject. It was based on this book and he did show me the manuscript he was using. I present this as evidence that he devoted considerable time and energy on thinking how mathematics, at all levels, should be presented and motivated. He had strong opinions on this subject, and I learned much about this aspect of my profession from him as well.

His gentleness, his generosity, his broad interest (in politics and literature, as well mathematics), and his sense of humor have been well described in other articles. I cannot ignore, in this article, the tragedy that we experienced during the meeting in Miraflores: the sudden death of Stélios Pichorides, the last of Zygmund's students. This meeting is also dedicated to him. I would like to echo an observation made by Jean-Pierre Kahane in the article he wrote about Stélios that also appears in these proceedings. To those of you who did not know Zygmund personally, but did have a chance to get to know Stélios, let me say that the latter presented an image that was similar to that of the former. It is remarkable how these two individuals shared the same gentleness, generosity, general interest, and sense of humor.

2

Stylianos Pichorides 1940–1992

Jean-Pierre Kahane

This article consists of two parts. The first, written in English, is a brief account of the life and work of Pichorides. It is intended to provide basic information for those who did not know him. The second is more personal. It is an attempt to revive part of his ideas, style, and human behavior through writings, testimonies, and remembrances. It is written in French, the language which he and I used to communicate with each other. It may help those who knew him to conjure up his warm and charming personality.

2.1 Life and works

Pichorides was born in Athens on October 18, 1940. He attended primary school in Athens from 1946 to 1953. He graduated from the Varvakion Model School, Athens, in 1958, then entered the Athens Polytechnic, and graduated from the School of Electrical Engineering in June 1963.

From June 1963 to October 1968 he worked as an electrical engineer, with the exception of military service in 1964. In the meantime he enrolled in the mathematical department of the University of Athens. In 1967 he took an examination in mathematics given by the State Scholarship Foundation of Greece and received a scholarship to be used abroad from September 1968.

He was then interested in set theory and mathematical logic. From October 1968 to June 1969 he studied at the University of Chicago and obtained the M.Sc. From June 1969 to June 1971 he turned to trigonometric series, under the guidance of Zygmund, and earned his Ph.D. on the best constants in Riesz's inequalities and related questions.

Then he returned to Athens and obtained a research position in the Nuclear Center Democritos—his permanent position from June 1972 to April 1983. Meanwhile, he occupied a series of temporary positions abroad. For two years (October

1974–October 1976) he was *Attaché de Recherches* at the *Centre National de la Recherche Scientifique* (CNRS), France, and worked at Orsay with *Equipe d'Analyse Harmonique*. From September 1979 to May 1981 he was a visiting professor at Orsay (*Université de Paris-Sud*) (September 1979–April 1980) and UCLA (September 1980–May 1981). In 1980 he was awarded the Salem Prize for his work on the Littlewood conjecture.

After April 1983 his permanent position was professor of mathematics at the University of Crete. The University had already existed, but he can be considered as one of its main founders. Before moving there he organized a most successful conference in harmonic analysis in Iraklion, July 1978, together with Petridis and Varopoulos.

He initiated a number of new and important activities at the University. In particular he supervised the first doctoral thesis in mathematics by Katsoprinakis.

He also visited a number of places. From September 1989 to July 1990 he was a visiting professor at Orsay, Caltech, and the University of Chicago. From September 1991 to May 1992 he remained a professor at the University of Cyprus. Moreover, from 1971 to 1992, he was invited in a number of places for shorter visits: Cambridge University, Brown University, Institut Mittag-Feffler, Courant Institute, and the University of Chicago among others.

He was taking part in a conference in Spain, when he died suddenly, on June 18, 1992.

His scientific work focuses on a few difficult questions in the theory of Fourier series; essentially, inequalities in Fourier series, or, equivalently, inequalities for trigonometric polynomials. Usually the questions are elementary, the statements easy to understand, and the proofs very ingenious.

His dissertation thesis [1] and the subsequent papers [2] and [3] deal with two conjugate functions, f, $\tilde{f}$, either on the circle or on the line. The most spectacular result in [1] is the computation of the best constants A_p in the Riesz inequalities between L^p-norms:

$$\|\tilde{f}\|_p \leq A_p \|f\|_p, \qquad (1 < p < \infty),$$

namely $\mathrm{tg}\left(\frac{\pi}{2p}\right)$ if $1 < p \leq 2$, $\mathrm{cotg}\left(\frac{\pi}{2p}\right)$ if $2 \leq p < \infty$. The result was conjectured and proved in particular cases ($p = 2^n$, $n = 1, 2, \ldots$) by Gokhberg and Krupnik a few years before. The charm of Pichorides's proof is that it relies in a clever way on the device used in Zygmund's Trigonometric Series for proving the Riesz inequalities, a device discovered by Calderón. It provides optimal results also for the inequality of Zygmund

$$\|\tilde{f}\|_1 \leq A \frac{1}{2\pi} \int_{-\pi}^{\pi} |f| \log^+ |f| + B$$

where $B = B(A)$, namely the best possible value for $A \left(= \frac{2}{\pi}\right)$, and for the inequalities of Kolmogorov

$$\|\tilde{f}\|_p \leq B_p \|f\|_1, \qquad (0 < p < 1)$$

(where the best B_p for positive functions f is computed).

Conjugate functions is another term for Hilbert transforms on the circle. Hilbert transforms on the line can be treated as well and sometimes in a simpler way. For example, if $f = 1_E$ (the indicator function of a set E), the distribution of the Hilbert transform Hf depends only on $|E|$, the Lebesgue measure of E: this is a theorem of Stein and Weiss. This theorem is a motivation for [2], which gives the best possible upper estimate for $\frac{\|sh\,\tilde{f}\|_2}{\|f\|^2}$ when $|f| \leq k < \frac{\pi}{4}$ as a function of $k \left(= (\cos 2k)^{-1/2}\right)$.

[3] is closely related to the Riesz theorem on the line. It states that

$$\int_{\mathbb{R}} |Hf|^a |f|^{p-a} \leq C \int_{\mathbb{R}} |Hf|^b |f|^{p-b}$$

$(0 \leq a, b \leq p, \; C = C(a, b, p))$ with $C < \infty$ if $p > 2$, and with $C = \infty$ only if $1 < p \leq 2$ and $a \neq b$).

[12] goes back to a corollary of the Riesz theorem on the circle, namely inequalities

$$\|S_n\|_p \leq C_p \|f\|_p$$

for partial sums of the Fourier series of f. Actually inequalities for $\|\tilde{f}\|_p$ and uniform inequalities for $\|S_n\|_p$ are equivalent. However the best C_p is not known. The proof of Pichorides is elementary, complete, 13 lines long, with a good estimate of C_p. This is incredible.

[13] and [14] are inspired by works of Bourgain (1985 and 1988) on the Littlewood–Paley function

$$\gamma(x) = \left(\sum_j \left| \sum_{2^{j-1} \leq k < 2^j} \hat{f}(k) e^{ikx} \right|^2 + \left| \sum_{2^{j-1} \leq k < 2^j} \hat{f}(-k) e^{-ikx} \right|^2 \right)^{1/2}.$$

The Littlewood–Paley inequalities are

$$A_p \|f\|_p \leq \|\gamma\|_p \leq B_p \|f\|_p$$

and Bourgain proved $B_p = O(p) \; (p \to \infty)$ and $B_p = O\left((p-1)^{-3/2}\right) (p \to 1)$. The exponent $-\frac{3}{2}$ may look strange. Pichorides shows first, that it can be replaced by -1 for Taylor-type series $(\hat{f}(-k) = 0$ for $k \geq 0)$ [13]; then, that it can be understood by exploring the behavior of A_p as $p \to \infty$. I shall say a few words about this in the second part.

In the works of Pichorides L^p inequalities $(1 < p < \infty)$ correspond to the first and last period: 1972–1975 and 1990–1992. The topic got a new impetus in 1975 with the work of Beckner on the Hausdorff–Young and Young inequalities for Fourier transforms and convolutions on the line

$$\|\widehat{f}\|_{p'} \leq C_p \|f\|_p, \qquad \left(1 \leq p \leq 2, \frac{1}{p} + \frac{1}{p'} = 1\right)$$

$$\|f * g\|_r \leq C_{p.q.r} \|f\|_p \|g\|_q, \qquad \left(\frac{1}{r} = \frac{1}{p} + \frac{1}{q} - 1\right)$$

Beckner was able to compute the best constants and was awarded the Salem Prize for this achievement. The methods of Beckner have no relation to those of Pichorides, but the simple fact that Pichorides was able to compute best constants in the Riesz inequalities stimulated similar investigations.

A few years after Beckner, in 1980, Pichorides obtained the Salem Prize for this contribution to the solution of the Littlewood conjecture. Here, only the L^1-norm is concerned. The Littlewood conjecture was

$$\int \left|e^{i\lambda_1 x} + \cdots + e^{i\lambda_n x}\right| dx \geq C \log n$$

whenever the λ_j are distinct integers. I shall try to explain the point of view of Pichorides in the second part. There were contributions by Salem, Cohen, and Davenport (see Davenport, Mathematika 7, 1960). Cohen obtained a lower estimate $\left(\frac{\log n}{\log \log n}\right)^{1/8}$, Davenport $\left(\frac{\log n}{\log \log n}\right)^{1/4}$. Then Pichorides improved this lower bound in a series of steps: $\left(\frac{\log n}{\log \log n}\right)^{1/2}$ ([4], 1974), $(\log n)^{1/2}$ ([7], 1977; this was obtained independently by Fournier in 1978 with a quite different method), $\frac{\log n}{(\log \log n)^2}$ (October 1979, [10]). The method in [10], a very original dichotomy device already introduced in [7], attracted much attention and the Salem Prize both expressed and reinforced the feeling that for the first time the Littlewood conjecture could be reached. Actually Konjagin was able to use the dichotomy device in a stronger way to solve the conjecture in November 1980 (Izvestia Acad. Nauk SSSR, April 1981). Independently, McGehee, Pigno, and Smith used a completely new idea and also got a lower estimate $\log n$ (January 1981). Again, the breakthrough achieved by Pichorides proved very successful in stimulating other bright people.

[9] contains a nice result: if f is a trigonometric polynomial, $|f| \leq 1$, and all roots of f are real, then

$$\text{meas}\{t \in (0, 2\pi) : |f(t)| > \cos x\} \leq 4x, \qquad 0 \leq x \leq \frac{\pi}{2};$$

a best possible result (take $f(t) = \cos t$).

So far I have not mentioned the expository papers [6], [11], and [15]. [6] is almost a book and contains the motivations of Pichorides for studying the Littlewood conjecture. [11] contains personal ideas and conjectures on trigonometric polynomials. [15] is the best introduction to the thesis of Katsoprinakis and the further works of Nestoridis and Katsoprinakis on limit sets of partial sums of Taylor series. I shall take advantage of these papers in the second part.

2.2 Style et personnalité

La table où j'écris ce rapport est celle-là même qu'utilisait Pichorides durant ses derniers séjours à Orsay. C'est donc ici le lieu d'évoquer des souvenirs. Je mélangerai souvenirs personnels, souvenirs d'amis et collègues à Orsay, et témoignages venus d'ailleurs, en particulier de ses élèves, amis et collègues à l'Université de Crète.

Je commencerai par une image: Stélios au tableau noir, la main au menton, évoquant irrésistiblement Zygmund par sa façon de se tenir, son élocution, son écriture. Aucun des autres élèves de Zygmund ne m'a donné ce sentiment d'être en présence du maître lui-même, à la fois posé et détendu, le langage mesuré et l'esprit vif, tout juste rajeuni de quarante ans.

Une autre image: Stélios en Grec de l'Antiquité, curieux comme Ulysse, disert comme Platon, le goût d'Euclide pour la perfection géométrique. Sur les plages de Crète, en sa compagnie, on renouait avec trente siècles d'histoire.

Une autre encore: Stélios bavardant avec Mme Dumas, égrenant des souvenirs d'enfance, d'une enfance marquée par la gène, la guerre mondiale, la guerre civile. Avec finesse et humour, il constatait, que, grâce à son père maçon, il était sans doute le seul mathématicien à savoir bâtir un mur.

Avec lui, je n'avais guère de conversations intimes. Nous parlions de mathématiques, d'enseignement, quelquefois un peu de politique, et là encore je retrouvais Zygmund: une connaissance profonde de sujets variés en mathématique, une vaste culture, le souci des élèves à tous les niveaux, et une ouverture d'esprit à toutes les grandes questions du monde. A l'Université de Crète, il semble avoir été un modèle admiré et respecté bien au delà du cercle des mathématiciens. Il avait une haute conscience du métier universitaire. Dès son arrivée en Crète en 1983, il organisa un séminaire d'analyse régulier, mit au point un nouveau cursus, initia un système tutorial. Ensuite, il eut la charge de superviser les programmes d'études avancées de toute l'université. Il dirigea la première thèse de doctorat de l'Université de Crète en mathématiques (Katsoprinakis) et était en train de diriger la seconde. Et pendant ce temps, sur 14 semestres d'enseignement qu'il assura à Iraklion, il enseigna 14 sujets différents.

A Orsay, il donna des séminaires et des cours de différents niveaux. Ses cours avancés ("de troisième cycle") étaient très appréciés, les cours les plus élémentaires (mathématiques pour biologistes en première année d'université, où nous fonctionnions dans deux sections parallèles) également. Heureusement nous avons gardé trace de certains exposés, dans les actes du Séminaire d'Analyse Harmonique d'Orsay. Pour comprendre son style en mathématiques, il faut à la fois se reporter aux produits achevés (les treize lignes de démonstration de l'inégalité $\|S_n\|_p \leq C_p \|f\|_p$ dans [12]), et à la lente génèse des idées. A cet égard, la lecture de [6], [11], et [15], dont je viens de dire un mot est essentiel.

Les idées de base de Stélios étaient toujours simples, il avait d'ailleurs une manière bien à lui de définir l'analyse harmonique: on appelle analyse harmonique la partie de la trigonométrie élémentaire qui s'occupe des sommes de sinus et de cosinus ("Harmonic analysis is that part of high school trigonometry which deals with sums of sines and cosines"). Cette boutade m'a été communiquée par Saffari. On peut apprécier sa portée dans [6], où se trouve indiqué dès le début le rôle des sommes d'exponentielles $e^{i\lambda_1 x} + \cdots + e^{i\lambda_n x}$ en analyse harmonique et en théorie des nombres. Comme Calderón l'avait montré (cf. Zygmund, *Trigonometric Series,* chap. XIII), la convergence presque partout des séries de Fourier des fonctions de carré sommable équivaut à une propriété des noyaux de Dirichlet

$$D_n(x) = \sum_{-n}^{n} e^{ikx},$$

à savoir la majoration uniforme de l'intégrale double

$$\int \int D_{\min(n(x),n(y))}(x-y)\, dx\, dy$$

pour toutes les fonctions $n(\cdot)$ à valeurs entières positives (et c'est la voie par laquelle Fefferman a démontré en 1973 le théorème de Carleson de 1966). Quant à la théorie des nombres, le théorème de Vinogradov de 1937, disant que tout nombre impair assez grand est la somme de trois nombres premiers, c'est une propriété des sommes

$$f(x) = \sum_{p \text{ premier. } p<n} e^{ipx},$$

à savoir

$$\int f^3(x) e^{-inx}\, dx > 0.$$

Ainsi, deux des théorèmes les plus difficiles de la théorie des nombres et de l'analyse harmonique concernent les sommes d'exponentielles.

Quand q est un entier positif, on a

$$n^{1/2} \leq \left\| \sum_1^n \exp(i\lambda_k x) \right\|_{2q} \leq \left\| \sum_1^n \exp(ikx) \right\|_{2q}.$$

La première inégalité est inaméliorable comme le montrent les suites λ_k lacunaires, et la seconde inégalité, due à Hardy et Littlewood, montre le rôle extrémal du noyau de Dirichlet; on en tire

$$\left\| \sum_1^n \exp(i\lambda_k x) \right\|_p \leq C \left\| \sum_1^n \exp(ikx) \right\|_p$$

pour tout $p \geq 2$, et pour une constante C convenable (inconnue d'ailleurs). Vu la convexité de ces normes comme fonctions de p, et leur égalité quand $p = 2$, il est naturel de penser à une inégalité en sens opposé pour $p < 2$: c'est la signification de la conjecture de Littlewood, relative au cas $p = 1$.

On trouve dans [6] et dans [11] l'histoire de la conjecture de Littlewood et de sa solution. Je voudrais donner ici une idée de la décomposition dyadique de Pichorides. On part d'une suite $\Lambda = (\lambda_1, \ldots, \lambda_n)$, et le problème est invariant par translation. On décomposera Λ en Λ_1 (entiers pairs) et Λ_0' (entiers impairs), et, quitte à translater, on suppose qu'il y a au moins $\frac{n}{2}$ termes dans Λ_1; ainsi

$$\Lambda = \Lambda_1 \cup \Lambda_0', \ \Lambda_1 \subset 2\mathbb{Z}, \quad \Lambda_0' \subset 2\mathbb{Z} + 1, \quad |\Lambda_1| \geq |\Lambda_0'|.$$

On poursuit la décomposition, en supposant (quitte à translater à chaque stade) pour $j = 1, 2, \ldots, \sigma - 1$,

$$\Lambda_j = \Lambda_{j+1} \cup \Lambda_j', \ \Lambda_{j+1} \subset 2^{j+1}\mathbb{Z}, \ \Lambda_j' \subset 2^j(2\mathbb{Z} + 1), \ |\Lambda_{j+1}| \geq |\Lambda_j'|$$

jusqu'à ce que, pour $j = \sigma$, Λ_j n'ait plus qu'un élément; on pose alors $\Lambda_\sigma' = \Lambda_\sigma$. On a ainsi partagé la somme

$$F(x) = \sum_{\lambda \in \Lambda} e^{i\lambda x}$$

sous la forme

$$F(x) = F_0(x) + F_1(x) + \cdots + F_\sigma(x),$$

avec

$$F_j(x) = \sum_{\lambda \in \Lambda_j'} e^{i\lambda x}, \qquad (\Lambda_j' \subset 2^j(2\mathbb{Z} + 1)).$$

Un argument simple permet de conclure quand le nombre des $F_j \not\equiv 0$ est assez grand ($\geq (\log n)^3$). Dans le cas contraire, il y a un nombre significatif ($\geq \log \log n$) de j pour lesquels F_j contient beaucoup de termes (plus de $\frac{n}{(\log n)^4}$). A ces F_j on associe le signe de leurs parties réelles, g_j. Ainsi le spectre de g_j est contenu dans $2^j(2\mathbb{Z} + 1)$, et $g_j = \pm 1$; donc la distribution des g_j est celle de fonctions de Rademacher indépendantes. Des techniques de martingales dyadiques permettent d'obtenir

$$\|F\|_1 > C_\xi \frac{\log n}{(\log \log n)^\xi}$$

pour tout $\xi > \frac{1}{2}$. Il faut encore une autre idée, de Konjagin, pour passer de là à $\|F\|_1 > C \log n$. C'est le terme d'une laborieuse ascension, entreprise par Pichorides dès 1974.

Il y avait une face de la montagne très cachée, mais beaucoup plus accessible: c'est ce qu'ont découvert McGehee, Pigno, et Smith. Je me souviens encore de mon incrédulité puis de ma surprise en recevant, début 1981, une lettre circulaire de Smith donnant la solution complète en quelques pages.

Cela n'épuise pas le sujet de la distribution des sommes d'exponentielles F. En particulier, est-il vrai que

$$\int_E |F| \geq C > 0$$

pour tout ensemble E de mesure π contenu dans $(0, 2\pi)$, C étant une constante absolue? La question—avec les conséquences d'une réponse positive—se trouve posée dans [11]. [11] contient également des idées très intéressantes sur les polynomes dont tous les zéros se trouvent sur le cercle unité $|z| = 1$, et sur les polynomes à coefficients unimodulaires dont le module est presque constant sur le cercle unité.

Voici une autre question de Pichorides, relative à la fonction γ de Littlewood–Paley (voir la première partie). Est-il vrai que

$$\|f\|_p \leq \sqrt{p}\|\gamma\|_p?$$

L'exemple des séries lacunaires montre que la majoration en $\sqrt{p}$ est la meilleure qu'on puisse souhaiter. Dans [13] Pichorides obtient une majoration en $p \log p$. La majoration en $\sqrt{p}$ aurait comme conséquence intéressante que la formule de Bourgain la plus surprenante $(B_p = 0((p-1)^{-3/2}) \, (p \to 1)$ apparaîtrait comme corollaire de l'autre $(B_p = 0(p) \, (p \to \infty))$.

Lors de son dernier séjour à Orsay Stélios avait fait deux exposés de séminaire [15], [15'], qui montraient la variété de ses intérêts, et l'influence qu'il pouvait avoir sur le développement des mathématiques en Grèce. Dans ces deux cas, il s'agit de problèmes d'aspect élémentaire, difficiles, alliant l'analyse et la géométrie. Ce sont pourtant des problèmes bien dissemblables.

Le premier est une présentation du contexte de la thèse de son élève Katsoprinakis et des travaux qui l'ont suivi (Nestoridis). Le point de départ est un théorème de Marcinkiewicz et Zygmund qui est lui même l'aboutissement de leurs recherches sur les limites supérieure et inférieure des sommes partielles S_n et $\tilde{S}_n$ d'une fonction f et de sa conjuguée $\tilde{f}$. Le bon cadre est celui des ensembles limites des sommes partielles d'une série de Taylor: sous la condition que la série soit commable $(C, 1)$, ces ensembles limites ont presque partout la "structure circulaire". Ce théorème, pour être bien compris, requiert une collection d'exemples, amorcée dans le livre de Zygmund. Le point de vue de Pichorides est de faire apparaître ces exemples comme conséquences d'hypothèses très simples sur le comportement des sommes partielles. Pour une bonne part, c'est de la géométrie d'Euclide, mais inspirée par l'analyse contemporaine.

Le second traite d'un problème fameux sur les sections des corps convexes et des travaux de Ball et Giannopoulos: l'exposé n'a que trois pages, et naturellement il est loin d'être complet. Mais il donne la saveur du sujet, l'état de la question (ouverte alors pour les dimensions n entre 3 et 6), et surtout un aperçu suggestif sur les techniques. Depuis, les cas $n = 5$ et 6 ont été traités par Papadimitrakis, dans un article qui se termine par des remerciements à Pichorides "for the many discussions we had about this problem."

Je reviens pour finir à deux témoignages d'amis proches de Stélios, Bahman Saffari (Babar), chez qui il avait coutume de loger pendant ses séjours en France, et Nestoridis, son collaborateur à l'Université de Crète.

Babar se souvient de ce que Stélios lui avait raconté de ses débuts. Après avoir obtenu le diplôme d'ingénieur électricien qui lui donnait le droit et la responsabilité de signer l'autorisation de fonctionnement des usines électriques nouvellement créées, son premier travail d'enseignement—pour améliorer ses fins de mois—fut de donner des cours d'électricité dans une école de coiffure. Babar observe aussi que, plus tard, dans l'atmosphère assez tendue des universités parisiennes ("dog eat dog") Stélios est l'un des mathématiciens étrangers qui avait le plus d'amis et le moins d'ennemis.

Nestoridis m'a raconté une histoire de Crète. Stélios avait cours le matin, à 8 heures. Un jour, il y eut passage à l'heure d'hiver. Tout le monde le savait, sauf Stélios. Il se présente donc en classe à 7 heures croyant qu'il était 8 heures. La classe était vide. Un autre serait peut-être parti. Stélios resta et fit son cours en l'écrivant soigneusement au tableau. Au moment de partir, il vit les étudiants arriver, et tout le monde fut de bonne humeur.

La mort de Stélios, brutale, imprévue, choquante, a bouleversé tous ses amis, ses collègues et ses étudiants. Mais il n'est pas mauvais de garder de lui cette image de professeur obstiné à faire son cours devant la classe vide, puis éclatant de rire, inflexible sur l'essentiel, souriant de lui-même à l'occasion et respectueux des autres jusque dans la plaisanterie.

References

Publications by Pichorides

(References [1], [2],... [15] are taken from a list prepared by Pichorides himself. He observes that [6], [11], and [15] are partly expository. [6'], [8'],... [15'] were added by J.-P. Kahane.)

[1]   On the best values of the constants in the theorems of M. Riesz, A. Zygmund and Kolmogorov. *Studia Math.* 64 (1972), 165–179.

[2]   On the conjugate of bounded functions. *Bull. AMS* 81 (1975), 143–145.

[3]   Une propriété de la transformée de Hilbert. *C. R. Acad. Sci. Paris* 280 (1975), 1197–1199.

[4]   A lower bound for the L^1-norm of exponential sums. *Mathematika* 21 (1974), 155–159.

[5]   A remark on exponential sums. *Bull. AMS* 83 (1977), 283–285.

[6] L^p norms of exponential sums. Sém. Anal. Harm., Publ. Math. d'Orsay 77-73 (1976), 1–65.

[6'] Norms of exponential sums (supplement) (Sept. 1977). Sém. Anal. Harm., Publ. Math. d'Orsay 78-12 (1977–78), 65–79.

[7] On a conjecture of Littlewood concerning exponential sums I. *Bull. Greek Math. Soc.* 18 (1977), 8–16.

[8] On a conjecture of Littlewood concerning exponential sums II. *Bull. Greek Math. Soc.* 19 (1978), 274–277.

[8'] Quelques remarques sur les sommes exponentielles. CNRS. Talence 1978, Sém. Théorie des nombres (1977–78), exposé n° 2.

[9] Une remarque sur les polynômes trigonométriques réels dont toutes les racines sont réelles. *C. R. Acad. Sci. Paris* 286 (1978), 17–19.

[9'] A theorem of J.-F. Fournier on Littlewood's conjecture (Sept. 1978). Sém. Anal. Harm. Orsay (1978–79), 56–63.

[9''] On the L^1-norm of exponential sums. Harmonic Analysis, Iraklion 1978. Lecture Notes in Math. 781, Springer-Verlag 1980, 171–176.

[10] On the L^1-norm of exponential sums. *Ann. Inst. Fourier* 30 (1980), 79–89.

[10'] Une inégalité de H. Montgomery sur les sommes d'exponentielles. Sém. Anal. Harm. Orsay (1980–81), 88–72.

[11] Notes on trigonometric polynomials. Conference on Harmonic Analysis, Chicago, 1983, vol. 1, 84–94.

[12] Une remarque sur le théorème de M. Riesz concernant la norme L^p de sommes partielles de séries de Taylor. *C. R. Acad. Sci. Paris* 310 (1990), 549–551.

[13] A note on the Littlewood–Paley square function inequality. *Colloquium Math.* 60/61 (1990), 687–691.

[14] A remark on the constants of the Littlewood–Paley inequality. (Univ. of Crete, preprint n° 20). *Proc. AMS* 114 (1992), 787–789.

[15] (with V. Nestoridis) The circular structure of the set of limit points of partial sums of Taylor series. Sém. Anal. Harm. Orsay (1989–90), 71–77.

[15'] On a recent result of A. Giannopoulos on sections of convex symmetric bodies in $\mathbb{R}^n$. Sém. Anal. Harm. Orsay (1989–90), 83–86.

Other references

W. Beckner: Inequalities in Fourier analysis. *Ann. Math.* 102 (1975), 159–182.

J. Bourgain: On square functions on the trigonometric system. *Bull. Soc. Math. Belge,* ser. B 37 (1985), 20–26.

J. Bourgain: On the behaviour of the constant in the Littlewood–Paley inequality, in Geometric aspects of functional analysis, Israel Sem. 1987–1988, Lecture Notes in Math. 1376, Springer-Verlag 1989, 202–208.

H. Davenport: On a theorem of P. J. Cohen. *Mathematica* 7 (1960), 93–97.

J.-F. Fournier: On a theorem of Paley and the Littlewood conjecture. *Arkiv för Mat.* 17 (1979), 199–216.

E. Katsoprinakis: On a theorem of Marcinkiewicz and Zygmund for Taylor series. *Arkiv för Mat.* 27 (1989), 105–126.

E. Katsoprinakis: Taylor series with limit-points on a finite number of circles. Preprint n° 23, University of Crete, Mathematics (1991).

E. Katsoprinakis and V. Nestoridis: Partial sums of Taylor series on a circle. *Ann. Inst. Fourier* 39 (1989), 715–736.

E. Katsoprinakis: Taylor series with limit-points on a finite number of circles. Preprint n° 23, University of Crete, Mathematics (1991).

S. B. Konjagin: O probleme Littlewood'a. *Izv. A.N. SSSR,* ser. mat. 45, 2 (1981), 243–265.

O. C. McGehee, L. Pigno, and B. Smith: Hardy's inequality and L^1-norms of exponential sums. *Ann. Math.* (2) 113 (1981), 3, 613–618.

V. Nestoridis: Limit points of partial sums of Taylor series. *Mathematika* 38 (1991), 2, 239–249.

M. Papadimitrakis: On the Busemann-Petty problem about convex, centrally symmetric bodies in $\mathbb{R}^n$. *Mathematika* 39, 2 (1992), 258–266.

A. Zygmund: *Trigonometric series,* vol. I, CUP, 1959.

Part II

Main Lectures

3

Band-Limited Wavelets

Aline Bonami

Fernando Soria

Guido Weiss[1]

ABSTRACT   This manuscript represents a series of lectures, a "mini course," presented by Guido Weiss during the Conference in Harmonic Analysis and Partial Differential Equations that was held in Miraflores de la Sierra in June 1992. This presentation can be regarded as an introduction to "wavelets," and this material is, for the most part, self-contained. The lectures, however, are aimed at an audience that knows the fundamentals of harmonic analysis. We are considering those wavelets whose Fourier transforms are compactly supported since many of their properties are more easily formulated and derived than those of the more general wavelets. As is pointed out in the text, however, much of the material presented here is true more generally. Much of this development appears, perhaps in different form, elsewhere; nevertheless, we do introduce some new results. All this is carefully explained in the text.

3.1   Introduction

An *orthonormal wavelet* is a function $\psi \in L^2(\mathbb{R})$ such that the system $\{\psi_{j,k}\} = \{2^{j/2}\psi(2^j x + k)\}$, $j, k \in \mathbb{Z}$, is an orthonormal basis of $L^2(\mathbb{R})$. We claim that there are two classes of equalities, involving the Fourier transform, $\widehat{\psi}$, of ψ that characterize orthonormal wavelets. The first one is

$$\sum_{\ell \in \mathbb{Z}} \widehat{\psi}(2^j[\xi + 2\ell\pi]) \overline{\widehat{\psi}(\xi + 2\ell\pi)} = \delta_{0,j} \tag{I}$$

[1]The research of the last named author was supported by the DARPA grant administered through AFSOR Grant 90-0323.

for a.e. $\xi \in \mathbb{R}$ whenever $j \geq 0$ (see equation (1.8) below). The second class of equations is

$$\text{(i)} \quad \sum_{j \in \mathbb{Z}} |\widehat{\psi}(2^j \xi)|^2 = 1,$$

$$\text{(ii)} \quad \sum_{j=0}^{\infty} \widehat{\psi}(2^j \xi)\overline{\widehat{\psi}(2^j(\xi + (2m+1)2\pi))} = 0 \qquad\qquad \text{(II)}$$

for a.e. $\xi \in \mathbb{R}$ whenever $m \in \mathbb{Z}$ (see equations (9) and (16) below). As we shall see, (I) is equivalent to the orthonormality of the system $\{\psi_{jk}\}$ and (II) is equivalent to the completeness of this system in $L^2(\mathbb{R})$.

Although these facts are quite general, we shall concentrate our study on the case when ψ is band-limited, in which case (as we shall show) the sums in (I) and (II) are finite for each ξ. More precisely, a function in $L^2(\mathbb{R})$ is said to be *band-limited* if its Fourier transform has compact support. Lemarié and Meyer [LM] have constructed a band-limited wavelet having a Fourier transform that is infinitely differentiable. The function ψ they produced satisfied

$$\widehat{\psi}(\xi) = e^{i\xi/2} b(\xi), \tag{1}$$

where b is an even, non-negative "bell shaped" function whose support is the union

$$\left[\frac{-8\pi}{3}, -\frac{2\pi}{3}\right] \cup \left[\frac{2\pi}{3}, \frac{8\pi}{3}\right].$$

Our study will examine general band-limited wavelets. Also, we shall examine in depth a class of more specialized wavelets, which includes the one introduced by Lemarié and Meyer, and show how they are connected with the local sine/cosine bases of Coifman and Meyer.

We would like to thank Auscher, Berkson, Cohen, Paluszynski, and Wang, with whom we discussed these matters; each gave us very helpful suggestions.

Let us first make some simple observations about general wavelets. It is a well-known fact that $\{\psi(\cdot + k)\}$, $k \in \mathbb{Z}$, is an orthonormal system in $L^2(\mathbb{R})$ if and only if

$$\sum_{\ell \in \mathbb{Z}} |\widehat{\psi}(\xi + 2\ell\pi)|^2 = 1 \tag{2}$$

for a.e. $\xi \in \mathbb{R}$. The definition of the Fourier transform we are using is given by the equality

$$\widehat{f}(\xi) = \int_{\mathbb{R}} f(x) e^{-i\xi \cdot x} \, dx.$$

Thus, the Plancherel theorem involves the identity

$$\langle f, g \rangle = \int_{\mathbb{R}} f(x)\overline{g(x)} \, dx = \frac{1}{2\pi} \langle \widehat{f}, \widehat{g} \rangle. \tag{3}$$

The simple argument that establishes (2) involves a "periodization" idea (associated with the Poisson summation formula) that we shall use repeatedly: $\{\psi(\cdot + k)\}$, $k \in \mathbb{Z}$, is orthonormal if and only if

$$
2\pi \delta_{k,m} = 2\pi \int_{\mathbb{R}} \psi(x+k)\overline{\psi(x+m)}\, dx
$$

$$
= \int_{\mathbb{R}} \widehat{\psi}(\xi)e^{ik\xi}\,\overline{\widehat{\psi}(\xi)}e^{-im\xi}\, d\xi
$$

$$
= \sum_{\ell \in \mathbb{Z}} \int_{2\ell\pi}^{2(\ell+1)\pi} |\widehat{\psi}(\xi)|^2 e^{-i(m-k)\xi}\, d\xi
$$

$$
= \int_{0}^{2\pi} \sum_{\ell \in \mathbb{Z}} |\widehat{\psi}(\xi + 2\ell\pi)|^2 e^{-i(m-k)\xi}\, d\xi.
$$

But this means that the Fourier coefficients of the 2π-periodic function appearing on the left in equality (2) are all zero with the exception of the constant coefficient, which equals 1. This establishes (2), from which we immediately see that the support of $\widehat{\psi}$, when ψ is a band-limited wavelet, must measure at least 2π (with equality only possible if $|\widehat{\psi}(\xi)|$ is a.e. either 0 or 1). The wavelet constructed by Lemarié and Meyer has a Fourier transform with support of measure 4π and, thus, certainly satisfies the above necessary condition.

Let

$$
\gamma_{j,k}(\xi) = \frac{2^{-j/2}}{\sqrt{2\pi}} e^{i2^{-j}k\xi}\, \widehat{\psi}(2^{-j}\xi),
$$

for $j, k \in \mathbb{Z}$. Let us assume that $\{\psi_{j,k}\}$ is an orthonormal system; then, by the Plancherel theorem, $\{\gamma_{j,k}\}$ is also an orthonormal system. We shall denote $\gamma_{0,0}$ simply by γ, so that $\gamma = \frac{\widehat{\psi}}{\sqrt{2\pi}}$. Let W_j be the closure of the span of $\{\psi_{j,k} : k \in \mathbf{Z}\}$, $\widehat{W}_j$ be the closure of the span of $\{\gamma_{j,k} : k \in \mathbb{Z}\}$, and Q_j the projection onto $\widehat{W}_j$. Then, for an $f \in L^2(\mathbb{R})$,

$$
(Q_j f)(\xi) = \sum_{k \in \mathbb{Z}} \langle f, \gamma_{j,k} \rangle \gamma_{j,k}(\xi). \tag{4}
$$

Let $g(\eta) \equiv f(\eta)\overline{\widehat{\psi}(2^{-j}\eta)}$ and $F(\xi) = \sum_{\ell \in \mathbb{Z}} g(\xi + 2^{j+1}\ell\pi)$. The last function is $2^{j+1}\pi$-periodic, and so are the functions $E_k^{(j)}(\xi) = \frac{2^{-j/2}}{\sqrt{2\pi}} e^{i2^{-j}k\xi}$, which form an orthonormal basis for $L^2([0, 2^{j+1}\pi))$, $k = 0, \pm 1, \pm 2, \ldots$. A simple "periodization" argument, like the one we just used to establish (2), shows that

$$
\langle f, \gamma_{j,k} \rangle = \langle F, E_k^{(j)} \rangle
$$

where the first inner product is the one associated with $L^2(\mathbb{R})$ and the second is the one associated with $L^2([0, 2^{j+1}\pi))$. Hence,

$$
F(\xi) = \sum_{k \in \mathbb{Z}} \langle f, \gamma_{j,k} \rangle E_k^{(j)}(\xi).
$$

More explicitly, this equality is

$$\sum_{\ell \in \mathbb{Z}} f(\xi + 2^{j+1}\ell\pi)\overline{\widehat{\psi}(2^{-j}\xi + 2\ell\pi)} = \sum_{k \in \mathbb{Z}} \langle f, \gamma_{j,k} \rangle E_k^{(j)}(\xi). \tag{5}$$

The following is a natural consequence of (5):

THEOREM 1

(A) $f \perp \widehat{W}_j$ if and only if $\displaystyle\sum_{\ell \in \mathbb{Z}} f(\xi + 2^{j+1}\ell\pi)\overline{\widehat{\psi}(2^{-j}\xi + 2\ell\pi)} = 0$ a.e.;

(B) $(Q_j f)(\xi) = \widehat{\psi}(2^{-j}\xi)\displaystyle\sum_{\ell \in \mathbb{Z}} f(\xi + 2^{j+1}\ell\pi)\overline{\widehat{\psi}(2^{-j}\xi + 2\ell\pi)}.$

PROOF   If $f \perp \widehat{W}_j$ all the inner products $\langle f, \gamma_{j,k} \rangle$, $k \in \mathbb{Z}$, are 0; thus, the expression on the left in (5) is 0 almost everywhere. On the other hand, if this last assertion is true, then all the coefficients $\langle f, \gamma_{j,k} \rangle$ in the Fourier expansion of $F(\xi)$ must be 0 (see the equality preceding (5) as well as (4)). But this tells us that $f \perp \widehat{W}_j$ and part (A) is proven.

If we multiply both sides of (5) by $\widehat{\psi}(2^{-j}\xi)$, observe that $\gamma_{j,k}(\xi) = \widehat{\psi}(2^{-j}\xi)$ $\cdot E_k^{(j)}(\xi)$, and compare with (3), we obtain the equality in part (B). $\blacksquare$

REMARK 1   An examination of the argument we gave to establish (B) shows that we do not use the orthogonality of the system $\{\widehat{\psi}_{j,k}\}$. That is, this formula is valid whenever Q_j is the operator defined by equality (4) with the functions $\gamma_{j,k}$ defined as we did by using an arbitrary function $\psi \in L^2(\mathbb{R})$. $\blacksquare$

It is a natural consequence of these considerations to obtain the following characterization of those functions $\psi \in L^2(\mathbb{R})$ that generate an orthonormal system $\{\psi_{j,k}\}$, $j, k \in \mathbb{Z}$:

THEOREM 2

If $\psi \in L^2(\mathbb{R})$ then the functions $\psi_{j,k}(x) = 2^{j/2}\psi(2^j x + k)$, $j, k \in \mathbb{Z}$, form an orthonormal system if and only if

(A) $\displaystyle\sum_{\ell \in \mathbb{Z}} |\widehat{\psi}(\xi + 2\ell\pi)|^2 = 1$

for a. e. ξ, *and*

(B) $\displaystyle\sum_{\ell \in \mathbb{Z}} \widehat{\psi}(\xi + 2^{j+1}\ell\pi)\overline{\widehat{\psi}(2^{-j}\xi + 2\ell\pi)} = 0$

for a.e. ξ *whenever* $j > 0$.

PROOF   We have already noted that condition (A) is equivalent to the orthonormality of the system $\{\psi_{0,k}\}, k \in \mathbb{Z}$ (see equality (2)). If we write out the corresponding orthonormality relations and perform the change of variables $\xi \to 2^j \xi$, we obtain the orthonormality relations for the system $\{\psi_{j,k}\}, k \in \mathbb{Z}$. This shows that condition (A) is equivalent to the orthonormality of each of the systems $\{\psi_{j,k}\}, k \in \mathbb{Z}$, that span the subspaces W_j. We shall show that condition (B) is equivalent to the fact that the spaces $\widehat{W}_j$ are mutually orthogonal.

We first observe that these spaces are mutually orthogonal if and only if $\widehat{W}_0 \perp \widehat{W}_j$ for each $j > 0$. In fact, we have $\rho^j W_q = W_{q+j}$, where ρ is the unitary operator $(\rho f)(x) = \sqrt{2} f(2x)$. If $n = m + j, j \neq 0$, applying ρ^m to W_q when $q = 0$ and $q = j$, we see that $W_m \perp W_n$ if and only if $W_0 \perp W_j$. There is no loss in generality if we assume that, say, $m < n$; thus $n = m + j$ with $j > 0$. To obtain a condition equivalent to this last orthogonality relation, we can apply condition (A) of Theorem 1 to each of the functions $f(\xi) = \sqrt{2\pi} \gamma_{0,h}(\xi) = e^{ih\xi} \widehat{\psi}(\xi), h \in \mathbb{Z}$, that span $\widehat{W}_0$ and obtain

$$\sum_{\ell \in \mathbb{Z}} e^{ik(\xi + 2^{j+1}\ell\pi)} \widehat{\psi}(\xi + 2^{j+1}\ell\pi)\overline{\widehat{\psi}(2^{-j}\xi + 2\ell\pi)}$$

$$= e^{ik\xi} \sum_{\ell \in \mathbb{Z}} \widehat{\psi}(\xi + 2^{j+1}\ell\pi)\overline{\widehat{\psi}(2^{-j}\xi + 2\ell\pi)} = 0$$

for almost every $\xi \in \mathbb{R}$ (the fact that j is a positive integer allows us to use the 2π-periodicity of $e^{ik\xi}$). Since $e^{ik\xi}$ is never 0 we obtain condition (B).   ∎

REMARK 2   If we replace ξ in equality (B) of Theorem 2 by $2^j \eta$ we can consolidate the two conditions in this theorem into the one equality

$$\sum_{\ell \in \mathbb{Z}} \widehat{\psi}(2^j[\eta + 2\ell\pi])\overline{\widehat{\psi}(\eta + 2\ell\pi)} = \delta_{0,j} \tag{6}$$

for a.e. η whenever $j \geq 0$.   ∎

Under certain hypotheses on the support of $\widehat{\psi}$, we shall see that the orthonormality of the system $\{\psi_{j,k}\}, j, k \in \mathbb{Z}$, implies the completeness of the system. This is easily recognized as generally false. For example, if $\{\psi_{j,k}\}$ is orthonormal, and $\alpha(x) \equiv \sqrt{2}\psi(2x)$, then a simple calculation shows that

$$\alpha_{j,k} = \psi_{j+1,2k}.$$

Thus, $\{\alpha_{j,k}\}$ is an orthonormal system that cannot be complete. In the next section we shall give necessary and sufficient conditions for the completeness of such systems when ψ is band-limited. In the third section we shall consider a class of wavelets that includes the one introduced by Lemarié and Meyer and show how it is connected with the local sine/cosine bases we mentioned above. We shall also show that any such system in easily seen to be derived from a *multiresolution analysis* (MRA). Many of the equalities we develop in these first two sections have appeared elsewhere; in particular, in the work of Lemarié one can find much of this material, including the important formula (14) (see [L₁]). Perhaps our treatment is more direct than what is found in the literature, and some of the material we develop probably appears here explicitly for the first time: the formula for the projection in Theorem 1 and the existence of an interval about the origin that is disjoint from the support of the Fourier transform of a band-limited wavelet are examples of "new" results. The material in the third section, however, does contain some novel features: this applies to the conditions in Theorem 7 that are equivalent to the orthonormality and completeness of the system discussed, and we make some observations concerning the connection of these bases with the local sine/cosine bases. In the fourth section we give further applications of the results obtained, a proof, in the band-limited case, of Auscher's theorem that shows that $H^2(\mathbb{R})$ cannot have a (smooth) wavelet basis, and we show a connection between the local cosine/sine bases of Coifman and Meyer and the Wilson bases. At the end of this article we shall present a more extensive discussion of the existing literature and its relation to this presentation. We have chosen to concentrate on the band-limited case since this presentation is intended to have an expository aspect. Since many of our results involve the Fourier transform of wavelets, the band-limited case avoids some difficulties (often concerning the meaning of certain infinite sums) that one encounters in the general case. We repeat, however, that many of the results obtained are valid more generally, and we indicate when this is the case.

3.2 Necessary and sufficient conditions for the completeness of a band-limited wavelet system

Hereafter we shall assume that $\widehat{\psi}$ is compactly supported. Thus, there exists a positive integer J such that sup $\widehat{\psi} \subset [-2^J\pi, 2^J\pi]$. All the results we shall consider are valid when ψ is not band-limited; however, as we shall see this assumption does simplify many of the arguments and formulae that we will present.

We begin by proving the following theorem, in which we shall find one of the two conditions characterizing completeness:

THEOREM 3
If $\widehat{\psi}$ has compact support and $\{\psi_{j,k}\}$ is an orthonormal basis of $L^2(\mathbb{R})$, then

$$\sum_{j\in\mathbb{Z}} |\widehat{\psi}(2^j\xi)|^2 = 1 \tag{7}$$

for all $\xi \neq 0$.

We begin by establishing a lemma from which the above theorem will follow rather easily:

LEMMA 1

Suppose $f \in L^2(\mathbb{R})$ is supported in the interval $I = (a, b)$, where $b - a \leq 2^{-J}\pi$ and $I \cap [-\pi, \pi] = \emptyset$, then

$$(Q_j f)(\xi) = f(\xi)|\widehat{\psi}(2^{-j}\xi)|^2 \tag{8}$$

whenever $\xi \in I$.

PROOF We first observe that if $-j \leq J, \ell \neq 0$ and $\xi \in I$, then $\xi + 2^{j+1}\ell\pi$ lies outside the support of f. Thus, in this case, equality (B) in Theorem 1 reduces to (8). If $-j > J$ and $\xi \in I$, then $|2^{-j}\xi| \geq 2^J|\xi| \geq 2^J\pi$ and $\widehat{\psi}(2^{-j}\xi) = 0$. Hence, again using equality (B) in Theorem 1, we have $(Q_j f)(\xi) = 0 = f(\xi)|\widehat{\psi}(2^{-j}\xi)|^2$, which is the desired equality (8). ∎

PROOF OF THEOREM 3 We first observe that $w(\xi) \equiv \sum_{j \in \mathbb{Z}} |\widehat{\psi}(2^{-j}\xi)|^2$ satisfies $w(2^n\xi) = w(\xi)$ for each $n \in \mathbb{Z}$. Thus, if we show that (7) holds when $0 < \pi < |\xi| \leq 2\pi$, then, by this observation, we can deduce that (7) holds for a.e. $\xi \in \mathbb{R}$. We shall, in fact, show that (7) holds for a.e. $\xi \in I$, where I satisfies the hypotheses of Lemma 1; the set $\{\xi \in \mathbb{R} : \pi \leq |\xi| \leq 2\pi\}$ is the union of 2^{J+1} such intervals (up to a finite number of points). From Lemma 1 we see that for such an interval I

$$\int_I \left| \sum_{j=-M}^{M} (Q_j f)(\xi) \right|^2 d\xi = \int_I |f(\xi)|^2 \left(\sum_{j=-M}^{M} |\widehat{\psi}(2^{-j}\xi)|^2 \right)^2 d\xi \tag{9}$$

when $\text{sup } f \subset I$. But the Q_j's are mutually orthogonal projections; thus, the left side of (9) is dominated by

$$\left\| \sum_{j=-M}^{M} Q_j f \right\|_{L^2(\mathbb{R})}^2 \leq \|f\|_{L^2(\mathbb{R})}^2 = \int_I |f(\xi)|^2 d\xi.$$

Hence,

$$\int_I |f(\xi)|^2 \left(\sum_{j=-M}^{M} |\widehat{\psi}(2^{-j}\xi)|^2 \right)^2 d\xi \leq \int_I |f(\xi)|^2 d\xi$$

whenever $\text{sup } f \subset I$. It follows that

$$\sum_{j=-M}^{M} |\widehat{\psi}(2^{-j}\xi)|^2 \leq 1 \qquad a.e. \tag{10}$$

when $\xi \in I$. On the other hand, since $\{\gamma_{j,k}\}$ is an orthonormal basis

$$\lim_{M \to \infty} \left\| \sum_{j=-M}^{M} Q_j f - f \right\|_2 = 0.$$

But this implies

$$\int_I |f(\xi)|^2 \left(1 - \sum_{j=-M}^{M} |\widehat{\psi}(2^{-j}\xi)|^2 \right)^2 d\xi = \int_I \left| f(\xi) - \sum_{j=-M}^{M} (Q_j f)(\xi) \right|^2 d\xi$$

$$\leq \left\| \sum_{j=-M}^{M} Q_j f - f \right\|_2^2 \to 0 \text{ as } M \to \infty$$

whenever sup $f \subset I$. This and (10) give us equality (7) for a.e. $\xi \in I$. As indicated at the beginning of this proof, this gives us (7) for a.e. $\xi \in \mathbb{R}$. ∎

One can adapt the above proof to show that the conclusion of Theorem 3 holds under the weaker hypothesis that there exists $\beta > 0$ such that

$$\int_{-\infty}^{\infty} |\widehat{\psi}(\xi)|^2 |\xi|^{\beta} d\xi < \infty.$$

We shall now study the behavior of $\widehat{\psi}$ near the origin.

THEOREM 4
If $\widehat{\psi}$ is continuous and $\{\psi_{j,k}\}$ is an o.n. basis, then $\widehat{\psi}(\xi) = 0$ in a neighborhood of the origin.

We begin by showing $\widehat{\psi}(0) = 0$:

LEMMA 2
If $\widehat{\psi}$ is continuous at 0 and $\{\psi_{j,k}\}$ is an orthonormal system, then $\widehat{\psi}(0) = 0$.

PROOF   Since $\widehat{\psi} \perp \widehat{W}_j$, $j \neq 0$, part (A) of Theorem 1 tells us that

$$\sum_{\ell \in \mathbb{Z}} \widehat{\psi}(\xi + 2^{j+1}\ell\pi)\overline{\widehat{\psi}(2^{-j}\xi + 2\ell\pi)} = 0 \tag{11}$$

a.e. when $j \neq 0$. Since sup $\widehat{\psi} \subset [-2^J\pi, 2^J\pi]$, $\xi + 2^{j+1}\ell\pi$ lies outside of the support of $\widehat{\psi}$ when $\ell \neq 0, \xi \in \text{sup } \widehat{\psi}$ and $J \leq j$. Thus, (11) reduces   to

$\widehat{\psi}(\xi)\overline{\widehat{\psi}(2^{-j}\xi)} = 0$ when $J \leq j$ and $\xi \in \sup \widehat{\psi}$. Obviously this equality is also true when $\xi \notin \sup \widehat{\psi}$; thus,

$$\widehat{\psi}(\xi)\overline{\widehat{\psi}(2^{-j}\xi)} = 0 \tag{12}$$

for all $\xi \in \mathbb{R}$ when $j \geq J$. Letting $j \to \infty$ we obtain $\widehat{\psi}(\xi)\overline{\widehat{\psi}(0)} = 0$, for all $\xi \in \mathbb{R}$. Since $\widehat{\psi}$ is not identically 0, this means $\widehat{\psi}(0) = 0$. ∎

PROOF OF THEOREM 4   (12) implies that $\widehat{\psi}(2^j\xi) = 0$ when $|j| \geq J$ and $\widehat{\psi}(\xi) \neq 0$. Thus, from Theorem 3 we must have

$$\sum_{|j|<J} |\widehat{\psi}(2^j\xi)|^2 = 1,$$

and it follows that there exists an integer $j_0 \in (-J, J)$ such that

$$|\widehat{\psi}(2^{j_0}\xi)| \geq \left(\frac{1}{2J-1}\right)^{1/2}.$$

By Lemma 2, there exists $\epsilon > 0$ such that

$$|\widehat{\psi}(\eta)| \leq \frac{1}{2}\left(\frac{1}{2J-1}\right)^{1/2}$$

if $|\eta| < \epsilon$. Thus, $\epsilon \leq |2^{j_0}\xi| \leq 2^J|\xi|$. This shows that, if $\widehat{\psi}(\xi) \neq 0$, then $|\xi| \geq \epsilon 2^{-J}$ and, consequently, $\widehat{\psi}(\xi) = 0$ for $\xi \in (-\epsilon 2^{-J}, \epsilon 2^{-J})$. ∎

REMARK 3   The conclusion of Lemma 2 holds when $\widehat{\psi}$ is much more general. For example, if, for an $\alpha > \frac{1}{2}$

$$|\widehat{\psi}(\xi)| \leq (1 + |\xi|)^{-\alpha},$$

then, instead of (12), one can show that $\widehat{\psi}(\xi)\overline{\widehat{\psi}(2^{-j}\xi)} = O(2^{-j\alpha})$ as $j \to \infty$ and the rest of the proof is valid. In fact, a somewhat more complicated argument gives this conclusion if $\widehat{\psi}$ is assumed only to be square integrable, continuous at 0, and $\{\psi_{j,k}\}$ is an orthonormal system. We shall present this argument elsewhere. ∎

By Theorem 4 we see that under these assumptions we can find a positive integer J such that

$$\sup \widehat{\psi} \subset \{\xi : 2^{-J}\pi \leq |\xi| \leq 2^J\pi\}. \tag{13}$$

It follows that the series in (7) is finitely non-zero under these assumptions.

In the following lemma we introduce a second condition, that, together with (7), will be shown to be equivalent to the completeness of the system $\{\psi_{j,k}\}$.

LEMMA 3

If $\{\psi_{j,k}\}$ is a complete orthonormal system satisfying (13), then for each $m \in \mathbb{Z}$ and a.e. ξ,

$$\sum_{j=0}^{\infty} \widehat{\psi}(2^j \xi) \overline{\widehat{\psi}(2^j(\xi + (2m+1)2\pi))} = 0. \tag{14}$$

(observe that this last sum has only finitely many non-zero terms).

PROOF   By Theorem 1, part (B),

$$(Q_j f)(\xi) = \widehat{\psi}(2^{-j}\xi) \sum_{\ell \in \mathbb{Z}} f(\xi + 2^j \ell 2\pi) \overline{\widehat{\psi}(2^{-j}\xi + 2\ell\pi)}.$$

Suppose $0 < a < |\xi| < b$, then by (13) we see that there are only a finite number of integers j (say, $|j| \le M$) such that $\widehat{\psi}(2^{-j}\xi) \ne 0$. For each such j there are at most a finite number of indices ℓ such that $\widehat{\psi}(2^{-j}\xi + 2\ell\pi) \ne 0$. Thus, there are only a finite number of values of the form $v = 2^j \ell$ that involve non-zero terms in the above sum: $v = 0$ could be one such term, the others must have the form $2^n(2m+1)$ for a finite number of $m, n \in \mathbb{Z}, n \ge j$. Thus,

$$(Q_j f)(\xi) = f(\xi)|\widehat{\psi}(2^{-j}\xi)|^2 + \widehat{\psi}(2^{-j}\xi) \sum_{n \ge j}$$

$$\cdot \sum_{m \in \mathbb{Z}} f(\xi + 2^n(2m+1)2\pi) \overline{\widehat{\psi}(2^{-j}[\xi + 2^n(2m+1)2\pi])}.$$

Thus, summing over j, we have, since $\{\psi_{j,k}\}, j, k \in \mathbb{Z}$ is a complete system,

$$f(\xi) = \sum_{j \in \mathbb{Z}} (Q_j f)(\xi)$$

$$= \sum_{|j| \le M} (Q_j f)(\xi)$$

$$= f(\xi) \sum_{|j| \le M} |\widehat{\psi}(2^{-j}\xi)|^2 + \sum_{m,n \in \mathbb{Z}} f(\xi + 2^n(2m+1)2\pi)$$

$$\cdot \sum_{j \le n} \widehat{\psi}(2^{-j}\xi) \overline{\widehat{\psi}(2^{-j}[\xi + 2^n(2m+1)2\pi])}$$

$$= f(\xi) + \sum_{m,n \in \mathbb{Z}} f(\xi + 2^n(2m+1)2\pi)$$

$$\cdot \sum_{j \le n} \widehat{\psi}(2^{-j}\xi) \overline{\widehat{\psi}(2^{-j}[\xi + 2^n(2m+1)2\pi])}$$

since $\sum_{|j| \le M} |\widehat{\psi}(2^{-j}\xi)|^2 = \sum_{j \in \mathbb{Z}} |\widehat{\psi}(2^{-j}\xi)|^2 = 1$ (all these sums being finite).

Thus,

$$\sum_{m,n\in\mathbb{Z}} f(\xi + 2^n(2m+1)2\pi)$$

$$\cdot \sum_{j\le n} \widehat{\psi}(2^{-j}\xi)\overline{\widehat{\psi}(2^{-j}[\xi + 2^n(2m+1)2\pi])} = 0 \qquad (15)$$

when $\{\xi : a < |\xi| < b\}$. Since the finite number of values v of the form $2^n(2m+1)$ in the above sums are distinct, for each $\xi_0 \in \{\xi : a \le |\xi| \le b\}$ we can find a small neighborhood $\aleph$ of ξ_0 such that $U \equiv \aleph + v$ contains no point of the form $\xi_0 + v'$ if $v' \ne v$. If we let $f = \chi_U$, the characteristic function of U, in equality (15), we obtain

$$\sum_{j\le n} \widehat{\psi}(2^{-j}\xi)\overline{\widehat{\psi}(2^{-j}[\xi + 2^n(2m+1)2\pi])} = 0, \qquad (16)$$

whenever $\xi \in U$. Since ξ_0 is an arbitrary point of $(-b, -a) \cup (a, b)$, the last equality must hold whenever $a < |\xi| < b$. Since the positive numbers a and b are arbitrary, this gives us equality (16) for all $\xi \ne 0$ and $m, n \in \mathbb{Z}$. In particular, taking $n = 0$ and changing $-j$ to j we obtain equality (14) for all ξ. When $\xi = 0$ this equality is an immediate consequence of (13). ∎

We shall now show that equalities (7) and (14) are not only consequences of the completeness of the system $\{\psi_{j,k}\}$, $j, k \in \mathbb{Z}$, but their validity implies that the system is complete.

THEOREM 5

Suppose ψ is a band-limited function satisfying (13), equality (14) and $\{\psi_{j,k}\}$ is an o.n. system, then $\{\psi_{j,k}\}$ is a complete orthonormal system.

PROOF Observe that $\widehat{\psi}$ satisfies (13). Then, as in the proof of Lemma 3, for $f \in L^2(\mathbb{R})$,

$$(Q_j f)(\xi) = f(\xi)|\widehat{\psi}(2^{-j}\xi)|^2 + \widehat{\psi}(2^{-j}\xi) \sum_{m\in\mathbb{Z}}$$

$$\cdot \overline{\sum_{n\ge j} f(\xi + 2^n(2m+1)2\pi)\widehat{\psi}(2^{-j}[\xi + 2^n(2m+1)2\pi])}$$

let us restrict this equality to a compact set E. The fact that $\widehat{\psi}$ satisfies (13), guarantees, for $\xi \in E$, that $|\widehat{\psi}(2^{-j}\xi)|^2$ and, thus, $(Q_j f)(\xi)$ is non-zero for only a finite number of indices j (say, $|j| \leq M$) and, for each such j, only a finite number of non-zero terms appear in the last sums. But, if we sum over j, and apply (7)

$$\sum_{j\in\mathbb{Z}}(Q_j f)(\xi) = \sum_{|j|\leq M}(Q_j f)(\xi)$$

$$= f(\xi)\sum_{|j|\leq M}|\widehat{\psi}(2^{-j}\xi)|^2 + \sum_{m,n\in\mathbb{Z}}f(\xi + 2^n(2m+1)2\pi)$$

$$\cdot\sum_{j\leq n}\widehat{\psi}(2^{-j}\xi)\overline{\widehat{\psi}(2^{-j}[\xi + 2^n(2m+1)2\pi])}$$

$$= f(\xi)\cdot 1 + \sum_{m,n\in\mathbb{Z}}f(\xi + 2^n(2m+1)2\pi)$$

$$\cdot\sum_{j\leq n}\widehat{\psi}(2^{-j}\xi)\overline{\widehat{\psi}(2^{-j}[\xi + 2^n(2m+1)2\pi])}.$$

But, by (14), the last sum is 0 (write $\xi = 2^n\eta$ and $j+ = n - j$); thus, $\sum_{j\in\mathbb{Z}}(Q_j f)(\xi) = f(\xi)$ for all $\xi \in E$. Since E is an arbitrary compact subset of $\mathbb{R} - \{0\}$, we immediately obtain the desired completeness. ∎

If we unite Theorem 3, Lemma 3, and Theorem 5, we obtain the following characterization of the band-limited orthonormal systems $\{\psi_{j,k}\}$ that form an orthonormal basis for $L^2(\mathbb{R})$:

THEOREM 6
If ψ is a band-limited function in $L^2(\mathbb{R})$ such that $\{\psi_{j,k}\}$ is an orthonormal system and $|\widehat{\psi}(\xi)|^2 \leq C|\xi|$ for $|\xi| \leq 1$, then this system is complete if and only if

$$\sum_{j\in\mathbb{Z}}|\widehat{\psi}(2^j\xi)|^2 = 1$$

for all $\xi \neq 0$ and, for each $m \in \mathbb{Z}$,

$$\sum_{j\geq 0}\widehat{\psi}(2^j\xi)\overline{\widehat{\psi}(2^j(\xi + (2m+1)2\pi))} = 0.$$

Each of these sums involves only a finite number of non-zero terms for each $\xi \in \mathbb{R}$.

To complete our discussion of completeness and orthonormality, let us make a rather simple observation concerning the operator

$$Qf = \sum_{j,k\in\mathbb{Z}}\langle f, \gamma_{j,k}\rangle\gamma_{j,k}$$

where $\psi \in L^2(\mathbb{R})$ and γ is defined in terms of ψ as in Section 1 (of this chapter). Such operators are sometimes called "frame operators." And it is defined in the

L^2 sense if $\{\gamma_{j,k}\}$ is a frame (see [D]). We remarked (after the proof of Theorem 1) that representation (B) in Theorem 1 for each of the operators

$$Q_j f = \sum_{k \in \mathbb{Z}} \langle f, \gamma_{j,k} \rangle \gamma_{j,k}$$

did not require that the system $\{\psi_{j,k}\}$ be orthonormal. If, for example, ψ satisfies (13), the last series is defined as a finitely convergent pointwise series for each ξ (and the same is true for the series defining Q). Let us point out that if $\|\psi\|_2 = 1$ and Q is the identity operator, then the system $\{\psi_{j,k}\}$, or equivalently, $\{\gamma_{j,k}\}$, is orthonormal. This is a consequence of the following computation:

$$1 = \langle \gamma_{m,n}, \gamma_{m,n} \rangle = \left\langle \sum_{j,k \in \mathbb{Z}} \langle \gamma_{m,n}, \gamma_{j,k} \rangle \langle \gamma_{j,k}, \gamma_{m,n} \rangle \right\rangle = \sum_{j,k \in \mathbb{Z}} |\langle \gamma_{m,n}, \gamma_{j,k} \rangle|^2.$$

Since the last sum contains the term $\langle \gamma_{m,n}, \gamma_{m,n} \rangle = 1$, the other non-negative terms must all be zero; but this means that our system is orthonormal.

A careful inspection of the argument we gave to establish Theorem 5 shows that we do not need to use the orthonormality of the system $\{\gamma_{j,k}\}$ to conclude that the operator Q equals the identity. Thus, equalities (7) and (14) and the simple hypothesis that $\|\psi\|_2 = 1$, not only imply completeness, but they also give us orthonormality.

3.3 The systems of Lemarié and Meyer

In this section we shall study wavelet systems that are "strongly" band-limited. We shall also show how they are connected with the one introduced by Lemarié and Meyer that we described at the beginning of this chapter. We shall give a complete characterization of them and show how they are related to the local sine/cosine bases of Coifman and Meyer.

Let us begin by proving the following theorem:

THEOREM 7
Suppose $\psi \in L^2(\mathbb{R})$ and $b = |\widehat{\psi}|$ is an even continuous function with support equal to $[-2\pi - \epsilon', -\pi + \epsilon] \cup [\pi - \epsilon, 2\pi + \epsilon']$, where $0 < \epsilon, \epsilon'$ and $\epsilon + \epsilon' \leq \pi$. Then $\{\psi_{j,k}\} \equiv \{2^{-j/2}\psi(2^{-j}x - k)\}$, $j, k \in \mathbb{Z}$, is an orthonormal system if and only if

(i) $2\epsilon = \epsilon'$;
(ii) $b(\pi + \xi) = b(2(\pi - \xi))$ *if* $\xi \in [-\epsilon, \epsilon]$;
(iii) $b^2(\pi + \xi) + b^2(\pi - \xi) = 1$ *if* $\xi \in [-\epsilon, \epsilon]$;
(iv) $b(\xi) = 1$ *if* $\xi \in [\pi + \epsilon, 2(\pi - \epsilon)]$;
(v) $\widehat{\psi}(\xi) = e^{i\alpha(\xi)}b(\xi)$,

where the phase function α satisfies

$$\alpha(\xi + \pi) - \alpha(2(\xi + \pi)) - \alpha(\xi - \pi) + \alpha(2(\xi - \pi)) = (2m + 1)\pi$$

for some $m = m(\xi) \in \mathbb{Z}$ for all ξ in $[-\epsilon, \epsilon]$.

Moreover, if $\{\psi_{j,k}\}$ is such an orthonormal system, then it is a basis for $L^2(\mathbb{R})$.

REMARK 4 The "simplest" solution of the functional equation in part (v) of Theorem 7 is $\alpha(\xi) = \frac{\xi}{2}$, which is the phase used by Lemarié and Meyer (see equality (1)). Other solutions are given by $\alpha(\xi) = \left(\frac{\xi}{2}\right) + \beta(\xi)$, where β is 2π-periodic (all of which, clearly, give rise to the same spaces W_j). Another solution is $\alpha(\xi) = \left(\frac{\xi}{2}\right) - \frac{\pi(sgn\xi)}{2}$. This gives us the wavelets ψ whose Fourier transform is $-i(sgn\xi)e^{i\xi/2}b(\xi)$, the Hilbert transform of the wavelet ψ such that $\widehat{\psi} = e^{i\xi/2}b(\xi)$. The most general solution is easy to find: choose any function α_0 on the complement of $[-2\pi - 2\epsilon, -2\pi] \cup [2\pi, 2\pi + 2\epsilon]$, then the functional equation leads us to define $\alpha(\xi) \equiv \alpha_0\left(\frac{\xi}{2}\right) - \alpha_0\left(\left[\frac{\xi}{2}\right] - 2\pi\right) + \alpha_0(\xi - 4\pi) - (2m + 1)\pi$ on $[2\pi, 2\pi + 2\epsilon]$ and $\alpha(\xi) \equiv \alpha_0\left(\frac{\xi}{2}\right) - \alpha_0\left(\left[\frac{\xi}{2}\right] + 2\pi\right) + \alpha_0(\xi + 4\pi) + (2m + 1)\pi$ on $[-2\pi - 2\epsilon, -2\pi]$. De Michele pointed out to us that this functional equation was studied by Paganoni [P]. ∎

PROOF Let us rewrite condition (B) of Theorem 2 when $j = 1$, taking into account the extra assumptions about $\widehat{\psi}$ that we are making in Theorem 7. Replacing ξ by 2ξ, this condition becomes

$$\sum_{\ell \in \mathbb{Z}} \widehat{\psi}(2(\xi + 2\ell\pi))\overline{\widehat{\psi}(\xi + 2\ell\pi)} = 0$$

for a.e. ξ. Since the left side of this equality represents a 2π-periodic function, it suffices to consider this equation when $\xi \in [0, 2\pi]$. In this case $2(\xi + 2\ell\pi)$ is outside the support of $\widehat{\psi}$ when $\ell < -1$ and when $\ell > 0$. Thus, we have, using the continuity of $\widehat{\psi}$,

$$\widehat{\psi}(\xi)\overline{\widehat{\psi}(2\xi)} + \widehat{\psi}(\xi - 2\pi)\overline{\widehat{\psi}(2(\xi - 2\pi))} = 0 \qquad (17)$$

for all $\xi \in [0, 2\pi]$. Let $g(\xi) = \widehat{\psi}(\xi)\overline{\widehat{\psi}(2\xi)}$. Then Supp $g = \left\{\xi : \pi - \epsilon \leq |\xi| \leq \pi + \left(\frac{\epsilon'}{2}\right)\right\}$ and, as a consequence of (17), we have $|g(\xi)| = |g(2\pi - \xi)|$ since $|g|$ is an even function. But this last equality tells us that $|g(\xi)|$ is an even function with respect to π (observe that ξ and $2\pi - \xi$ are equidistant from π and on opposite sides of π). Hence, $\epsilon = \frac{\epsilon'}{2}$ and (i) is proved.

Condition (A) of Theorem 2 asserts that the 2π-periodic function $\sum_{\ell \in \mathbb{Z}} b^2 \cdot (\xi + 2\ell\pi)$ equals 1. On the period $[-2\epsilon, 2\pi - 2\epsilon]$ this assertion is equivalent to the four equalities:

(a) $b^2(\xi - 2\pi) + b^2(\xi + 2\pi) = 1$ when $\xi \in [-2\epsilon, 2\epsilon]$,

(b) $b(\xi - 2\pi) = 1$ when $\xi \in [2\epsilon, \pi - \epsilon]$,

(c) $b^2(\xi - 2\pi) + b^2(\xi) = 1$ when $\xi \in [\pi - \epsilon, \pi + \epsilon]$,

(d) $b(\xi) = 1$ when $\xi \in [\pi + \epsilon, 2\pi - 2\epsilon]$.

Each of (b) or (d) is, of course, equality (iv). Equality (iii) is equivalent to (c) (after the change of variables $\xi = \eta + \pi$). If we rewrite (a) in the form

(a') $b^2(2(\xi - \pi)) + b^2(2(\xi + \pi)) = 1$ when $\xi \in [-\epsilon, \epsilon]$,

then (a'), (iii) and the equality

(e) $b(\xi + \pi)b(2(\xi + \pi)) = b(\xi - \pi)b(2(\xi - \pi))$ for $\xi \in [-\pi, \pi]$

immediately imply (ii) ((e), of course, is equivalent to $|g(\xi)| = |g(2\pi - \xi)|$ when $\xi \in [0, 2\pi]$, which was established immediately following (17)).

The only remaining problem is to derive the functional equation (v). We claim that the system of two vectors $(\widehat{\psi}(\xi + \pi), \widehat{\psi}(\xi - \pi))$ and $(\widehat{\psi}(2[\xi + \pi])$, $\widehat{\psi} \cdot (2[\xi - \pi]))$ is orthonormal for $\xi \in [-\epsilon, \epsilon]$: equalities (a') and (iii) give us the normality and (17) implies the orthogonality. Thus, there exists $\delta(\xi)$ such that

$$e^{i\delta(\xi)}(b(\xi + \pi)e^{i\alpha(\xi + \pi)}, \ b(\xi - \pi)e^{i\alpha(\xi - \pi)})$$
$$= (-b(2[\xi - \pi])e^{i\alpha(2[\xi - \pi])}, \ b(2[\xi + \pi])e^{i\alpha(2[\xi + \pi])})$$

for $|\xi| \leq \epsilon$. Since $b(\xi + \pi) = b(2[\xi - \pi])$, $b(\xi - \pi) = b(2[\xi + \pi])$ (by (ii)) and $[\pi - \epsilon, 2\pi + 2\epsilon]$ lies in the support of b, we must have

$$e^{i\{\alpha(\xi + \pi) - \alpha(2[(\xi - \pi)]) - \pi\}} = e^{i\{\alpha(\xi - \pi) - \alpha(2[\xi + \pi])\}}$$

and (v) follows (more precisely, the above equality must hold when the cited values of b are not 0; if they are 0, we might as well choose α at these points so that (v) is satisfied).

Let us now turn to the proof of the sufficiency of these conditions. We remark that the equality of Supp $\widehat{\psi}$ and the set $[-2\pi - 2\epsilon, -\pi + \epsilon] \cup [\pi - \epsilon, 2\pi + 2\epsilon]$ was only used in the very last argument, and the extremality of ϵ, ϵ' is used in the proof of (i). For the sufficiency we only need the inclusion Supp $\widehat{\psi} \subset [-2\pi - 2\epsilon, -\pi + \epsilon] \cup [\pi - \epsilon, 2\pi + 2\epsilon]$. Let us then suppose that this last relation is true and $\widehat{\psi} = e^{i\alpha}b$ satisfies (ii), (iii), (iv), and (v). We shall show that $\{\widehat{\psi}_{j,k}\}$ is an orthonormal system. Using Theorem 2, the orthonormality of the system $\{\psi_j k\}$ will be demonstrated if we can establish equality (6). An easy calculation shows that if $j \geq 2$ then either $2^j(\eta + 2\ell\pi)$ or $\eta + 2\ell\pi$ lies outside the support of $\widehat{\psi}$ if, for instance, η is in the period interval $[2\epsilon, 2\pi + 2\epsilon]$. Thus, we need to establish only the cases $j = 0$ and $j = 1$ of (6). But we have already shown that the case $j = 1$ is equivalent to (17) for ξ in the period interval $[0, 2\pi]$. This equality is equivalent to

$$\widehat{\psi}(\xi + \pi)\overline{\widehat{\psi}(2[\xi + \pi])} + \widehat{\psi}(\xi - \pi)\overline{\widehat{\psi}(2[\xi - \pi])} = 0, \tag{18}$$

for $\xi \in [-\pi, \pi]$. But if $\xi \in [-\pi, -\epsilon] \cup [\epsilon, \pi]$, an immediate calculation shows that one of the factors in each of two products is 0. For $\xi \in [-\epsilon, \epsilon]$, we can reverse the argument we just used to establish (v); it is easily seen that (v), together with (ii), (iii), and (iv), give us (18) on $[-\epsilon, \epsilon]$.

The case $j = 0$ is also a consequence of an argument that involves retracing the steps of an argument we have already given: we need only show that equalities (a), (b), (c), and (d) hold in the intervals $[-2\epsilon, 2\epsilon]$, $[2\epsilon, \pi - \epsilon]$, $[\pi - \epsilon, \pi + \epsilon]$, and $[\pi + \epsilon, 2\pi - 2\epsilon]$. But this was essentially done when we established (ii), (iii), and (iv).

Finally, we must show that any orthonormal system $\{\psi_{j,k}\}$, $j, k \in \mathbb{Z}$, with Supp $\widehat{\psi} = [-2\pi - 2\epsilon, -\pi + \epsilon] \cup [\pi - \epsilon, 2\pi + 2\epsilon]$ and $|\widehat{\psi}|$ even must be complete. Suppose, then, that $f \in L^2(\mathbb{R})$ is orthogonal to all $\widehat{\psi}_{j,k}$. We claim that $f(\xi) = 0$ for a.e. $\xi \in \mathbb{R}$. Since $\langle f, \widehat{\psi}_{j,k} \rangle = 0$ is equivalent to $\langle \widetilde{f}, \widehat{\psi}_{j,k} \rangle = 0$, where $\widetilde{g}(\xi) \equiv g(-\xi)$, it follows that it suffices to show that $f(\xi) = 0$ for a.e. $\xi > 0$ (note that $\widetilde{\widehat{\psi}}$ is an orthonormal wavelet if and only if ψ is an orthonormal wavelet and $\widetilde{\widehat{\psi}} = \widehat{\widetilde{\psi}}$). Since $g(\xi) \equiv f(2^m \xi)$ also satisfies $\langle g, \widehat{\psi}_{j,k} \rangle = 0$ for all $j, k \in \mathbb{Z}$, if we show that $f(\xi) = 0$ for a.e. $\xi \in [\pi - \epsilon, 2(\pi - \epsilon)]$, then $f(\xi) = 0$ for a.e. $\xi \in [2^m(\pi - \epsilon), 2^{m+1}(\pi - \epsilon)]$ for all $m \in \mathbb{Z}$. Thus we have reduced the problem to showing $f(\xi) = 0$ for $\xi \in [\pi - \epsilon, 2(\pi - \epsilon)]$. The orthogonality $f \perp \widehat{W}_0$ can be expressed by equality (A) in Theorem 1 with $j = 0$. When $\xi \in [\pi + \epsilon, 2(\pi - \epsilon)]$, all points of the form $\xi + 2\ell\pi$, $\ell \neq 0$, lie outside the support of $\widehat{\psi}$ and, by (iv), $\widehat{\psi}(\xi) = 1$. This shows that $f(\xi) = 0$ for $\xi \in [\pi + \epsilon, 2(\pi - \epsilon)]$. For $\xi \in [\pi - \epsilon, (\pi + \epsilon)]$ equality (A) for $j = 0$ becomes $f(\xi - 2\pi)\overline{\widehat{\psi}(\xi - 2\pi)} + f(\xi)\overline{\widehat{\psi}(\xi)} = 0$ which is equivalent to

$$f(\xi - \pi)\overline{\widehat{\psi}(\xi - \pi)} + f(\xi + \pi)\overline{\widehat{\psi}(\xi + \pi)} = 0 \tag{19}$$

when $\xi \in [-\epsilon, \epsilon]$. But in this range, the orthogonality $f \perp \widehat{W}_{-1}$, which can be expressed by equality (A) of Theorem 1 when $j = -1$, is equivalent to

$$f(\xi - \pi)\overline{\widehat{\psi}(2[\xi - \pi])} + f(\xi + \pi)\overline{\widehat{\psi}(2[\xi + \pi])} = 0. \tag{19'}$$

But the determinant of the system of two linear equations (19) and (19') is

$$e^{-i\alpha(\xi - \pi)}b(\xi - \pi)e^{-i\alpha(2(\xi + \pi))}b(2(\xi + \pi))$$

$$- e^{-i\alpha(\xi + \pi)}b(\xi + \pi)e^{-i\alpha(2(\xi - \pi))}b(2(\xi - \pi))$$

$$= \{b^2(\xi - \pi) + b^2(\xi + \pi)\} e^{-i[\alpha(\xi - \pi) + \alpha(2[\xi + \pi])]}$$

$$= e^{-i\alpha(\xi - \pi) + \alpha(2[\xi + \pi])]} \neq 0$$

for $\xi \in [-\epsilon, \epsilon]$ because of (ii), (iii), and (iv). But this means that $f(\xi + \pi) = 0$ in this range or, equivalently, that $f(\xi) = 0$ for $\xi \in [\pi - \epsilon, \pi + \epsilon]$. This gives us the desired conclusion: $f(\xi) = 0$ for $\xi \in [\pi - \epsilon, 2(\pi - \epsilon)]$. The proof of Theorem 7 is complete. ∎

REMARK 5 The two linear equations (19) and (19′) are a consequence of conditions (ii), (iii), and (iv) of Theorem 7 (by the usual periodization argument and the support hypothesis on $\widehat{\psi}$); they do not involve (v). If the determinant of the system were 0, then we must have $b^2(\xi + \pi) = b^2(\xi - \pi)$. But this equality and condition (iii) imply $b(\xi - \pi) = \frac{1}{\sqrt{2}} = b(\xi + \pi)$. Thus, if b does not assume the value $\frac{1}{\sqrt{2}}$ on $[\pi - \epsilon, \pi + \epsilon]$ in a set of positive measure, we can still conclude that the system $\{\psi_{j,k}\}$ is complete, even if (v) fails. More specifically, let b be a non-negative even function satisfying the hypotheses of Theorem 7 except (v); then, if $\widehat{\psi}(\xi) = b(\xi)$, and b assumes the value $\frac{1}{\sqrt{2}}$ on a set of measure zero, it follows that $\{\psi_{j,k}\}$, $j, k \in \mathbb{Z}$, spans a subspace that is dense in $L^2(\mathbb{R})$. ∎

REMARK 6 The conditions $\epsilon' = 2\epsilon$ and $\epsilon + \epsilon' \leq \pi$ imply $\epsilon \leq \frac{2\pi}{3}$. Thus, $\sup b \subset \left[-\frac{8\pi}{3}, -\frac{2\pi}{3}\right] \cup \left[\frac{2\pi}{3}, \frac{8\pi}{3}\right]$, and we see that the bell functions of Theorem 7 are the ones introduced by Lemarié and Meyer. The condition that Supp b coincides with $[-2\pi - \epsilon', -\pi + \epsilon] \cup [\pi - \epsilon, 2\pi + \epsilon']$ was used to establish the necessity of $\epsilon' = 2\epsilon$. In fact, as is indicated at the end of this chapter (see the first of the last two remarks in this paper), the only requirement is the above intervals are the minimal ones containing Supp b. ∎

3.4 The local bases of Coifman and Meyer and their connection with the Lemarié–Meyer basis and Wilson bases

Let us now examine more carefully the function $b(\xi) = |\widehat{\psi}(\xi)|$ in Theorem 7. We claim that conditions (i), (ii), (iii), and (iv) are precisely the properties that define the "bell functions" associated with the local cosine/sine bases of Coifman and Meyer (see [AWW], [CM]). Let us describe these bases. Suppose $I = [\alpha, \beta]$ is a finite closed interval in $\mathbb{R}$. It is often useful to focus on the (local) properties of a function f (or, in engineering terms, "signal" f) when it is restricted to I. To achieve this, one can expand this restriction of f in terms of a complete orthonormal system of $L^2(I)$. An example of such a system is

$$\left\{ \sqrt{\frac{2}{|I|}} \chi_I(\xi) \cos \left\{ \frac{(2k+1)\pi(\xi - \alpha)}{2|I|} \right\} \right\}, k = 0, 1, 2, \ldots . \qquad (20)$$

Expansions in terms of such bases are known as "windowed" or "short time" Fourier transforms. Systems such as the one presented in (20) (the cosine can be replaced by the sine and the "odd" half-integer frequencies $\frac{(2k+1)}{2}$ can be replaced by k) are appropriate for focusing on local properties (i.e., what happens on the interval I); however, the abrupt "cutoff" effected by the multiplication by the characteristic function χ_I involves some undesirable artifacts. Coifman and Meyer (see [CM] or [AWW]) constructed a projection P_I that has many similarities to the one obtained by multiplication by χ_I, but it is smoother—these artifacts are therefore minimized. Let us describe P_I in terms of the notation and in terminology used in [AWW].

Let ϵ, ϵ' be positive numbers such that $\epsilon + \epsilon' \leq \beta - \alpha$. A function $b = b_I = b_{I,\epsilon,\epsilon'}$ is called a *bell over* $I = [\alpha, \beta]$ if Supp $b \subset [\alpha - \epsilon, \beta + \epsilon']$ and satisfies the properties

(a) $b^2(\xi) + b^2(2\alpha - \xi) = 1$ if $\xi \in [\alpha - \epsilon, \alpha + \epsilon]$;

(b) $b(\xi) = 1$ if $\xi \in [\alpha + \epsilon, \beta - \epsilon']$;

(c) $b^2(\xi) + b^2(2\beta - \xi) = 1$ if $\xi \in [\beta - \epsilon', \beta + \epsilon']$.

It is very easy to construct all such bell functions that are infinitely differentiable. For example, we can begin by choosing an even non-negative C^∞ function τ with Supp $\tau \subset [-\epsilon, \epsilon]$ so normalized that $\int_{\mathbb{R}} \tau(t)\,dt = \frac{\pi}{2}$ and let $\theta(\xi) = \int_{-\infty}^{\xi} \tau(t)\,dt$. An immediate consequence of the fact that τ is even is that $\theta(\xi) + \theta(-\xi) = \frac{\pi}{2}$. Thus, if we put $s_\epsilon(\xi) \equiv \sin\theta(\xi)$ and $c_\epsilon(\xi) \equiv \cos\theta(\xi)$, it follows that $c_\epsilon(\xi) = \cos\left[\frac{\pi}{2} - \theta(-\xi)\right] = s_\epsilon(-\xi)$. Thus, the graph of c_ϵ is the mirror image, through the vertical axis $\xi = 0$, of the graph of s_ϵ. It is also clear that

$$s_\epsilon^2(\xi) + c_\epsilon^2(\xi) = 1.$$

It is now easy to check that the function

$$b(\xi) = b_I(\xi) = s_\epsilon(\xi - \alpha)\, c_{\epsilon'}(\xi - \beta) \tag{21}$$

satisfies the above defining conditions for a bell over I. There are four "natural" projections associated with the function b. Each is determined by the pair of choices of signs in the expression

$$P_I f(\xi) = b_I(\xi)\{b_I(\xi)f(\xi) \pm b_I(2\alpha - \xi)f(2\alpha - \xi)$$
$$\pm (b_I(2\beta - \xi)f(2\beta - \xi)\}. \tag{22}$$

That is, when ϵ and ϵ' are fixed, the operator P_I is well defined if we choose the two *polarities* (that is, the choice of signs) at each end point α and β. These polarities are particularly important when we want to study the properties of P_I and P_J when I and J are adjacent intervals. Of course, the choice of ϵ and ϵ' is also important in these considerations. More precisely, we say that two adjacent intervals $I = [\alpha, \beta]$ and $J = [\beta, \gamma]$ are said to be *compatible* and have associated *bells that are compatible* if

$$\alpha - \epsilon < \alpha < \alpha + \epsilon \leq \beta - \epsilon' < \beta < \beta + \epsilon' \leq \gamma - \epsilon'' < \gamma < \gamma + \epsilon''$$

and

$$b_I(\xi) = b_J(2\beta - \xi)$$

when $\xi \in [\beta - \epsilon', \beta + \epsilon']$. From this equality and the properties (a), (b), and (c) we obtain the following summation formula for compatible adjacent bells:

$$b_I^2 + b_J^2 = b_{I \cup J}^2.$$

The important role played by the polarities of two adjacent intervals becomes clear when we want to have a corresponding additive property for projections:

PROPOSITION 1

Suppose $I = [\alpha, \beta]$ and $J = [\beta, \gamma]$ are adjacent compatible intervals and P_I, P_J have opposite polarity at β. Then $P_I + P_J$ is the orthogonal projection $P_{I \cup J}$

$$P_I + P_J = P_{I \cup J}.$$

Moreover, P_I and P_J are orthogonal to each other

$$P_I P_J = P_J P_I = 0.$$

All the above assertions are proved in [AWW], but let us point out that Proposition 1 follows readily from (22) since the terms involving the end point β cancel each other. This proposition allows us to decompose $L^2(\mathbb{R})$ into a direct sum of mutually orthogonal subspaces that are images under such projections:

$$L^2(\mathbb{R}) = \bigoplus_{k \in \mathbb{Z}} \mathcal{H}_k. \tag{23}$$

We do this as follows: choose a sequence $\{\alpha_k\}$, $k \in \mathbb{Z}$, of reals and accompanying positive numbers $\{\epsilon_k\}$ such that

$$\alpha_k + \epsilon_k < \alpha_{k+1} - \epsilon_{k+1}$$

for all k. Thus, each pair of adjacent intervals $I_{k-1} = [\alpha_{k-1}, \alpha_k]$ and $I_k = [\alpha_k, \alpha_{k+1}]$ is a compatible pair. Let b_k be the bell over I_k and $P_k = P_{I_k}$. If $\lim_{k \to \pm\infty} \alpha_k = \pm\infty$, we then have $\mathbb{R} = \cup_{k \in \mathbb{Z}} I_k$. It follows from Proposition 1 that, as long as P_{k-1} and P_k have opposite polarity at α_k

$$\sum_{-N}^{N} P_k = P_{[\alpha_{-N}, \alpha_{N+1}]}.$$

Letting $N \to \infty$ in this equality, we obtain the decomposition (23) with $\mathcal{H}_k = P_k L^2(\mathbb{R})$.

Theorem 7 shows that the wavelets being considered produce, on the Fourier transform side, such an orthogonal decomposition of $L^2(\mathbb{R})$. Indeed, the dilates by 2^j on the interval $[\pi, 2\pi]$ make up a disjoint covering of the right half of $\mathbb{R}$ (similarly, the same dilations of $[-2\pi, 2\pi]$ cover the negative reals). Properties (ii), (iii), and (iv) show that the functions $b(2^{-j}\xi), \xi > 0$, are supported on $2^j \cdot [\pi - \epsilon, \pi + 2\epsilon]$ and are bells over $[2^j\pi, 2^{j+1}\pi], j \in \mathbb{Z}$, (compare with properties (a), (b), and (c)). In fact, it follows easily from our formulation of these three properties that any two of these adjoining bells are compatible. Thus, the corresponding projections P_j^+ give us a direct sum decomposition, which we will call Proposition 1'

$$L^2(\mathbb{R}^+) = \bigoplus_{j \in \mathbb{Z}} \mathcal{H}_j,$$

where $\mathcal{H}_j = P_j^+ L^2(\mathbb{R})$ and $\mathbb{R}^+$ is the positive real axis. The same considerations apply to the negative reals, $\mathbb{R}^-$, and we obtain an orthogonal decomposition of $L^2(\mathbb{R}^-)$, which, together with Proposition (1'), yields the desired decomposition of $L^2(\mathbb{R})$.

We shall now describe the local sine/cosine bases of Coifman and Meyer and show how they are related to the bases $\Psi_j = \{\widehat{\psi}_{j,k}\}_{k \in \mathbb{Z}}$, $j = 0, \pm 1, \pm 2, \ldots$. Let us return to the general case of the interval $I = [\alpha, \beta]$ and its associated projections P_I. It turns out that if we replace $\chi_I(\xi)$ in (20) by $b_I(\xi)$ we obtain the system

$$\left\{ \sqrt{\frac{2}{|I|}} \chi_I b_I(\xi) \cos \left\{ \frac{(2k+1)\pi(\xi - \alpha)}{2|I|} \right\} \right\}, k = 0, 1, 2, \ldots . \tag{24}$$

It is shown in [CM] (also in [AWW]) that this system is an orthonormal basis for the space $P_I L^2(\mathbb{R})$, where the polarity of P_I at α is $+$ and, at β, is $-$. Let us say that a function f is *even (odd) about the point* γ if $f(\xi) = f(2\gamma - \xi)$, $(f(\xi) = -f(2\gamma - \xi))$, for ξ in a neighborhood of γ (recall that we introduced this notion of evenness in the sentence immediately following (17)). It is easy to check that the elements in the image of the projection P_I are of the form bS where S is even about α and odd about β, a property enjoyed by the above cosine functions. It is precisely this property that is crucial for obtaining the orthonormality of the system in (24). If we replace the cosine in (24) by the sine, we obtain functions that are odd about α and even about β, and the system

$$\left\{ \sqrt{\frac{2}{|I|}} b_I(\xi) \sin \left\{ \frac{(2k+1)\pi(\xi - \alpha)}{2|I|} \right\} \right\}, k = 0, 1, 2, \ldots . \tag{25}$$

is an orthonormal basis for $P_I L^2(\mathbb{R})$ when P_I has negative polarity at α and positive polarity at β. In Section 4 of [AWW] it is shown that when $I = [\pi, 2\pi]$ or $[-2\pi, -\pi]$, these two bases can be combined to obtain the orthonormal system $\{\gamma_{0,k}(\xi)\} = \{e^{ik\xi} \gamma(\xi)\}$, $k \in \mathbb{Z}$, that forms an orthonormal basis of the space $\widehat{W}_0$, where $\gamma(\xi) = \frac{e^{i\xi/2} b(\xi)}{\sqrt{2\pi}}$. Let us present this situation in terms that are appropriate to the material we have discussed. We write $b^-(\xi)$ for the restriction of $b(\xi)$ to the interval $[-2\pi - 2\epsilon, -\pi + \epsilon]$ and let $b^+ = b - b^-$. Then the functions

$$c_k^-(\xi) = \sqrt{\frac{2}{\pi}} b^-(\xi) \cos \frac{2k+1}{2} (\xi + 2\pi),$$

$$s_k^-(\xi) = \sqrt{\frac{2}{\pi}} b^-(\xi) \sin \frac{2k+1}{2} (\xi + 2\pi),$$

$$c_k^+(\xi) = \sqrt{\frac{2}{\pi}} \, b^+(\xi) \cos \frac{2k+1}{2}(\xi - \pi),$$

$$s_k^+(\xi) = \sqrt{\frac{2}{\pi}} \, b^+(\xi) \sin \frac{2k+1}{2}(\xi - \pi),$$

$k = 0, 1, 2, \ldots$, are the local cosine and sine bases described in (24) and (25) when $I = [-2\pi, -\pi]$ and when $I = [\pi, -\pi]$, respectively. But a simple computation gives us

$$
\begin{aligned}
\sqrt{2\pi} \, \gamma_{0.k}(\xi) &= b^+(\xi) \cos \frac{2k+1}{2}\xi + b^-(\xi) \cos \frac{2k+1}{2}\xi \\
&\quad + i \left\{ b^+(\xi) \sin \frac{2k+1}{2}\xi + b^-(\xi) \sin \frac{2k+1}{2}\xi \right\} \\
&= \sqrt{\frac{\pi}{2}} \left\{ (-1)^{k+1} s_k^+(\xi) - c_k^-(\xi) + i \left\{ (-1)^k c_k^+(\xi) - s_k^-(\xi) \right\} \right\}.
\end{aligned}
\tag{26}
$$

Since the real part of this expression is even and the imaginary part is odd, we clearly have

$$\langle (-1)^{k+1} s_k^+ - c_k^-, \ (-1)^{\ell} c_\ell^+ - s_\ell^- \rangle = 0$$

for all $k, \ell \in \mathbb{Z}$. Moreover, $c_k^- \perp c_\ell^+, s_\ell^+$ and $s_k^- \perp c_\ell^+, s_\ell^+$ because of the disjointedness of supports. Thus, the orthonormality relations $\langle \gamma_{0.k}, \gamma_{0.\ell} \rangle = \delta_{k.\ell}$, give us

$$\langle s_k^-, \ s_\ell^- \rangle + \langle s_k^+, \ s_\ell^+ \rangle + \langle c_k^-, \ c_\ell^- \rangle + \langle c_k^+, \ c_\ell^+ \rangle = 4\delta_{k.\ell},$$

for all $k, \ell \in \mathbb{Z}$. Of course, each of the terms in the last sum equals $\delta_{k.\ell}$. The easy "unfolding" argument showing this is given in [AWW], and we shall not reproduce it here. The moral of this story is:

The wavelet bases we are considering in this and the third section not only can be constructed from these local sine/cosine bases (a fact that is known and is explained in detail in Section 4 of [AWW]); but, in addition, we have the converse result that wavelet bases lead directly to the concept of local bases. We say this, in particular, because Theorem 7 shows that the bell function b is a very special feature of these wavelets. Furthermore, the explicit formula for the bell function presented in (21) gives a simple construction of these wavelet bases.

Let us make another observation that ties this discussion to the result by Auscher that shows there are no smooth wavelet bases of the Hardy space H^2. Formula (26) can be rewritten in the form

$$\gamma_{0.k}(\xi) = \sqrt{\frac{1}{2\pi}} \left\{ (-1)^k i b^+(\xi) e^{i\frac{2k+1}{2}(\xi - \pi)} - i b^-(\xi) e^{i\frac{2k+1}{2}(\xi + 2\pi)} \right\}. \tag{27}$$

In view of this formula, one is tempted to conjecture that the system

$$\left\{ \frac{1}{\sqrt{\pi}} b^+(\xi) e^{i\frac{2k+1}{2}(\xi-\pi)} \right\}, \qquad k \in \mathbb{Z}, \tag{28}$$

is an orthonormal basis associated with $L^2([\pi - \epsilon,\ 2\pi + 2\epsilon])$, which is clearly false (similarly for the other term in (27)). Moreover, it is well known that "local exponential bases" of this type cannot be constructed (this being a consequence of the ideas involved in the Balian–Low theorem). Let us observe that if the system (28) and its dyadic dilates did span $L^2(\mathbb{R}^+)$, the inverse Fourier transforms of this system would be a smooth wavelet basis of H^2. But this is precisely what Auscher has shown not to be possible.

Our entire development does not involve the notion of a *multiresolution analysis* (MRA). Perhaps it is appropriate to point out that the wavelets we are considering can be very easily constructed from a scaling function φ by the quadrature mirror filter methods developed by Mallat and Meyer (see [M], [Ma] and [D]). In fact, it is easy to exhibit the function φ in terms of the wavelet ψ. Let us recall (see [D] for the standard notation and the relevant definitions) that if we do have an MRA with an associated scaling function φ,

$$|\widehat{\varphi}(\xi)|^2 = \sum_{j=1}^{\infty} |\widehat{\psi}(2^j \xi)|^2. \tag{29}$$

This equality and the first four properties in Theorem 7 then give us

$$|\widehat{\varphi}(\xi)| = \begin{cases} 1 & \text{when } -\pi + \epsilon \leq \xi \leq \pi - \epsilon \\ |\widehat{\psi}(2\xi)| & \text{when } \pi - \epsilon \leq |\xi| \leq \pi + \epsilon \\ 0 & \text{when } \pi + \epsilon \leq |\xi|. \end{cases} \tag{30}$$

The following graphs show how an MRA associated with systems in Theorem 7 can be constructed.

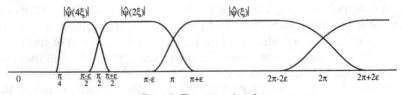

Figure 1. The construction of φ.

For simplicity we consider only the case when $\alpha(\xi) = \frac{\xi}{2}$. Then we can choose $\widehat{\varphi} = |\widehat{\varphi}|$ and from the relation

$$\widehat{\varphi}(2\xi) = \widehat{\varphi}(\xi) m_0(\xi) \tag{31}$$

we see that $m_0(\xi) = \widehat{\varphi}(2\xi)$ when $\xi \in [-\pi, \pi]$:

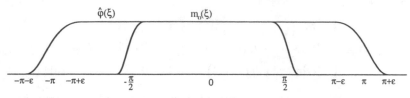

Figure 2. The construction of m_0.

We extend m_0 so that it is 2π-periodic and obtain ψ from the well known formula

$$\widehat{\psi}(2\xi) = e^{i\xi}\widehat{\varphi}(\xi)\overline{m_0(\xi + \pi)}. \tag{32}$$

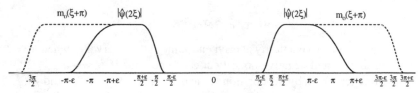

Figure 3. The construction of ψ.

Before considering other topics involving wavelets, let us show how the local sine/cosine bases can be used to obtain a class of bases that are related to wavelet bases, the *Wilson bases*. As we have seen, the wavelet bases are generated by performing certain dilations and translations of the "time" variable x; these, of course, are converted by the Fourier transform into corresponding dilations of the "frequency" variable ξ and multiplications by appropriate exponentials. Suppose we seek a basis that is also generated by a single function b (often called a *window*) by performing translations in x and translations in ξ. More precisely, we ask if one can construct an orthonormal basis of $L^2(\mathbb{R})$ of the form

$$\psi_{k,n}(x) = e^{ik\alpha x}b(x - n\beta), \tag{33}$$

$k, n \in \mathbb{Z}$, where α and β are positive numbers. Such basis is known as a *Gabor basis*. A theorem of Balian and Low ([B], [L₁]) tells us that such a basis can only occur if $\alpha\beta = 2\pi$ and either $xb(x) \notin L^2(\mathbb{R})$ or $\xi\widehat{b}(\xi) \notin L^2(\mathbb{R})$. For example, letting the window b be the characteristic function of a finite interval furnishes us with a basis, for appropriate α and β, but a smooth version (in x and ξ) of this example cannot exist. Several efforts have been made to obtain a wider class of bases having these features (see the introduction in [DJJ]). Perhaps the most satisfactory result in this direction (obtained before the local bases of Coifman and Meyer were introduced) is the one in [DJJ], which is a variant of a construction by Wilson [Wi], in which they show that smooth windows that vanish rapidly at infinity can be used if one replaces the exponential functions in (33) by sines and cosines. More precisely, they showed that there exist orthonormal bases for $L^2(\mathbb{R})$

having the form

$$\psi_{k,n}(x) = \begin{cases} \sqrt{2}\,b(x-n), & \text{if } k = 0 \\ 2(\cos 2\pi kx)b\left(x - \frac{n}{2}\right), & \text{if } k > 0 \text{ and } n + k \text{ is even} \\ 2(\sin 2\pi kx)b\left(x - \frac{n}{2}\right), & \text{if } k > 0 \text{ and } n + k \text{ is odd.} \end{cases} \tag{34}$$

An example of such a system is provided by a system of local cosine and sine bases. This observation was made independently by Auscher [A₂] and Laeng [La]. All we need to do is to choose a decomposition of the type described in (23) in the following way. Let $\{\alpha_n\} \equiv \left\{\frac{2n-1}{4}\right\}$, $n \in \mathbb{Z}$, $I_n = [\alpha_n, \alpha_{n+1}]$ and $\{\epsilon_n\}$ be a constant sequence of positive numbers such that $\epsilon_n + \epsilon_{n+1} < \frac{1}{2}$ (for example, $\epsilon_n = \frac{1}{5}$ for all $n \in \mathbb{Z}$). Then each pair I_n, I_{n+1} of adjacent intervals is compatible since $\alpha_{n+1} - \alpha_n = \frac{1}{2}$ and

$$\alpha_n + \epsilon_n < \alpha_n - \epsilon_{n+1}.$$

Let us assign positive polarity at each endpoint of the interval I_n when $n = 2m$ is an even integer and negative polarity at each endpoint of I_n when $n = 2^m + 1$ is an odd integer. Then, by Proposition 1, the projections $P_{I_{2m}} \equiv P_{2m}$ and $P_{I_{2m+1}} \equiv P_{2m+1}$ are orthogonal to each other and furnish us with the decomposition (23) of $L^2(\mathbb{R})$. Instead of the local bases (24) and (25), however, we now must use

$$\left\{\sqrt{\frac{1}{|I|}}\,b_I(x), \left\{\sqrt{\frac{2}{|I|}}\,b_I(x)\cos\left\{\frac{k\pi(x-\alpha)}{|I|}\right\}\right\}\right\}, \quad k = 1, 2, \ldots, \tag{35}$$

when $I = [\alpha, \beta]$ has positive polarity at each endpoint and

$$\left\{\sqrt{\frac{2}{|I|}}\,b_I(x)\sin\left\{\frac{k\pi(x-\alpha)}{|I|}\right\}\right\}, \quad k = 1, 2, \ldots \tag{36}$$

when $I = [\alpha, \beta]$ has negative polarity at each endpoint. Let us write down explicitly the form that these bases assume in our case. When $n = 2m$ is an even integer, the intervals I_n form the class

$$\{\ldots, I_0, I_2, \ldots, I_{2m}, \ldots\}$$
$$= \left\{\ldots, \left[-\frac{1}{4}, \frac{1}{4}\right], \left[\frac{3}{4}, \frac{5}{4}\right], \ldots, \left[m - \frac{1}{4}, m + \frac{1}{4}\right], \ldots\right\},$$

while the odd integers $n = 2m + 1$ give us the class

$$\{\ldots, I_1, I_3, \ldots, I_{2m+1}, \ldots\}$$
$$= \left\{\ldots, \left[\frac{1}{4}, \frac{3}{4}\right], \left[\frac{5}{4}, \frac{7}{4}\right], \ldots, \left[m + \frac{1}{4}, m + \frac{3}{4}\right], \ldots\right\}.$$

It is easy to check that the functions $\cos 2k\pi\left(x - \left[m - \frac{1}{4}\right]\right) = \cos 2k\pi\left(x + \frac{1}{4}\right)$

are even at each of the endpoints of I_{2m}, and the functions $\sin 2k\pi \left(x - \left[m + \frac{1}{4}\right]\right)$ $= \sin 2k\pi \left(x - \frac{1}{4}\right)$ are odd at each of the points of I_{2m+1}. The bell functions $b_n = b_{I_n}$ can be chosen in the following way: b_0 can be constructed by using formula (21) with $I = \left[-\frac{1}{4}, \frac{1}{4}\right]$, $\epsilon = \frac{1}{5} = \epsilon'$ and the rest of the bell functions are, simply, translates by half-integers of b_0. Finally, let us choose the basis of $L^2(\mathbb{R})$ obtained by taking all the functions in (35) and (36) with I ranging through the intervals I_{2m} in the first case and through the intervals I_{2m+1} in the latter case. This gives us the collection

$$\tilde{\psi}_{k,n}(x) = \begin{cases} \sqrt{2}\, b\left(x - \frac{n}{2}\right), & \text{if } n \text{ is even and } k = 0 \\ 2b\left(x - \frac{n}{2}\right)\left(\cos 2k\pi \left(x + \frac{1}{4}\right)\right), & \text{if } n \text{ is even and } k > 0 \quad (37) \\ 2b\left(x - \frac{n}{2}\right)\left(\cos 2k\pi \left(x - \frac{1}{4}\right)\right), & \text{if } n \text{ is odd and } k > 0. \end{cases}$$

But, except for factors that are integral powers of -1, this is the basis (34). This follows from

$$\cos 2k\pi \left(x + \frac{1}{4}\right) = \begin{cases} \cos 2k\pi x & \text{if } k = 4q, \\ -\cos 2k\pi x & \text{if } k = 2(2q + 1), \\ \sin 2k\pi x & \text{if } k = 2(2q + 1) + 1, \\ -\sin 2k\pi x & \text{if } k = 4q + 1 \end{cases} \quad \text{for } q \in \mathbb{Z}$$

and

$$\cos 2k\pi \left(x - \frac{1}{4}\right) = \begin{cases} \sin 2k\pi x & \text{if } k = 4q, \\ -\sin 2k\pi x & \text{if } k = 2(2q + 1), \\ \cos 2k\pi x & \text{if } k = 2(2q + 1) + 1, \\ -\cos 2k\pi x & \text{if } k = 4q + 1. \end{cases} \quad \text{for } q \in \mathbb{Z}$$

3.5 The non-existence of smooth wavelets bases of $H^2(\mathbb{R})$ and other observations about the support of band-limited wavelets

Let $H^2 = H^2(\mathbb{R})$ be the subspace of $L^2(\mathbb{R})$ of all those functions f such that $\sup \hat{f} \subset [0, \infty)$. The simplest example of a wavelet basis of H^2 is the one generated by the function ψ whose Fourier transform is the characteristic function of $[2\pi, 4\pi]$. It has been an open problem for some time whether there are other such orthonormal wavelets for H^2. Auscher has recently shown (see [A_3]) that if one assumes some smoothness for the function $\hat{\psi}$, then ψ cannot generate an orthonormal system that forms a basis for H^2. A precise version of this theorem is the following:

THEOREM 8
There does not exist an orthogonal wavelet $\psi \in H^2$ that generates a basis of this space if $\hat{\psi} \in C^1(\mathbb{R})$ and satisfies $|\hat{\psi}(\xi)| + |\hat{\psi}'(\xi)| \leq C|\xi|^{-\alpha}$ for $\xi \geq 1$ and $\alpha > \frac{1}{2}$.

We present a particularly elementary argument (described by Auscher) that establishes this theorem in the band-limited case. Let us suppose that $\psi \in L^2(\mathbb{R})$, $\widehat{\psi}$ has compact support that lies within $[0, \infty)$, and that $\{\psi_{j,k}\}$, $j, k \in \mathbb{Z}$, is an orthonormal system. Let $H(\xi) \equiv \sum_{j=1}^{\infty} |\widehat{\psi}(2^j \xi)|^2$ and assume that $H(0+) = 1$ and both H and H' belong to $L^1(\mathbb{R})$ (we will show that these properties follow from the assumptions made in Theorem 8).

LEMMA 4

$\sum_{k \in \mathbb{Z}} H(\xi + 2k\pi) = \frac{3\pi - \xi}{2\pi}$ when $0 < \xi < 2\pi$.

PROOF   Since $1 + \sum_{\ell \neq 0} \frac{e^{i\ell\xi}}{2\pi i\ell} = \frac{3\pi - \xi}{2\pi}$ for $0 < \xi < 2\pi$, it suffices to show that the Fourier coefficients c_ℓ of the 2π-periodic function $\sum_{k \in \mathbb{Z}} H(\xi + 2k\pi)$ satisfy

$$c_\ell = \begin{cases} 1 & \text{if } \ell = 0 \\ \frac{1}{2\pi i\ell} & \text{if } \ell \neq 0. \end{cases} \tag{38}$$

We have

$$\int_0^{2\pi} e^{-i\ell\xi} \sum_{k \in \mathbb{Z}} H(\xi + 2k\pi) \, d\xi = \int_{\mathbb{R}} e^{-i\ell\xi} H(\xi) \, d\xi. \tag{39}$$

Thus,

$$c_0 = \left(\frac{1}{2\pi}\right) \int_{\mathbb{R}} H(\xi) \, d\xi$$

$$= \left(\frac{1}{2\pi}\right) \sum_{j=1}^{\infty} \int_0^{\infty} |\widehat{\psi}(2^j \xi)|^2 \, d\xi = \left(\frac{1}{2\pi}\right) \|\widehat{\psi}\|_2^2 \sum_{j=1}^{\infty} 2^{-j} = 1.$$

We shall use (39) to calculate c_ℓ when $\ell \neq 0$ as well; to do this, we first observe that it follows from (2) that, when $\ell \neq 0$,

$$\int_0^{\infty} |\widehat{\psi}(\xi)|^2 e^{-i\ell\xi} \, d\xi = \int_0^{2\pi} e^{-i\ell\xi} \sum_{k \in \mathbb{Z}} |\widehat{\psi}(\xi + 2k\pi)|^2 \, d\xi = 0. \tag{40}$$

Since $H\left(\frac{\xi}{2}\right) = |\widehat{\psi}(\xi)|^2 + H(\xi)$, an application of (40) gives us the equality $c_\ell = 2c_{2\ell}$. On the other hand, an integration by parts gives us

$$c_\ell = \left(\frac{1}{2\pi}\right) \int_0^{\infty} H(\xi) e^{-i\ell\xi} \, d\xi$$

$$= -\frac{i}{2\pi\ell} + \left(\frac{1}{2\pi}\right) \int_0^{\infty} H'(\xi) \frac{e^{-i\ell\xi}}{i\ell} \, d\xi \equiv -\frac{i}{2\pi\ell} + \frac{d_\ell}{i\ell}.$$

Thus, $-\frac{i}{2\pi\ell} + \frac{d_\ell}{i\ell} = c_\ell = 2c_{2\ell} = 2\left\{-\frac{i}{4\pi\ell} + \frac{d_{2\ell}}{i2\ell}\right\} = -\frac{i}{2\pi\ell} + \frac{d_{2\ell}}{i\ell}$, and we obtain $d_\ell = d_{2\ell}$. But the Riemann–Lebesgue theorem then implies $d_\ell = \lim_{n \to \infty} d_{2^n\ell} = 0$. This completes the proof of (38) and of Lemma 4.   ∎

An immediate consequence of Lemma 4 is Corollary 1.

COROLLARY 1

$\lim_{\xi \to 2\pi-} \sum_{k \in \mathbb{Z}} H(\xi + 2k\pi) = \frac{1}{2}$.

Since $H(\xi + 2k\pi) = \sum_{j=1}^{\infty} |\widehat{\psi}(2^j(\xi + 2k\pi))|^2 \to \sum_{j=1}^{\infty} |\widehat{\psi}(2^{j+1}(k+1)\pi)|^2$ as $\xi \to 2\pi-$, we also have

$$\lim_{\xi \to 2\pi-} \sum_{k \in \mathbb{Z}} H(\xi + 2k\pi) = \sum_{k \in \mathbb{Z}} \sum_{j=1}^{\infty} |\widehat{\psi}(4 \cdot 2^{j-1}(k+1)\pi)|^2$$

$$= \sum_{m=1}^{\infty} \epsilon(m) |\widehat{\psi}(4m\pi)|^2,$$

where $\epsilon(m)$ is the number of representations of $m = 2^{j-1}(k+1)$ when $k \geq 0$ and $j > 0$ (since $\xi \to 2\pi-$ we can assume that $0 \leq \xi < 2\pi$ and, thus, the above sums only involve $k \geq 0$). This observation and Corollary 1 give us the equality

$$\sum_{m=1}^{\infty} \epsilon(m) |\widehat{\psi}(4m\pi)|^2 = \frac{1}{2}. \tag{41}$$

But, using (2) with $\xi = 0$, we have, since Supp $\widehat{\psi} \subset [0, \infty)$,

$$1 = \sum_{\ell \in \mathbb{Z}} |\widehat{\psi}(2\pi\ell)|^2 = \sum_{m=0}^{\infty} |\widehat{\psi}(2(2m+1)\pi)|^2 + \sum_{m=1}^{\infty} |\widehat{\psi}(2 \cdot 2m\pi)|^2.$$

Thus,

$$\sum_{m=0}^{\infty} |\widehat{\psi}(2(2m+1)\pi)|^2 = 1 - \sum_{m=1}^{\infty} |\widehat{\psi}(4m\pi)|^2 \geq 1 - \sum_{m=1}^{\infty} \epsilon(m) |\widehat{\psi}(4m\pi)|^2 = \frac{1}{2}.$$

This means (a) that there exists $m \geq 1$ such that $\widehat{\psi}(4m\pi) \neq 0$, and, also, (b) there exists $m \geq 0$ such that $\widehat{\psi}(2(2m+1)\pi) \neq 0$. Let us also observe that our proof of Theorem 3 can be adapted to show that equality (7) is valid for $\xi > 0$; from this it is also easy to see that $\widehat{\psi}$ satisfies Lemma 3 and, in particular, we have equality (14) for $\widehat{\psi}$. From this last equality we can make the following claim:

If $\widehat{\psi}(\xi) \neq 0$ for some $\xi > 0$ and $\widehat{\psi}(2^j\xi) = 0$ whenever $j \geq 1$, then $\widehat{\psi}(\xi + 2(2\ell + 1)\pi) = 0$ for all $\ell \in \mathbb{Z}$. $\tag{42}$

Since the support of $\widehat{\psi}$ is compact, a consequence of (a) and (b) above is the fact that there exists a maximal positive integer k such that $\widehat{\psi}(2k\pi) \neq 0$. Thus applying (42) to $\xi = 2k\pi > 0$, we obtain

$$0 = \widehat{\psi}(2k\pi + 2(2\ell + 1)\pi) = \widehat{\psi}(2\pi(k + 2\ell + 1)) \tag{43}$$

for all $\ell \in \mathbb{Z}$. If k is even, let m satisfy (b) and choose the integer $\ell = \frac{(2m-k)}{2}$; this gives us the contradiction

$$0 = \widehat{\psi}(2\pi(k + 2\ell + 1)) = \widehat{\psi}(2\pi(2m + 1)) \neq 0.$$

If k is odd, let m satisfy (a) and choose the integer $\ell = \frac{(2m-k-1)}{2}$; this gives us the contradiction

$$0 = \widehat{\psi}(2\pi(k + 2\ell + 1)) = \widehat{\psi}(4\pi m) \neq 0.$$

We have shown that if ψ generates an orthonormal system $\{\psi_{j,k}\}$, $j, k \in \mathbb{Z}$, that is a basis for H^2, $H(0+) = 1$ and H, H' belong to $L^1(\mathbb{R})$, then we obtain the above contradictions. Let us see how these last properties are related to the assumptions made in the announcement of Theorem 8. We have already observed that equality (7) is valid provided $\xi > 0$. Thus,

$$H(\xi) = \sum_{j=1}^{\infty} |\widehat{\psi}(2^j \xi)|^2 = 1 - \sum_{j=0}^{\infty} |\widehat{\psi}(2^{-j}\xi)|^2.$$

This equality and the assumptions in Theorem 8 imply

$$|1 - H(\xi)| = \sum_{j=0}^{\infty} |\widehat{\psi}(2^{-j}\xi)|^2 \leq C|\xi|^2 \sum_{j=0}^{\infty} 2^{-j} \tag{44}$$

and we obtain $H(0+) = 1$. We also have the estimate

$$H(\xi) = \sum_{j=1}^{\infty} |\widehat{\psi}(2^j \xi)|^2 \leq C \sum_{j=1}^{\infty} 2^{-2j\alpha} \xi^{-2\alpha}$$

when $\xi \geq 1$. Since we are assuming $\alpha > \frac{1}{2}$, this estimate gives us $\int_1^{\infty} H(\xi)\, d\xi < \infty$. This last inequality and (44) imply that $H \in L^1(\mathbb{R})$. The fact that $H' \in L^1(\mathbb{R})$ is a consequence of the estimate

$$|H'(\xi)| \leq C \sum_{j=1}^{\infty} 2^j 2^{-j\alpha} |\xi|^{-\alpha} 2^{-j\alpha} |\xi|^{-\alpha} = C_\alpha |\xi|^{-2\alpha}$$

when $\xi \geq 1$, where $C_\alpha < \infty$ since $\alpha > \frac{1}{2}$ explains the role played by the hypothesis $\alpha > \frac{1}{2}$ in Auscher's theorem. In the band-limited case, as we have seen repeatedly, the situation is much simpler, since the support of H is compact and the integrability of H and H' holds under weaker assumptions.

If ψ arises from an MRA, the function H equals $|\widehat{\varphi}(\xi)|^2$, where φ is the scaling function of the MRA (this is a well-known fact; see, for example, [D], page 136). When ψ arises from a "smooth" scaling function, however, it cannot generate a basis for H^2 (see the discussion on page 551 of [JL]).

We shall end this presentation with some observations concerning the interval about the origin that lies outside the support of $\widehat{\psi}$, when ψ is band-limited (see Theorem 4). In the class of wavelets we discussed in Section 3 of this chapter, the minimal interval about 0 that avoids the support of $\widehat{\psi}$ is $\left(-\frac{2\pi}{3}, \frac{2\pi}{3}\right)$. Thus, the minimal length of these intervals is $\frac{4\pi}{3}$. One may wonder whether this is a general phenomenon. There are more general results along these lines: for example, if the support of $\widehat{\psi}$ is made up of two intervals lying on the two sides of the origin (but is not necessarily symmetric about 0), it is not hard to show that the length of the interval about 0 that is disjoint from this support is at least $\frac{4\pi}{3}$. In general, however, the length of this interval can be arbitrarily small.

PROPOSITION 2
Let $\epsilon > 0$. There exists a band-limited orthonormal wavelet ψ such that $\widehat{\psi} \in C^\infty$ and $|\widehat{\psi}|$ is an even function that does not vanish identically on $[-\epsilon, \epsilon]$.

Before giving the proof of this proposition, let us make the following observation that shows that the points $-\frac{2\pi}{3}$ and $\frac{2\pi}{3}$ have a "special" relation to the support of a band-limited wavelet.

PROPOSITION 3
Suppose ψ is an orthonormal wavelet that is constructed from a scaling function φ such that $\widehat{\varphi}$ is continuous and has compact support, then either $\widehat{\psi}\left(\frac{2\pi}{3}\right)$ or $\widehat{\psi}\left(-\frac{2\pi}{3}\right)$ must be 0.

PROOF By equality (32) we see that it suffices to show that either $m_0\left(\frac{2\pi}{3}\right)$ or $m_0\left(-\frac{2\pi}{3}\right)$ must be 0. From (2) with $\xi = \frac{2\pi}{3}$ we have

$$\sum_{\ell \in \mathbb{Z}} \left|\widehat{\varphi}\left(\frac{2\ell\pi + 2\pi}{3}\right)\right|^2 = 1.$$

This equality and the fact that $\widehat{\varphi}$ has compact support guarantee the existence of a largest $k \in \mathbb{Z}$ such that

$$\widehat{\varphi}\left(\frac{2k\pi + 2\pi}{3}\right) \neq 0. \tag{45}$$

Without loss of generality we can assume that $k \geq 0$. Thus, using the 2π-periodicity of m_0 and (31), we have

$$0 = \widehat{\varphi}\left([8k + 2]\pi + \frac{2\pi}{3}\right)$$

$$= \widehat{\varphi}\left(2\left[4k\pi + 2 \cdot \frac{2\pi}{3}\right]\right)$$

$$= \widehat{\varphi}\left(4k\pi + 2 \cdot \frac{2\pi}{3}\right) m_0\left(4k\pi + 2 \cdot \frac{2\pi}{3}\right)$$

$$= \widehat{\varphi}\left(2k\pi + \frac{2\pi}{3}\right) m_0\left(2k\pi + \frac{2\pi}{3}\right) m_0\left(4k\pi + \frac{4\pi}{3}\right)$$

$$= \widehat{\varphi}\left(2k\pi + \frac{2\pi}{3}\right) m_0\left(\frac{2\pi}{3}\right) m_0\left(-\frac{2\pi}{3}\right).$$

Thus, (45) and this last calculation imply $m_0\left(\frac{2\pi}{3}\right) m_0\left(-\frac{2\pi}{3}\right) = 0$ and the proposition follows. ∎

We shall produce a wavelet having the properties announced in Proposition 2 by constructing an appropriate MRA. The following result asserts that such an MRA exists:

LEMMA 5

There exists an MRA generated by an even band-limited scaling function φ and associated 2π-periodic function m_0 such that

(1) $\widehat{\varphi} \in C^\infty$ and $\widehat{\varphi}(\xi)$ is never 0 when $\xi \in \left(-\frac{4\pi}{3}, \frac{4\pi}{3}\right)$,

(2) $m_0 \in C^\infty$, m_0 is even, $m_0(\xi)$ is never 0 when $\xi \in \left(-\frac{2\pi}{3}, \frac{2\pi}{3}\right)$ and m_0 is not identically 0 on $\left[\pi - \frac{\epsilon}{2}, \pi + \frac{\epsilon}{2}\right]$.

Observe that Lemma 5 does imply Proposition 2. Recall (see either [D] or [M]) that $\widehat{\psi}$ is constructed from $\widehat{\varphi}$ and m_0 by formula (32) which can be written in the form

$$\widehat{\psi}(\xi) = e^{-i\xi/2}\overline{m_0\left(\pi + \frac{\xi}{2}\right)}\widehat{\varphi}\left(\frac{\xi}{2}\right).$$

In particular, then, the support of $\widehat{\psi}$ is compact since $\widehat{\varphi}$ has compact support. Also $\widehat{\psi} \in C^\infty$ since $\widehat{\varphi}, m_0 \in C^\infty$. Since m_0 is 2π-periodic and $m_0, \widehat{\varphi}$ are even, $|\widehat{\psi}|$ is even

$$|\widehat{\psi}(-\xi)| = \left|m_0\left(\pi - \frac{\xi}{2}\right)\right| \left|\widehat{\varphi}\left(-\frac{\xi}{2}\right)\right|$$

$$= \left|m_0\left(\frac{\xi}{2} - \pi\right)\right| \left|\widehat{\varphi}\left(\frac{\xi}{2}\right)\right|$$

$$= \left|m_0\left(\frac{\xi}{2} + \pi\right)\right| \left|\widehat{\varphi}\left(\frac{\xi}{2}\right)\right| = |\widehat{\psi}(\xi)|.$$

Without loss of generality we can assume that $\epsilon < \frac{8\pi}{3}$. Suppose $\widehat{\psi}(\xi) \equiv 0$ on $[-\epsilon, \epsilon]$; then, we must have $m_0\left(\frac{\xi}{2} + \pi\right) \equiv 0$ on $[-\epsilon, \epsilon]$ since $\widehat{\varphi}\left(\frac{\xi}{2}\right)$ is never 0 for $\xi \in \left[-\frac{8\pi}{3}, \frac{8\pi}{3}\right] \supset [-\epsilon, \epsilon]$. Thus, $m_0(\xi) \equiv 0$ for $\xi \in \left[\pi - \frac{\epsilon}{2}, \pi + \frac{\epsilon}{2}\right]$, contradicting the last announced property of m_0 in Lemma 5.

Let us now turn to the proof of the lemma. The idea is first to construct a function m_0 having properties that imply that the function φ defined by

$$\widehat{\varphi}(\xi) = \prod_{j=1}^{\infty} m_0(2^{-j}\xi) \tag{46}$$

is a scaling function for a $\mathbb{C}^\infty$ MRA. We shall construct such a function m_0 that satisfies the following properties:

(a) $m_0 \in \mathbb{C}^\infty$ and is 2π-periodic,

(b) $m_0(\xi)$ is never 0 on $\left[-\frac{2\pi}{3}, \frac{2\pi}{3}\right]$,

(c) $|m_0(\xi)|^2 + |m_0(\xi + \pi)|^2 = 1$ for all ξ,

(d) $m_0(\xi) \equiv 1$ for $\xi \in [-\delta, \delta]$ for some $\delta > 0$.   (47)

The proof of Lemma 5 will be based on two lemmas. The first is Lemma 6.

LEMMA 6
If m_0 satisfies the properties in (47) then the function φ defined by equality (46) is a scaling function for an MRA.

To state the second lemma, we shall impose one more property on m_0 that will give us, in particular, property (2) in Lemma 5. Let I be an interval that is contained in $\left[\frac{2\pi}{3}, \pi\right]$ and denote by $2\pi - I$ the reflection of I with respect to π. The function m_0 that we are considering, in addition to the properties in (47), is even and satisfies

$$m_0 \equiv 0 \text{ on } \left[\tfrac{2\pi}{3}, \pi\right] \backslash (I \cup [2\pi - I]) \text{ and}$$

$$m_0(\xi) \neq 0 \text{ for all } \xi \in (I \cup [2\pi - I]).$$ (48)

It is not hard to construct a function m_0 satisfying (47) and (48). The following figure represents of $|m_0|^2$ on the period interval $[-\pi, \pi]$ for an arbitrary interval $I \subset \left[\frac{2\pi}{3}, \pi\right]$:

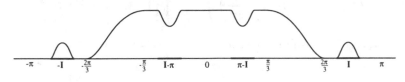

Figure 4

The conditions on the "bump" functions over the intervals I and $-I$, as well as the "valley" functions over $I - \pi$ and $\pi - I$, are determined by the fact that equality (47) (c) must be satisfied. The second lemma (Lemma 7) can now be stated.

LEMMA 7
If m_0 satisfies (47) and (48), and I is appropriately chosen, then φ is band-limited and m_0 assumes non-zero values in the interval $\left(\frac{\pi - \epsilon}{2}, \frac{\pi + \epsilon}{2}\right)$. An "appropriate" choice of I is the interval $[\pi - \pi 2^{-J} - \eta, \pi - \pi 2^{-J} + \eta]$, where J is large enough so that $\pi - \pi 2^{-J} > \pi - \frac{\epsilon}{2}$ and $\pi - 2\pi 2^{-J} > \frac{2\pi}{3}$, and $0 < \eta < \frac{\pi 2^{-J}}{(2^J + 1)}$.

PROOF OF LEMMA 5   Let $\widehat{\varphi}(\xi) = \prod_{j=1}^{\infty} m_0(2^{-j}\xi)$. Since $m_0(\xi) \equiv 1$ in a neighborhood of 0, this product converges for all ξ and defines a $\mathbb{C}^{\infty}$ function. Let

$$\widehat{f_k}(\xi) = \prod_{j=1}^{k} m_0(2^{-j}\xi) \chi_{[-2^k\pi, 2^k\pi]}(\xi).$$

Clearly $\widehat{f_k}(\xi) \to \widehat{\varphi}(\xi)$ for all $\xi \in \mathbb{R}$. We claim that

$$\langle f_k(\cdot), f_k(\cdot - n) \rangle = \delta_{o,n} \tag{49}$$

for all $k, n \in \mathbb{Z}$. To see this we first write

$$\langle f_k(\cdot), f_k(\cdot - n) \rangle = \left(\frac{1}{2\pi}\right) \int_{\mathbb{R}} \widehat{f_k}(\xi) \overline{\widehat{f_k}(\xi)} e^{in\xi} \, d\xi$$

$$= \left(\frac{1}{2\pi}\right) \int_{-2^k\pi}^{2^k\pi} \prod_{j=1}^{k} |m_0(2^{-j}\xi)|^2 e^{in\xi} \, d\xi.$$

But $\int_{-2^k\pi}^{2^k\pi} = \int_{-2^k\pi}^{0} + \int_{0}^{2^k\pi}$ can be expressed as an integral over $[0, 2^k\pi]$ by translating the variable in the first summand by 2^k. Then the 2π-periodicity of m_0 and (47) (c) can be used to show that

$$\int_{-2^k\pi}^{2^k\pi} \prod_{j=1}^{k} |m_0(2^{-j}\xi)|^2 e^{in\xi} \, d\xi = \int_{0}^{2\pi 2^{k-1}} \prod_{j=1}^{k-1} |m_0(2^{-j}\xi)|^2 e^{in\xi} \, d\xi.$$

An induction argument then gives us

$$\langle f_k(\cdot), f_k(\cdot - n) \rangle = \left(\frac{1}{2\pi}\right) \int_{0}^{2\pi 2^{k-i}} \prod_{j=1}^{k-i} |m_0(2^{-j}\xi)|^2 e^{in\xi} \, d\xi, \tag{50}$$

for $i = 1, \ldots, k - 1$ (this time the inductive step involves writing

$$\int_{0}^{2\pi 2^{k-i}} = \int_{0}^{2\pi 2^{k-i-1}} + \int_{2\pi 2^{k-i-1}}^{2\pi 2^{k-i}}$$

and using essentially the same argument as before). If we put $i = k - 1$ in (50) we have

$$\langle f_k(\cdot), f_k(\cdot - n) \rangle$$

$$= \left(\frac{1}{2\pi}\right) \int_{0}^{4\pi} |m_0(2^{-1}\xi)|^2 e^{in\xi} \, d\xi$$

$$= \left(\frac{1}{2\pi}\right) \int_{0}^{2\pi} \left|m_0\left(\frac{\xi}{2}\right)\right|^2 e^{in\xi} \, d\xi + \left(\frac{1}{2\pi}\right) \int_{0}^{2\pi} \left|m_0\left(\frac{\xi}{2} + \pi\right)\right|^2 e^{in\xi} \, d\xi$$

$$= \left(\frac{1}{2\pi}\right) \int_{0}^{2\pi} e^{in\xi} \, d\xi = \delta_{0,n}.$$

This established (49). In particular, $\| f_k \|_2 = 1$. Thus, by the Plancherel theorem and Fatou's lemma,

$$\| \varphi \|_2^2 = \frac{1}{2\pi} \| \widehat{\varphi} \|_2^2 = \frac{1}{2\pi} \int_{\mathbb{R}} \lim_{k \to \infty} |\widehat{f_k}|^2 \leq \frac{1}{2\pi} \liminf_{k \to \infty} \int_{\mathbb{R}} |\widehat{f_k}|^2 = 1.$$

Thus, φ belongs to $L^2(\mathbb{R})$. We claim that $\lim_{k \to \infty} f_k = \varphi$ in $L^2(\mathbb{R})$. To see this we first observe that the continuity of m_0 and (47) (b) imply that there exists $A > 0$ such that $|m_0(\xi)| \geq A$ for $\xi \in \left[-\frac{\pi}{2}, \frac{\pi}{2} \right]$. Thus, $|m_0(2^{-j}\xi)| \geq A$ when $\xi \in [-2^k \pi, 2^k \pi]$ and $j \geq k+1$, for any positive integer k. From (47) (d) we also have $|m_0(2^{-j}\xi)| = 1$ when $\xi \in [-2^k \pi, 2^k \pi]$ and $j \geq p \equiv k + 1 + \left[\log_2 \left(\frac{1}{\pi \delta} \right) \right]$. Consequently,

$$\prod_{j=k+1}^{\infty} |m_0(2^{-j}\xi)| = \prod_{j=k+1}^{p} |m_0(2^{-j}\xi)| \geq A^{\log_2(\frac{1}{\pi \delta})} \equiv B,$$

with B independent of k. Thus, we can apply the dominated convergence theorem to obtain

$$\lim_{k \to \infty} \int_{\mathbb{R}} |\widehat{f_k}(\xi) - \widehat{\varphi}(\xi)|^2 = 0$$

$\left(\text{since the last estimate implies } |\widehat{f_k}(\xi)|^2 \leq \frac{|\widehat{\varphi}(\xi)|^2}{B^2} \right)$. From (49), therefore, we have

$$\langle \varphi(\cdot), \varphi(\cdot - n) \rangle = \lim_{k \to \infty} \langle f_k(\cdot), f_k(\cdot - n) \rangle = \delta_{0,n}.$$

Since $\widehat{\varphi}(2\xi) = m_0(\xi)\widehat{\varphi}(\xi)$ and $\widehat{\varphi}(0) = 1$, we see that φ is a scaling function for an MRA. This completes the proof of this lemma. ∎

We now turn our attention to the proof of Lemma 7. Let I be the interval described in the second sentence of this lemma. Let $\xi \in [2^{j-1}\pi, 2^j \pi]$ for $j \geq J+2$ and let $t = 2^{-j}\xi$; thus $t \in \left[\frac{\pi}{2}, \pi \right]$. We have

$$\widehat{\varphi}(\xi) = m_0(t)m_0(2t) \ldots m_0(2^{j-1}t)\widehat{\varphi}(t). \tag{51}$$

Let us consider the case $t \in \left[\frac{2\pi}{3}, \pi \right]$ and take into account property (48). If $t \notin I$, then $m_0(t) = 0$ and, consequently, $\widehat{\varphi}(\xi) = 0$. If $t \in I$, then $2^J t + 2\pi - 2^J \pi \in [\pi - \eta 2^J, \pi + \eta 2^J]$. Our condition on η implies that this interval fall between I and $2\pi - I$ where $m_0(\xi) \equiv 0$. Thus, $m_0(2^J t) = m_0(2^J t + 2\pi - 2^J \pi) = 0$. But $m_0(2^J t)$ is one of the factors in the product in (51) and, thus, $\widehat{\varphi}(\xi) = 0$.

We now consider the case $t \in \left[\frac{\pi}{2}, \frac{2\pi}{3} \right]$. In this case $2t - 2\pi \in \left[-\pi, -\frac{2\pi}{3} \right]$. If $2t - 2\pi \notin -I$, then $m_0(2t) = m_0(2t - 2\pi) = 0$; hence by (51), $\widehat{\varphi}(\xi) = 0$. If $2t - 2\pi \in -I$, then $2^J(2t - 2\pi) \in -2^J I = [-2^J \pi + \pi - 2^J \eta, -2^J \pi + \pi + 2^J \eta]$ so that $2^{J+1}(t - \pi) + 2^J \pi \in [\pi - 2^J \eta, \pi + 2^J \eta]$. But we have already observed that $m_0(\xi) \equiv 0$ on the last interval. This gives us $m_0(2^{J+1}t) = 0$, and, by (51), $\widehat{\varphi}(\xi) = 0$ (recall that $j \geq J + 2$). We have shown that $\widehat{\varphi}(\xi) \equiv 0$ on $[2^{j-1}\pi, 2^j \pi]$ if $j \geq J + 2$; since $\widehat{\varphi}$ is even we conclude that φ is band-limited.

Recall that we chose J such that $\pi - \pi 2^{-J} > \frac{\pi - \epsilon}{2}$, so that

$$I \cap \left[\frac{\pi - \epsilon}{2}, \frac{\pi + \epsilon}{2} \right] \neq 0$$

and, thus m_0 does not vanish identically $\left[\frac{\pi - \epsilon}{2}, \frac{\pi + \epsilon}{2} \right]$. This finishes the proof of Lemma 7 and Lemma 5 as well.

3.6 Concluding remarks

As we have stated earlier, the purpose of these lectures was to present some of the basic properties of orthonormal wavelets through an elementary treatment. Although much of what has been discussed is true more generally, we have emphasized the band-limited case since it avoids some technical problems and some results have simpler statements and proofs. There are several facts about the band-limited case that we have not included. For example, it was pointed out to us by Paluszynski that if ψ is an orthonormal wavelet that satisfies sup $\widehat{\psi} = [-M, -N] \cup [K, L]$ then $N = 2\pi - \frac{L}{2}$, $K = 2\pi - \frac{M}{2}$, $4\pi < M + L < \frac{16\pi}{3}$, and $0 < L, M < 4\pi$. It follows that if $K = N$ and $L = M$, then $N = 2\pi - \frac{M}{2}$; hence, $N < \pi$ and $M > 2\pi$. We see, therefore, that if the support of $\widehat{\psi}$ consists of two disjoint intervals and $|\widehat{\psi}|$ is even, we are forced to make the assumptions about this support that are made in Theorem 7. Incidentally, we immediately see that Paluszynski's observation proves the assertion made in the paragraph preceding Proposition 2.

REMARK 7   To see that, in Theorem 7, we need only assume that the support of $\widehat{\psi} = e^{i\alpha} b$ is contained in $[-2\pi - \epsilon', -\pi + \epsilon] \cup [\pi - \epsilon, 2\pi + \epsilon']$ and ϵ, ϵ' are minimal in order to conclude that $\epsilon' = 2\epsilon$, we argue as follows: instead of using (B) of Theorem 2 when $j = 1$, we use Lemma 3 with $m = -1$. When $\xi \in [0, \pi]$, this reduces to a sum of two terms and then we have

$$b(\xi)b(2\pi - \xi) = b(2\xi)b(4\pi - 2\xi). \tag{52}$$

It is also important to write the conditions (a)–(d) (in the proof of Theorem 7) in terms of ϵ' instead of 2ϵ. More precisely, we shall use

(a') $b^2(\eta) + b^2(4\pi - \eta) = 1$ if $\eta \in [2\pi - \epsilon', 2\pi + \epsilon']$, and

(c') $b^2(\xi) + b^2(2\pi - \xi) = 1$ if $\xi \in [\pi - \epsilon, \pi + \epsilon]$.

First assume $\epsilon' > 2\epsilon$. Then $[2\pi - \epsilon', 2\pi - 2\epsilon]$ is a non-trivial interval and (a') is valid for η within it. If $\eta \in [2\pi - \epsilon', 2\pi - 2\epsilon]$ we also have $b\left(\frac{\eta}{2}\right) = 0$. Thus, using (52) with $\eta = 2\xi$, we must have

$$b(\eta)b(4\pi - \eta) = 0 \tag{53}$$

for $\eta \in [2\pi - \epsilon', 2\pi - 2\epsilon]$. By (a'), the continuity of b, the fact that $b(2\pi - \epsilon') = 1$ and (53), it follows that $b(4\pi - \eta) = 0$ for all such η. But $\eta \in [2\pi - \epsilon', 2\pi - 2\epsilon]$ if and only if $\xi = 4\pi - \eta \in [2\pi + 2\epsilon, 2\pi + \epsilon']$. Thus, $b(\xi) = 0$ if $\xi \geq 2\pi + 2\epsilon$. This contradicts the minimality of ϵ'. It follows that $\epsilon' \leq 2\epsilon$.

Now assume $\epsilon' < 2\epsilon$. Using (52) we have

$$b(\xi)b(2\pi - \xi) = 0 \tag{54}$$

on $\left[0, \pi - \frac{\epsilon}{2}\right]$ (since $b(2\xi) = 0$ for ξ in this interval). By our assumption, $\left[\pi - \epsilon, \pi - \frac{\epsilon'}{2}\right]$ is a non-trivial interval and, clearly, (54) and (c') are true on this interval. Since $b(\pi - \epsilon) = 0$, the continuity of b, (c') and (54) imply $b(\xi) = 0$ on $\left[\pi - \epsilon, \pi - \frac{\epsilon'}{2}\right]$. But this means $b(\xi) \equiv 0$ on $\left[0, \pi - \frac{\epsilon'}{2}\right] = [0, \pi - \epsilon] \cup \left[\pi - \epsilon, \pi - \frac{\epsilon'}{2}\right]$ and this contradicts the minimality of $\epsilon > 0$. Thus, $\epsilon' \geq 2\epsilon$. ∎

REMARK 8 We established conditions I, II (i), and II (ii) in the band-limited case. As we state below, these conditions were known to Meyer and Lemarié [L2]. A proof of these equalities under the simple assumption that $\psi \in L^2(\mathbb{R})$ can be found in [HKLS].

We end by making some further bibliographic comments. Formulations of equalities (I) and (II) do appear in the work of Lemarié (see [L1]). The recast connection with the formula for the projections Q_j, however, is not included in his paper; nor is the separate and explicit connection with the orthonormality of the system (equation (I)) and the completeness of the system (equation (II)). The last named author of this chapter, in a discussion with Meyer, found out that the latter was also aware of these formulae; in fact, most of the material on this subject that is known to this author is to Meyer's credit, who has continuously conveyed to him his many summaries of the most recent wavelet developments as they emerged from his remarkable school. The local bases (24) and (25) were also discovered by Malvar [Mal]. In his paper, however, there is no discussion of the projections P_l. ∎

References

[A₁] Auscher, P., *Il n'existe pas de bases d'ondelettes regulières dans l'espace de Hardy $H^2(\mathbb{R})$*, to appear in the C. R. Acad. Sci. Paris.

[A₂] Auscher, P., *Remarks on the local Fourier bases*, to appear in Wavelets: Mathematics and Applications, J. Benedetto and M. Frazier, eds.

[A₃] Auscher, P., *Solutions of two problems on wavelets*, submitted for publication.

[AWW]   Auscher, P., Weiss, G. and Wickerhauser, M. V., *Local Sine and Cosine Bases of Coifman and Meyer and the Construction of Smooth Wavelets,* Wavelets: A Tutorial in Theory and Applications, C. K. Chui ed., Acad. Press (1992), 237–56.

[B]   Balian, R., *Un principe d'incertitude fort en théorie du signal ou en mécanique quantique,* C. R. Acad. Sci. Paris, 292, série 2 (1981).

[CM]   Coifman, R. and Meyer, Y., *Remarques sur l'Analyse de Fourier à Fenêtre,* C. R. Acad. Sci. Paris, 312, série I (1991), 259–261.

[D]   Daubechies, I., *Ten Lectures on Wavelets,* SIAM-NSF Regional Conference, Series # 61, SIAM publ. (1992).

[DJJ]   Daubechies, I., Jaffard, S. and Journé, J. L., *A simple Wilson basis with exponential decay,* SIAM J. Math. Anal., 22 (1991), 554–72.

[HKLS] Ha, Y-H, Kang, H., Lee, J. and Seo, J.K., *Unimodular wavelets for L^2 and the Hardy space H^2,* preprint.

[JL]   Jaffard, S. and Laurencot, Ph., *Orthonormal Wavelets, Analysis of Operators and applications to Numerical Analysis,* Wavelets: A Tutorial in Theory and Applications, C. K. Chui ed., Acad. Press, (1992), 543–601.

[La]   Laeng, E., *Une base orthonormale de $L^2(\mathbb{R})$ dont les éléments sont bien localisés dans l'espace de phase et leurs supports adaptés à toute partition symétrique de l'espace des fréquences,* C.R. Acad. Sci. Paris, 311, série 2 (1990), 677–680.

[L_1] Lemarié, P.-G., *Ondelettes à Localisation Exponentielle,* J. Math. Pures et Appl., 67 (1988), 227–236.

[L_2] Lemarié, P.-G., *Analyse multi-échelle et ondelettes á support compact,* Springer Verlag, Lecture Notes in Math, 1438 (1990), 26–38.

[LM]   Lemarié, P.-G. and Meyer, Y., *Ondelettes et Bases Hilbertiennes,* Rev. Mat. Iberoamericana, 2 (1986), 1–18.

[M]   Meyer, Y., *Ondelettes et Opérateurs, I, II, and III,* Hermann Ed., Paris (1990).

[Ma]   Mallat, S., *A Theory for Multiresolution Signal Decomposition: the Wavelet Decompositions,* IEEE Transactions on Pattern Analysis and Machine Inteligence, 11 (1989), 674–693.

[Mal]   Malvar, H., *Lapped Transforms for Efficient Transform/Subband Coding,* IEEE Trans. Acoustics, Speech, and Signal Processing, 38 (1990), 969–978.

[P]   Paganoni, L., *Un Metodo per la Costruzione di Una Classe di Equazioni Funzionali,* Riv. Mat. Univ. Parma, 4 (11) (1985), 275–286.

[Wi]   Wilson, K. G., *Generalized Wannier Functions,* preprint, Cornell University.

4

A Family of Degenerate Differential Operators

Michael Christ[1]

4.1 Introduction

Consider the differential operator

$$D = \partial_x^2 + (\partial_y - x\partial_t)^2,$$

with coordinates (x, y, t) in $\mathbb{R}^3$. It is a prototypical example of a class of non-elliptic operators, with multiple characteristics, which are hypoelliptic. Recall that a differential operator L is said to be hypoelliptic in an open set V if, for any open $\Omega \subset V$, whenever $u \in [C_0^\infty(\Omega)]'$ is a distribution with $Lu \in C^\infty(\Omega)$, then necessarily $u \in C^\infty(\Omega)$. In contrast, L is said to be hypoelliptic at x_0 if whenever f is a distribution for which there exists a neighborhood ω of x_0 such that $Lf \in C^\infty(\omega)$, then there exists a possibly smaller neighborhood ω' of x_0 such that $f \in C^\infty(\omega')$. An example is the operator $\partial_x + i(\partial_y - x^2\partial_t)$ in $\mathbb{R}^3$, which is hypoelliptic at 0 (as follows from standard results concerning analytic extendibility of CR functions from non-pseudoconvex hypersurfaces), but is not hypoelliptic in any neighborhood of 0.

Consider a family of variants:

$$D_\theta = \partial_x^2 + (e^{i\theta}\partial_y - x^{m-1}\partial_t)^2$$

where $\theta \in \mathbb{R}$ and $m \in \{2, 4, 6, \ldots\}$; these are all hypoelliptic for $\theta = 0$. For $e^{i\theta} \notin \mathbb{R}$, they arise naturally in work of Pham The Lai and Robert [PR] (see Remark 3 below), but seem not to have been studied much.

PROPOSITION 1
Let $m \in \{2, 4, 6, \ldots\}$. Then D_θ is hypoelliptic at 0, if and only if, $e^{i\theta} \in \mathbb{R}$.

[1]Research supported by the National Science Foundation.

The positive result for $e^{i\theta} \in \mathbb{R}$ is well known. When $e^{i\theta} \notin \mathbb{R}$, then in any open set in which the x coordinate never vanishes, D_θ is of principal type and the well known results for that case imply that it is neither hypoelliptic nor locally solvable. But when $e^{i\theta}$ is neither real nor imaginary, D_θ has multiple characteristics where $x = 0$.

REMARK 1 $\partial_x^2 + (\partial_y - e^{i\theta} x^{m-1} \partial_t)^2$ is not hypoelliptic at 0, provided $\theta \neq 0$ is sufficiently small. This will follow from the same method.   ∎

REMARK 2   A number of results relate hypoellipticity of certain differential operators to the existence of "zeros" of an appropriate "symbol." A constant-coefficient operator, for example, is hypoelliptic if and only if the set Γ of zeros $z \in \mathbb{C}^n$ of its symbol has the property that $|\text{Im}(z)| \to \infty$ as $|\text{Re}(z)| \to \infty$, for $z \in \Gamma$. A second set of examples are homogeneous, left-invariant differential operators on graded nilpotent Lie groups; such an operator L is hypoelliptic, if and only if $\pi(L)$ is injective, for every irreducible unitary representation π of the group [RS], [HN1].

In our context a "symbol" is arrived at by separating variables and considering

$$u(x, y, t) = f(x)e^{i\eta y}e^{i\tau t}$$

where η, $\tau \in \mathbb{R}$. Then

$$-D_\theta u = e^{i\eta y}e^{i\tau t}\left[-\frac{d^2}{dx^2} + (\eta e^{i\theta} - x^{m-1}\tau)^2\right]f(x).$$

Thus arise the ordinary differential operators

$$L_\zeta = -\frac{d^2}{dx^2} + (\zeta - x^{m-1})^2 \tag{1}$$

obtained by setting $\tau = 1$ and $\zeta = \eta e^{i\theta}$. If there were to exist a value of $\eta \in \mathbb{R}$ and a function $f \in L^2(\mathbb{R})$, not identically vanishing, such that $L_\zeta f \equiv 0$, then it would follow immediately from an examination of the solutions $f(\tau^{1/m}x)e^{i\eta\tau^{1/m}y}e^{i\tau t}$, as $\tau \to +\infty$, that D_θ could not be hypoelliptic at 0. See Rothschild and Stein [RS] for this type of argument.   ∎

PROPOSITION 2
Let $m \in \{2, 3, 4, \ldots\}$. If $0 < |\arg(\zeta)| < \frac{\pi}{2}\frac{m-1}{m}$, if $f \in L^2(\mathbb{R})$ and $L_\zeta f \equiv 0$, then $f \equiv 0$.

Thus, for θ small but nonzero, there is no obstruction to hypoellipticity based upon "vanishing of the symbol," that is, noninjectivity of the L_ζ with $\zeta = \rho\, e^{i\theta}$ and ρ real, or nearly real.[2] The Lie bracket $[\partial_x, (e^{i\theta}\partial_y - x^{m-1}\partial_t)] = -(m-1)x^{m-2}\partial_t$

---

[2]For $m > 2$ and even, there do exist infinitely many $\zeta \in \mathbb{C}$ for which L_ζ has a nontrivial nullspace in L^2 [C1], [C2], so such an obstruction is present for some θ.

is independent of θ, so there is likewise no obstruction to hypoellipticity stemming from integrability considerations.

The crux of our analysis is that while there is no nullspace in L^2, it is a near thing:

PROPOSITION 3

Assume that $e^{i\alpha} \notin \mathbb{R}$ and write $\zeta = \rho e^{i\alpha}$. Then there exists $C < \infty$ and $\delta > 0$, and for each $\rho \geq 1$ a Schwartz function g, not identically vanishing, such that

$$\|L_\zeta g\|_{L^2(\mathbb{R})} \leq Ce^{-\delta\rho^{m/(m-1)}} \|g\|_{L^2(\mathbb{R})}.$$

REMARK 3 $\zeta \in \mathbb{C}$ is said [K], [PR] to be a nonlinear eigenvalue of the operator family $\{L_z : z \in \mathbb{C}\}$ if there exists $f \in L^2$, not identically zero, satisfying $L_\zeta f \equiv 0$. To each such ζ, one associates a generalized eigenspace Sp_ζ, which includes f but may be larger if ζ is, in an appropriate sense, of multiplicity greater than one; see Keldysh [K] or Pham The Lai and Robert [PR, Definition 1.3] for details. So far as we know, it remains an open question whether the linear span of all the Sp_ζ (as ζ ranges over the set of all nonlinear eigenvalues) is dense in L^2. With L_ζ defined as in (1), the span was proven to be dense for odd integers $m \geq 3$ [PR], [HH]. For $m = 2$ there are no nonlinear eigenvalues. For even $m \geq 4$ the set of nonlinear eigenvalues is infinite and discrete [C1] and [C2].

Proposition 2 implies that all the nonlinear eigenvalues lie in a narrow (for large m) conic neighborhood of the imaginary axis. This is of mild interest, for the following reason. According to the method of Pham The Lai and Robert [PR], to prove completeness it would suffice to partition $\mathbb{C}$ into finitely many sectors of aperture less than $\pi\frac{m-1}{m}$, bounded by rays of polynomial growth. A ray $\{\rho e^{i\theta} : \rho \geq 0\}$ is said to be of polynomial growth if there exists $r \in \mathbb{R}$, possibly negative, such that for all sufficiently large ρ and all $f \in C_0^\infty(\mathbb{R})$,

$$\|L_{\rho e^{i\theta}} f\|_{L^2} \geq \rho^r \|f\|_{L^2}.$$

The two real half-axes are such rays, as is every ray in $\{\zeta : Im(\zeta) < 0\}$ when m is odd. For m even, $\mathbb{R}^\pm$ are, according to Proposition 3, the only rays of polynomial growth; hence the method of [PR] cannot establish the existence of nonlinear eigenvalues, let alone completeness of the associated eigenspaces. ∎

REMARK 4 When θ is nonzero, L_θ has to some degree a hyperbolic character. For instance, its self-adjoint part is $\partial_x^2 + (\cos(\theta)\partial_y - x^{m-1}\partial_t)^2 - \sin^2(\theta)\partial_y^2$. This hyperbolic character becomes more apparent in those estimates of Section 4.3 that concern behavior near x_0. ∎

REMARK 5 As was suggested by Müller, our estimates lead naturally, in the spirit of [M], to a proof that for $e^{i\theta} \notin \mathbb{R}$, D_θ fails to satisfy a version of local solvability "at a point," which is related to the usual notion of local solvability in a neighborhood of that point in the same way that hypoellipticity at a point is related to hypoellipticity in an open set. However, the precise formulation of this notion is somewhat involved and seems to be of less intrinsic interest than that of hypoellipticity at a point, and the details of the proof are likewise more involved, so it will not be discussed here.[3]

Propositions 1 and 3 will be proven only for $0 < |\theta| < \frac{\pi}{2} \frac{m-1}{m}$, where the argument is simpler. The proofs of Propositions 2 and 3 rely on standard methods for the asymptotic analysis of solutions of ordinary differential equations with irregular singular points at infinity. The reader should be aware that while $C, c, \delta, \varepsilon$ denote always positive, finite constants, their values may change freely from one occurrence to the next. ∎

4.2 Reduction to a first-order system

Consider the ordinary differential equation

$$\frac{d^2}{dx^2} f = h^2 f$$

where $h \in C^\infty$ is given. In this section we reduce the equation to an approximately diagonalized first-order system; the WKB approximation $f \approx |\phi'|^{-1/2} \exp(\phi)$, where $\phi' = h$, will be implicit in the procedure. We proceed in a purely formal fashion, postponing issues such as division by zero, until particular instances of h are introduced.

Set

$$u = \begin{pmatrix} f \\ \frac{df}{dx} \end{pmatrix}$$

and

$$A = \begin{pmatrix} 0 & 1 \\ h^2 & 0 \end{pmatrix},$$

so that $\frac{du}{dx} = Au$. Introduce

$$S = \begin{pmatrix} 1 & 1 \\ h - \frac{1}{2}h'h^{-1} & -h - \frac{1}{2}h'h^{-1} \end{pmatrix}.$$

---

[3] Details may be found in a version of this paper circulated as a reprint.

Then

$$S^{-1} = -\frac{1}{2}h^{-1}\begin{pmatrix} -h - \frac{1}{2}h'h^{-1} & -1 \\ -h + \frac{1}{2}h'h^{-1} & 1 \end{pmatrix},$$

$$AS = \begin{pmatrix} h - \frac{1}{2}h'h^{-1} & -h - \frac{1}{2}h'h^{-1} \\ h^2 & h^2 \end{pmatrix}$$

and

$$S^{-1}AS = -\frac{1}{2}h^{-1}\begin{pmatrix} -2h^2 + \frac{1}{4}(h')^2h^{-2} & (-h - \frac{1}{2}h'h^{-1})^2 - h^2 \\ -(h - \frac{1}{2}h'h^{-1})^2 + h^2 & 2h^2 - \frac{1}{4}(h')^2h^{-2} \end{pmatrix}$$

$$= -\frac{1}{2}h^{-1}\begin{pmatrix} -2h^2 + \frac{1}{4}(h')^2h^{-2} & h' + \frac{1}{4}(h')^2h^{-2} \\ h' - \frac{1}{4}(h')^2h^{-2} & 2h^2 - \frac{1}{4}(h')^2h^{-2} \end{pmatrix}.$$

Rewrite the equation $u' = Au$ as

$$\frac{d}{dx}(S^{-1}u) = S^{-1}Au - S^{-1}S'S^{-1}u$$

$$= (S^{-1}AS - S^{-1}S')S^{-1}u.$$

Now

$$\frac{dS}{dx} = \begin{pmatrix} 0 & 0 \\ h' - \frac{1}{2}h''h^{-1} + \frac{1}{2}(h')^2h^{-2} & -h' - \frac{1}{2}h''h^{-1} + \frac{1}{2}(h')^2h^{-2} \end{pmatrix}$$

and

$$S^{-1}\frac{dS}{dx} =$$

$$-\frac{1}{2}h^{-1}\begin{pmatrix} -h' + \frac{1}{2}h''h^{-1} - \frac{1}{2}(h')^2h^{-2} & h' + \frac{1}{2}h''h^{-1} - \frac{1}{2}(h')^2h^{-2} \\ h' - \frac{1}{2}h''h^{-1} + \frac{1}{2}(h')^2h^{-2} & -h' - \frac{1}{2}h''h^{-1} + \frac{1}{2}(h')^2h^{-2} \end{pmatrix}.$$

Defining

$$V = S^{-1}u \quad \text{and} \quad B = S^{-1}AS - S^{-1}\frac{dS}{dx},$$

we obtain

$$\frac{dV}{dx} = Bv \tag{2}$$

with

$$B = -\frac{1}{2}h^{-1}\begin{pmatrix} -2h^2 + \frac{1}{4}(h')^2h^{-2} & h' + \frac{1}{4}(h')^2h^{-2} \\ + h' - \frac{1}{2}h''h^{-1} & -h' - \frac{1}{2}h''h^{-1} \\ + \frac{1}{2}(h')^2h^{-2} & + \frac{1}{2}(h')^2h^{-2} \\ & \\ h' - \frac{1}{4}(h')^2h^{-2} & 2h^2 - \frac{1}{4}(h')^2h^{-2} \\ -h' + \frac{1}{2}h''h^{-1} & + h' + \frac{1}{2}h''h^{-1} \\ - \frac{1}{2}(h')^2h^{-2} & - \frac{1}{2}(h')^2h^{-2} \end{pmatrix}.$$

The two terms $-\frac{1}{2}h'h^{-1}$ in the definition of S were introduced so that the cancellation $h' - h' = 0$ would occur in the off-diagonal entries of B; this amounts to the WKB approximation. Thus

$$B = \begin{pmatrix} h - \frac{1}{2}h'h^{-1} & 0 \\ 0 & -h - \frac{1}{2}h'h^{-1} \end{pmatrix} + E \tag{3}$$

where

$$E = \begin{pmatrix} -\frac{3}{8}(h')^2h^{-3} + \frac{1}{4}h''h^{-2} & -\frac{3}{8}(h')^2h^{-3} + \frac{1}{4}h''h^{-2} \\ \frac{3}{8}(h')^2h^{-3} - \frac{1}{4}h''h^{-2} & \frac{3}{8}(h')^2h^{-3} - \frac{1}{4}h''h^{-2} \end{pmatrix}. \tag{4}$$

4.3 Solutions of L_ζ

Let $\zeta = \rho e^{i\alpha}$ with $\rho \in \mathbb{R}^+$ and $0 < \alpha \le \frac{\pi}{2}$. The analysis of L_ζ for other values of ζ reduces to this case; for complex conjugation and reflection about $x = 0$ convert ζ to $\bar{\zeta}$ and to $-\zeta$, respectively. Set

$$\phi(x) = \zeta x - m^{-1}x^m \quad \text{and} \quad h(x) = \phi'(x) = \zeta - x^{m-1},$$

and

$$\Phi = \phi - \frac{1}{2}\log(-\phi')$$

where log denotes the natural logarithm. Since $\text{Im}(\phi'(x)) \equiv \rho\sin\alpha \ne 0$, $\log(-\phi'(x))$ may be uniquely defined as a continuous function of all $x \in \mathbb{R}$, so that its imaginary part tends to zero as $x \to +\infty$.

There exists a unique solution $\psi^+ = \psi_\zeta^+$ of $L_\zeta\psi^+ \equiv 0$ on all $\mathbb{R}$, such that

$$\psi^+(x) = e^{\Phi(x)}(1 + O(x^{-1})) \quad \text{as} \quad x \to +\infty.$$

See for instance Coddington and Levinson [CL, Chapter 5]. More precisely, there exists $C_0 < \infty$ such that

$$|\psi^+(x) - e^{\Phi(x)}| \leq C\rho^{1/(m-1)}x^{-1}|e^{\Phi(x)}| \tag{5}$$

for all $x \geq C_0\rho^{1/(m-1)}$, provided $\rho \geq 1$ [C2]. Also

$$\left|\frac{d}{dx}(\psi^+ - e^{\Phi})(x)\right| \leq C\rho^{1/(m-1)}x^{-1}\left|\Phi'e^{\Phi}\right|(x) \tag{6}$$

for the same range of x and ρ; this is implicit in the proof of (3.1) in [C2].

Define

$$x_0 = x_0(\rho, \alpha) = (\rho \cos \alpha)^{1/(m-1)}$$

to be the unique critical point of $\mathrm{Re}\,(\phi)$. Our present goal is to estimate ψ^+, from above and below, at x_0, by analyzing the equation (2) on the interval $[x_0, b]$, where $b = C_1\rho^{1/(m-1)}$ and $C_1 \geq C_0$. With $h(x) = \zeta - x^{m-1}$, and assuming always that $\rho \geq 1$ and $0 < \alpha < \frac{\pi}{2}$, the entries of E satisfy

$$|(h')^2h^{-3}(x)| \leq C\rho^{2(m-2)/(m-1)}\rho^{-3}$$

and

$$|h''h^{-2}(x)| \leq C\rho^{(m-3)/(m-1)}\rho^{-2},$$

for all $x \in [x_0, b]$ where C depends on α and on C_1. Thus

$$|E(x)| \leq C\rho^{-(m+1)/(m-1)} \qquad \forall x \in [x_0, b].$$

Let

$$u = \begin{pmatrix} \psi^+ \\ \frac{d}{dx}\psi^+ \end{pmatrix},$$

let $v = S^{-1}u$ as earlier, and set $w = e^{-\Phi}v$. Then

$$\frac{dw}{dx} = \begin{pmatrix} 0 & 0 \\ 0 & -2h \end{pmatrix} w + Ew.$$

Writing

$$w = \begin{pmatrix} w_1 \\ w_2 \end{pmatrix},$$

we have

$$\frac{d}{dx}|w(x)|^2 = 2\,\mathrm{Re}\,(-2\phi'(x))|w_2(x)|^2 + 2\,\mathrm{Re}\,\langle Ew, w\rangle(x).$$

For $x \in [x_0, b]$, $\mathrm{Re}\,(\phi'(x)) \leq 0$, so

$$\frac{d}{dx}|w(x)|^2 \geq -C\rho^{-(m+1)/(m-1)}|w(x)|^2.$$

Therefore

$$|w(x_0)|^2 \leq \exp\left(C \int_{x_0}^b \rho^{-(m+1)/(m-1)}\, dx\right) |w(b)|^2$$

$$\leq C|w(b)|^2$$

$$\leq C$$

for all $\rho \geq 1$. To justify the final step, recall that

$$w = e^{-\Phi} v$$

$$= e^{-\Phi} S^{-1} \begin{pmatrix} \psi^+ \\ \frac{d}{dx}\psi^+ \end{pmatrix}$$

$$= e^{-\Phi} \left(-\frac{1}{2}(\phi')^{-1}\right) \begin{pmatrix} -\phi' - \frac{1}{2}\phi''(\phi')^{-1} & -1 \\ -\phi' + \frac{1}{2}\phi''(\phi')^{-1} & 1 \end{pmatrix} \begin{pmatrix} \psi^+ \\ \frac{d}{dx}\psi^+ \end{pmatrix}$$

$$= \frac{1}{2}\begin{pmatrix} e^{-\Phi}(\psi^+ + [(\phi')^{-1} + \frac{1}{2}\phi''(\phi')^{-2}]\frac{d}{dx}\psi^+) \\ e^{-\Phi}(\psi^+ - (\phi')^{-1}\frac{d}{dx}\psi^+) - e^{-\Phi}\frac{1}{2}\phi''(\phi')^{-2}\frac{d}{dx}\psi^+ \end{pmatrix}.$$

At $x = b$, we have $e^{-\Phi}\psi^+ = 1 + O(C_1^{-1})$ and $e^{-\Phi}\frac{d}{dx}\psi^+ = \phi'(1 + O(C_1^{-1}))$, while $\phi''(\phi')^{-1}(b) = O(\rho^{-1/(m-1)})$ so that

$$|w_1(b)| \leq C$$

and

$$|w_2(b)| \leq C_2\rho^{-1/(m-1)} + CC_1^{-1}$$

where C_2 depends on C_1. In particular, once C_1 is fixed, $w(b) = O(1)$ as $\rho \to \infty$. Hence

$$|w(x)| \leq C \quad \text{for all} \quad x \in [x_0, C_1]. \tag{7}$$

We claim that

$$w_1(x_0) \to 1 \quad \text{and} \quad w_2(x_0) \to 0$$

as $\rho \to \infty$. Indeed,

$$\frac{dw_2}{dx}(x) = -2\phi'(x)w_2(x) + O(|E(x)| \cdot |w(x)|),$$

whence

$$\left|\frac{d}{dx}(e^{2\phi}w_2)(x)\right| \leq C\rho^{-(m+1)/(m-1)}|e^{2\phi}(x)|,$$

entailing

$$|w_2(x_0)| \le \left|e^{-2\phi(x_0)}e^{2\phi(b)}w_2(b)\right| + C\rho^{-(m+1)/(m-1)}\int_{x_0}^{b}\left|e^{2(\phi(s)-\phi(x_0))}\right|ds$$

$$\le Ce^{-\delta\rho^{m/(m-1)}}|w_2(b)| + \rho^{-(m+1)/(m-1)}|b - x_0|$$

$$\le Ce^{-\delta\rho^{m/(m-1)}} + C\rho^{-m/(m-1)}$$

$$\le C\rho^{-m/(m-1)}.$$

To obtain the second line, we have used the estimate $\operatorname{Re}(\phi(b) - \phi(x_0)) \le -\delta\rho^{m/(m-1)}$, valid for C_1 sufficiently large, and the fact that $\operatorname{Re}(\phi)$ has a global maximum at x_0. Thus $w_2(x_0) \to 0$.

Similarly

$$\left|\frac{d}{dx}w_1\right| \le |E| \cdot |w_1| \le C\rho^{-(m+1)/(m-1)}$$

so that

$$|w_1(x_0) - w_1(b)| \le C\rho^{-m/(m-1)},$$

where C depends again on C_1. But $w_1(b) = 1 + O(C_1^{-1})$, so

$$|w_1(x_0) - 1| \le C(C_1)\rho^{-m/(m-1)} + C \cdot C_1^{-1}.$$

Taking C_1 to be a sufficiently slowly increasing function of ρ, we find that $w_1(x_0) \to 1$.

To convert all this to information on $\psi^+(x_0)$,

$$\begin{pmatrix} \psi^+ \\ \frac{d}{dx}\psi^+ \end{pmatrix}(x_0) = Sv(x_0)$$

$$= e^{\Phi}\begin{pmatrix} 1 & 1 \\ \phi' - \frac{1}{2}\phi''(\phi')^{-1} & -\phi' - \frac{1}{2}\phi''(\phi')^{-1} \end{pmatrix}\begin{pmatrix} w_1 \\ w_2 \end{pmatrix}(x_0)$$

$$= \begin{pmatrix} e^{\Phi(x_0)}(1 + o(1)) \\ (e^{\Phi}\phi')(x_0)(1 + o(1)) \end{pmatrix} \tag{8}$$

as $\rho \to \infty$.

There exists a second solution, ψ^-, of $L_\xi\psi^- \equiv 0$, with

$$\psi^-(x) = e^{\Phi(x)}(1 + O(|x|^{-1})) \qquad \text{as} \qquad x \to -\infty.$$

All the preceding reasoning applies equally well to ψ^-.

Define

$$\gamma = \frac{1}{2}\frac{m-2}{m-1}. \tag{9}$$

Then $|\mathrm{Re}\,(\phi(x) - \phi(x_0))| \leq C$ whenever $|x - x_0| \leq C'\rho^{-\gamma}$. It follows as above that

$$\psi^+(x) = e^{\Phi(x)}(1 + o(1)) \qquad\qquad \forall\,|x - x_0| \leq C\rho^{-\gamma} \qquad (10)$$

and

$$0 < C^{-1} < \left|\frac{\psi^+(x)}{\psi^+(x_0)}\right| < C < \infty \qquad\qquad \forall\,|x - x_0| \leq C\rho^{-\gamma}. \qquad (11)$$

The same applies to ψ^-.

One has

$$\mathrm{Re}\,\phi(x_0) = \rho\cos\alpha \cdot (\rho\cos\alpha)^{1/(m-1)} - m^{-1} \cdot (\rho\cos\alpha)^{m/(m-1)}$$

$$= \frac{m-1}{m}(\cos\alpha)^{m/(m-1)}\rho^{m/(m-1)},$$

and

$$\left|\exp\left(-\frac{1}{2}\log(-\phi'(x_0))\right)\right| = |\phi'(x_0)|^{-1/2} = (\sin\alpha)^{-1/2}\rho^{-1/2},$$

so that

$$|\psi^+(x_0)|$$

$$= C\rho^{-1/2}\exp\left(\frac{m-1}{m}(\cos\alpha)^{m/(m-1)}\rho^{m/(m-1)}\right) \cdot (1 + o(1)) \qquad (12)$$

as $\rho \to \infty$. The same holds for ψ^-.

4.4   A variant

In this section, we repeat the preceding analysis for a related family of equations:

$$\tilde{L}_{\theta,\rho} = -\frac{d^2}{dx^2} + \tilde{h}^2$$

where

$$\tilde{h}(x) = e^{i\theta}(\rho - x^{m-1}) = \tilde{\phi}'(x)$$

and

$$\tilde{\phi}(x) = e^{i\theta}(\rho x - m^{-1}x^m),$$

with $0 < \theta < \frac{\pi}{2}$. In this case, the real and imaginary parts of $\tilde{\phi}'$ have a common zero, at

$$\tilde{x}_0 = \rho^{1/(m-1)}.$$

There exist unique solutions $\widetilde{\psi}^{\pm}$ of $L_{\theta,\rho}\widetilde{\psi}^{\pm} \equiv 0$, with

$$\left|\widetilde{\psi}^+(x) - e^{\widetilde{\Phi}(x)}\right| \le C\rho^{1(m-1)}|x|^{-1}|e^{\widetilde{\Phi}(x)}|$$

for all $x \ge C_0\rho^{1/(m-1)}$, for all $\rho \ge 1$. Here

$$\widetilde{\Phi} = \widetilde{\phi} - \frac{1}{2}\log(-\widetilde{\phi}') = \widetilde{\phi} - \frac{1}{2}\log(-(\rho - x^{m-1})) + \frac{i\theta}{2}$$

is defined for all $x > x_0$. Similarly for $\widetilde{\psi}^-$, with $\widetilde{\phi}$ replaced by $\widetilde{\phi} - \frac{1}{2}\log(\widetilde{\phi}')$.
Likewise

$$\left|\frac{d}{dx}(\widetilde{\psi}^+) - \widetilde{\phi}'e^{\widetilde{\Phi}}\right|(x) \le C\rho^{1/(m-1)}|x|^{-1}\left|\widetilde{\phi}'e^{\widetilde{\Phi}}\right|(x)$$

for all $x \ge C_0\rho^{1(m-1)}$ and $\rho \ge 1$, and analogously for $\frac{d}{dx}\widetilde{\psi}^-$. This follows directly from the proofs in [C2] of the corresponding statements for $\psi^{\pm}$.

Retain the definitions $\gamma = \frac{1}{2}\frac{m-2}{m-1}$ and $b = C_1\rho^{1/(m-1)}$, and set $a = \widetilde{x}_0 + \rho^{-\gamma}$.
Define

$$u = \begin{pmatrix} \widetilde{\psi}^+ \\ \frac{d}{dx}\widetilde{\psi}^+ \end{pmatrix},$$

reduce $L_{\theta,\rho}$ to a first-order system as in Section 2, and define v, w as before. Then

$$\frac{d}{dx}|w|^2 = 2\text{Re}\,(-2\widetilde{\phi}')|w|^2 + O(|E| \cdot |w|^2)$$

$$\ge -C|E| \cdot |w|^2$$

for $x > \widetilde{x}_0$.

The entries of E satisfy less favorable bounds than before:

$$|\widetilde{h}'(x)| \le C\rho^{(m-2)/(m-1)},$$

$$|\widetilde{h}''(x)| \le C\rho^{(m-3)/(m-1)},$$

and

$$|\widetilde{h}(x)| \ge C\rho^{2\gamma}|x - \widetilde{x}_0|$$

for all $x \in [a, b]$ and $\rho \ge 1$. Hence

$$|E(x)| \le C\rho^{2(m-2)/(m-1)}\rho^{-6\gamma}(x - \widetilde{x}_0)^{-3} + C\rho^{(m-3)/(m-1)}\rho^{-4\gamma}|x - \widetilde{x}_0|^{-2}$$

$$\le C\rho^{-2\gamma}(|x - \widetilde{x}_0|^{-3} + |x - \widetilde{x}_0|^{-2}).$$

Therefore

$$|w(a)|^2 \le \exp\left(C\int_a^b |E(x)|dx\right)|w(b)|^2$$

$$\le C\exp\left(C\rho^{-2\gamma}\int_a^b (|x - \widetilde{x}_0|^{-3} + |x - \widetilde{x}_0|^{-2})dx\right)$$

$$\leq C \exp\left(C\rho^{-2\gamma}\rho^{2\gamma}\right)$$

$$\leq C.$$

Consequently

$$\left|\widetilde{\psi}^{+}(a)\right| = \left|e^{\widetilde{\Phi}(a)}\right| \cdot |w_1(a) + w_2(a)|$$

$$\leq C \exp(\operatorname{Re}\widetilde{\phi}(a)) \, |\widetilde{\phi}'(a)|^{-1/2}.$$

Again

$$\left|\operatorname{Re}\left(\widetilde{\phi}(a) - \widetilde{\phi}(\widetilde{x}_0)\right)\right| \leq C\rho^{2\gamma}|a - \widetilde{x}_0|^2 \leq C,$$

while

$$|\widetilde{\phi}'(a)| \sim \rho^{2\gamma}|a - \widetilde{x}_0| \sim \rho^{\gamma}.$$

Since

$$\operatorname{Re}\widetilde{\phi}(\widetilde{x}_0) = \cos(\theta)(\rho \cdot \rho^{1/(m-1)} - m^{-1}\rho^{m/(m-1)})$$

$$= \frac{m-1}{m}\cos(\theta)\rho^{m/(m-1)},$$

we obtain

$$\left|\widetilde{\psi}^{+}(a)\right| \leq C\rho^{-\gamma/2}\exp\left(\frac{m-1}{m}\cos(\theta)\rho^{m/(m-1)}\right) \tag{13}$$

for all $\rho \geq 1$. In the same way

$$|(\widetilde{\psi}^{+})'(a)| \leq \left|\widetilde{\phi}' - \frac{1}{2}\widetilde{\phi}''(\widetilde{\phi}')^{-1}\right|(a) \cdot |w_1(a)| + \left|-\widetilde{\phi}' - \frac{1}{2}\widetilde{\phi}''(\widetilde{\phi}')^{-1}\right|(a) \cdot |w_2(a)|$$

$$\leq C\left(\rho^{\gamma} + \rho^{2\gamma}\rho^{-\gamma}\right)|e^{\widetilde{\Phi}(a)}|$$

$$\leq C\rho^{\gamma/2}\exp\left(\frac{m-1}{m}\cos(\theta)\rho^{m/(m-1)}\right). \tag{14}$$

To pass to bounds at $\widetilde{x}_0$,

$$\frac{d}{dx}\left(|\widetilde{\psi}^{+}(x)|^2 + s^{-2\gamma}|(\widetilde{\psi}^{+})'(x)|^2\right)$$

$$\leq 2\operatorname{Re}\left(\widetilde{\psi}^{+}(\widetilde{\psi}^{+})'(x)\right) + \rho^{-2\gamma} \cdot 2\operatorname{Re}\left((\widetilde{\psi}^{+})'(\widetilde{\phi}')^2\widetilde{\psi}^{+}\right)(x)$$

$$\leq C\rho^{\gamma}\left(|\widetilde{\psi}^{+}(x)|^2 + \rho^{-2\gamma}|(\widetilde{\psi}^{+})'(x)|^2\right)$$

for all $x \in [\widetilde{x}_0, a]$. Thus

$$|\widetilde{\psi}^{+}(\widetilde{x}_0)|^2 + \rho^{-2\gamma}|(\widetilde{\psi}^{+})'(\widetilde{x}_0)|^2$$

$$\leq \exp\left(\int_{\widetilde{x}_0}^{a} C\rho^{\gamma}dx\right) \cdot \left(|\widetilde{\psi}^{+}(a)|^2 + \rho^{-2\gamma}|(\widetilde{\psi}^{+})'(a)|^2\right)$$

$$\leq C\rho^{-\gamma}|e^{\widetilde{\Phi}(a)}|^2.$$

Therefore

$$|\widetilde{\psi}^+(\widetilde{x}_0)| + \rho^{-\gamma}|(\widetilde{\psi}^+)'(\widetilde{x}_0)| \le C\rho^{-\gamma/2} \exp\left(\frac{m-1}{m}\cos(\theta)\rho^{m/(m-1)}\right).$$

The same bounds hold for $\widetilde{\psi}^-$, by the same reasoning. Therefore we obtain the Wronskian bound

$$\left|\det\begin{pmatrix} \widetilde{\psi}^+ & \widetilde{\psi}^- \\ \frac{d}{dx}\widetilde{\psi}^+ & \frac{d}{dx}\widetilde{\psi}^- \end{pmatrix}(\widetilde{x}_0)\right| \le C\exp\left(2\frac{m-1}{m}\cos(\theta)\rho^{m/(m-1)}\right) \tag{15}$$

for all $\rho \ge 1$.

4.5 Wronskian estimate

Assume $\zeta = \rho e^{i\alpha}$ with $1 \le \rho \in \mathbb{R}^+$ and $0 < \alpha < \frac{\pi}{2}\frac{m-1}{m}$. Let $\psi^\pm$ be the solutions of L_ζ introduced in Section 4.3. Define

$$W = \det\begin{pmatrix} \psi^+ & \psi^- \\ \frac{d}{dx}\psi^+ & \frac{d}{dx}\psi^- \end{pmatrix}(x);$$

the determinant is independent of $x \in \mathbb{R}$. We will establish the bound

$$|W| \le C\exp\left(2\cos\left(\frac{m\alpha}{m-1}\right)\cdot\frac{m-1}{m}\rho^{m/(m-1)}\right). \tag{16}$$

First, the solutions $\psi^\pm$ extend to entire holomorphic functions of $z \in \mathbb{C}$, and satisfy

$$\psi^\pm(te^{i\theta}) = e^{\Phi(te^{i\theta})}(1 + O(|t|^{-1})) \tag{17}$$

as $t \to \pm\infty$ respectively, for all $|\theta| < \frac{\pi}{2}m^{-1}$, and

$$\left[-\frac{\partial^2}{\partial z^2} + (\zeta - z^{m-1})^2\right]\psi^\pm(z) \equiv 0,$$

where $\Phi(z) = \phi(z) - \frac{1}{2}\log(-\phi'(z))$ and $\phi(z) = \zeta z - m^{-1}z^m$. See, for instance, Coddington and Levinson [CL, Chapter 5]. The Wronskian becomes a constant function of $z \in \mathbb{C}$. To evaluate it, set

$$\theta = \frac{\alpha}{(m-1)}$$

and

$$g^\pm(t) = \psi^\pm(te^{i\theta}).$$

Then

$$\left[-e^{-2i\theta}\frac{d^2}{dt^2} + (\rho e^{i\alpha} - e^{i(m-1)\theta}t^{m-1})^2\right]g^{\pm} \equiv 0 \quad \text{for} \quad t \in \mathbb{R}.$$

The differential operator may be rewritten as

$$e^{i\alpha(m-2)/(m-1)}\left[-e^{-i\alpha m/(m-1)}\frac{d^2}{dt^2} + e^{i\alpha m/(m-1)}(\rho - t^{m-1})^2\right], \qquad (18)$$

or again as a constant of modulus one times

$$-\frac{d^2}{dt^2} + e^{2i\alpha m/(m-1)}(\rho - t^{m-1})^2 = \widetilde{L}_{\beta,\rho}$$

where $\beta = \frac{\alpha m}{m-1}$. Since

$$\phi(te^{i\theta}) = \rho e^{i\alpha} \cdot te^{i\theta} - m^{-1}(te^{i\theta})^m = e^{i\beta}(\rho t - m^{-1}t^m),$$

the asymptotics (17) imply that $g^{\pm}$ are the solutions $\widetilde{\psi}^{\pm}$ of $\widetilde{L}_{\beta,\rho}$ discussed in Section 4.4. Therefore by (15),

$$|W| = \left|\det\begin{pmatrix} g^+ & g^- \\ \frac{d}{dt}g^+ & \frac{d}{dt}g^- \end{pmatrix}(\widetilde{t}_0)\right|$$

$$\leq C\exp\left(2\frac{m-1}{m}\cos(\beta)\rho^{m/(m-1)}\right)$$

$$= C\exp\left(2\frac{m-1}{m}\cos\left(\frac{m\alpha}{m-1}\right)\rho^{m/(m-1)}\right).$$

Note that for any $y > 1$ and $\sigma \in \left(0, \frac{\pi}{2}y^{-1}\right)$,

$$\cos(y\sigma) < (\cos(\sigma))^y. \qquad (19)$$

Indeed, setting $f(\sigma) = y\log(\cos(\sigma)) - \log(\cos(y\sigma))$ we have

$$f'(\sigma) = y \cdot (\tan(y\sigma) - \tan(\sigma)) > 0.$$

Since $f(0) = 0$, $f(\sigma) > 0$ for $\sigma \in \left(0, \frac{\pi}{2}y^{-1}\right)$ and (19) follows.

The bound (16) now implies that great cancellation occurs in the Wronskian, for

$$|\psi^+(x_0) \cdot (\psi^-)'(x_0)| \sim \exp\left(2, \frac{m-1}{m}, [\cos(\alpha)]^{m/(m-1)}\rho^{m/(m-1)}\right)$$

$$= e^{\delta\rho^{m/(m-1)}}\exp\left(2, \frac{m-1}{m}\cos\left(\frac{m\alpha}{m-1}\right)\rho^{m/(m-1)}\right)$$

$$\geq ce^{\delta\rho^{m/(m-1)}}|W|$$

where $\delta = \left([\cos(\alpha)]^{m/(m-1)} - \cos\left(\frac{m\alpha}{m-1}\right)\right) \cdot 2\frac{m-1}{m} > 0.$

4.6 Proof of Proposition 3

Fix $\alpha \in (0, \frac{\pi}{2} \frac{m-1}{m})$, let $\rho \in \mathbb{R}^+$ and set $\zeta = \rho e^{i\alpha}$. Fix nonnegative functions $\eta^{\pm} \in C^{\infty}(\mathbb{R})$ satisfying $\eta^-(x) \equiv 1$ for $x \leq -1$, $\eta^+(x) \equiv 1$ for $x \geq 1$, and $\eta^+ + \eta^- \equiv 1$. Let $\psi^{\pm}$ be the solutions to L_{ζ} discussed previously, and recall the notation $x_0 = (\rho \cos(\alpha))^{1/(m-1)}$. To simplify the notation below, define

$$h^{\pm}(x) = \frac{\psi^{\pm}(x)}{\psi^{\pm}(x_0)}$$

and

$$\eta_{\rho}^{\pm}(x) = \eta^{\pm}(\rho(x - x_0)).$$

The approximate solution to L_{ζ} whose existence is asserted in Proposition 3 is

$$g = h^+ \eta_{\rho}^+ + h^- \eta_{\rho}^-.$$

Recall from Section 4.3 that

$$|h^{\pm}(x)| \leq C \qquad \forall x \in \mathbb{R}, \quad \rho \geq 1$$

and

$$\left| \frac{d}{dx} h^{\pm}(x_0) \right| \leq C\rho.$$

Hence $|g(x)| \leq C$, uniformly in ρ. The asymptotics (5) and (6) for $\psi^{\pm}$ imply that g and g' decay faster then any power of x as $|x| \to \infty$; but $L_{\zeta}g(x) \equiv 0$ for $|x - x_0| \geq \rho^{-1}$, whence g'' also decays rapidly. Differentiating the equation $g'' = (\zeta - x^{m-1})^2 g$ demonstrates the same for derivatives of higher order, therefore g is a Schwartz function.

We claim that

$$|h^+(x) - h^-(x)| \leq Ce^{\delta \rho^{m/(m-1)}} \qquad \forall |x - x_0| \leq \rho^{-1}. \tag{20}$$

Indeed, the Wronskian bound (16) together with (12) and (19) give

$$|(\psi^+(\psi^-)' - \psi^-(\psi^+)')(x_0)| \leq Ce^{-\delta \rho^{m/(m-1)}} |\psi^+(x_0)\psi^-(x_0)|,$$

which is to say

$$\left| \frac{d}{dx}(h^+ - h^-)(x_0) \right| \leq Ce^{-\delta \rho^{m/(m-1)}}.$$

Now $(h^+ - h^-)(x_0) = 1 - 1 = 0$ by definition, so since $(h^+ - h^-)'' = (\zeta - x^{m-1})^2(h^+ - h^-)$, there follows (20). This also gives $|(h^+ - h^-)'(x)| \leq Ce^{-\delta \rho^{m/(m-1)}}$ for $|x - x_0| \leq \rho^{-1}$.

For $|x - x_0| \geq \rho^{-\gamma}$, we have $L_\zeta g \equiv 0$, while in general

$$L_\zeta g = 2(h^+)'(\eta_\rho^+)' + 2(h^-)'(\eta_\rho^-)' + h^+(\eta_\rho^+)'' + h^-(\eta_\rho^-)''$$
$$= 2((h^+ - h^-)'(\eta_\rho^+)') + (h^+ - h^-)(\eta_\rho^+)''. \tag{21}$$

Thus

$$|L_\zeta g(x)| \leq Ce^{-\delta\rho^{m/(m-1)}} \cdot \rho^2 \leq Ce^{-\delta\rho^{m/(m-1)}}$$

with a smaller value of δ. On the other hand, $g(x_0) = 1$ and $|g'(x_0)| \leq C\rho$ for $|x - x_0| \leq \rho^{-1}$, so $|g(x)| \geq \frac{1}{2}$ for $|x - x_0| \geq C^{-1}\rho^{-1}$. Thus

$$\frac{\|L_\zeta g\|_{L^2}}{\|g\|_{L^2}} \leq Ce^{-\delta\rho^{m/(m-1)}}.$$

4.7   Proof of Proposition 1

Fix $\alpha \in \left(0, \frac{\pi}{2}\frac{m-1}{m}\right)$ and denote by g_ρ the approximate solution of L_ζ constructed in Section 4.6, where $\zeta = \rho e^{i\alpha}$. For large $\tau \in \mathbb{R}^+$, define

$$F_\tau(x, y, t) = g_\rho(\tau^{1/m}x)e^{i\tau^{1/m}\rho y}e^{i\tau t}$$

with $\rho = \rho(\tau) \in \mathbb{R}^+$ to be specified. Then

$$D_\alpha F_\tau(x, y, t) = \tau^{2/m}(L_\zeta g_\rho)(\tau^{1/m}x)e^{i\tau^{1/m}\rho y}e^{i\tau t}.$$

Because $g_\rho((\rho\cos(\alpha))^{1/(m-1)}) = 1$ and this is up to a constant factor the maximum value of g_ρ, we take

$$\rho = \tau^{[(m-1)/m](1-\varepsilon)}$$

where $\varepsilon \in (0, 1)$ is fixed. Then for any neighborhood Ω of 0,

$$0 < C^{-1} < \|F_\tau\|_{C^0(\Omega)} < C < \infty \tag{22}$$

uniformly as $\tau \to \infty$. Likewise

$$\left\|\frac{\partial}{\partial t}F_\tau\right\|_{C^0(\Omega)} \geq C\tau. \tag{23}$$

But

$$\|D_\alpha F_\tau\|_{C^0(\mathbb{R}^3)} \leq C\exp(-\delta\rho^{m/(m-1)}) \cdot \tau^{2/m}$$
$$\leq C\exp(-\delta\tau^{1-\varepsilon}).$$

Moreover, for any N there exist $C < \infty$, $\delta > 0$ such that

$$\|D_\alpha F_\tau\|_{C^N(\mathbb{R}^3)} \leq C\exp(-\delta\tau^{1-\varepsilon}). \tag{24}$$

Indeed, since finitely many powers of τ and $\rho(\tau)$ may always be absorbed at the expense of decreasing δ, it is sufficient to show that

$$\|L_\zeta g_\rho\|_{C^N(\mathbb{R})} \leq C \exp(-\delta \rho^{m/(m-1)}) \qquad \forall \rho \geq 1$$

for some $C < \infty$, $\delta > 0$ depending on N and on α. By (21), with the notation of Section 4.6, this would follow from

$$\left| \frac{d^k}{dx^k} (h^+ - h^-)(x) \right|$$
$$\leq C \exp(-\delta \rho^{m/(m-1)}) \qquad \forall k, \quad \forall |x - x_0| \leq \rho^{-1}, \tag{25}$$

where C, δ depend on k, α. But $(h^+ - h^-)'' = (\rho e^{i\alpha} - x^{m-1})^2 (h^+ - h^-)$, so that

$$\frac{d^k}{dx^k}(h^+ - h^-)(x) = p_k(x, \rho)(h^+ - h^-)(x) + q_k(x, \rho)(h^+ - h^-)'(x),$$

where p_k, q_k are polynomials, and (25) follows.

Combining (22), (23) and (24), we find that for any neighborhood Ω of 0, for any finite C or N, the inequality

$$\|f\|_{C^1(\Omega)} \leq C\|f\|_{C^0(\mathbb{R}^3)} + C\|D_\alpha f\|_{C^N(\mathbb{R}^3)}$$

is violated for $f = F_\tau$, for all sufficiently large τ. By the closed graph theorem, the Fréchet space

$$\{f \in C^0(\mathbb{R}^3) : D_\alpha f \in C^\infty(\mathbb{R}^3)\}$$

(with norms defined by suprema over $\mathbb{R}^3$) is therefore not contained in $C^1(\Omega)$ for any neighborhood Ω of 0, so D_α is not hypoelliptic at the origin.

4.8   Proof of Proposition 2

The space of solutions to L_ζ which do not grow rapidly as $x \to +\infty$ is one-dimensional and is spanned by ψ^+ (since there also exists a linearly independent solution that is asymptotic to $\exp\left(-\phi - \frac{1}{2}\log(-\phi')\right)$ [CL], and the space of all solutions has two dimensions). Likewise for ψ^-, as $x \to -\infty$. Thus to prove that L_ζ has no nullspace in L^2 is to prove ψ^+, ψ^- to be independent.

Assume $\alpha \in [0, \frac{\pi}{2} \frac{m-1}{m})$, write $\zeta = \rho e^{i\alpha}$ and proceed as in Section 4.4, setting $\theta = \frac{\alpha}{m-1}$. If ψ^+, ψ^- were dependent, the functions $\tilde{\psi}^\pm(t) = \psi^\pm(te^{i\theta})$ would be dependent. By (18),

$$\mathcal{L}\tilde{\psi}^\pm \equiv 0$$

where

$$\mathcal{L} = -e^{-i\alpha m/(m-1)} \frac{d^2}{dt^2} + e^{i\alpha m/(m-1)}(\rho - t^{m-1})^2.$$

For f in the Schwartz class,

$$\text{Re}\,\langle \mathcal{L}f,\, f\rangle = \cos\left(\frac{m}{m-1}\alpha\right)\left(\|f'\|_{L^2}^2 + \int |f(t)|^2(\rho - t^{m-1})^2\, dt\right).$$

Since

$$\left[\rho - t^{m-1},\, \frac{d}{dt}\right] = (m-1)t^{m-2},$$

and since $\cos\left(\frac{m}{m-1}\alpha\right) > 0$ by hypothesis, we obtain

$$\text{Re}\,\langle \mathcal{L}f,\, f\rangle \geq c\int |f(t)|^2 \cdot [(t^{m-2}) + (\rho - t^{m-1})^2]\, dt$$

$$\geq c\rho^{(m-2)/(m-1)}\|f\|_{L^2}^2.$$

But if $\widetilde{\psi}^+$ and $\widetilde{\psi}^-$ are linearly dependent then $\widetilde{\psi}^+$ is a Schwartz function, not identically zero, annihilated by $\mathcal{L}$—a contradiction.

REMARK 6 We believe that for any fixed α there exists $\delta > 0$ such that for $\zeta = \rho e^{i\alpha}$, for all $f \in C_0^2(\mathbb{R})$,

$$\|L_\zeta f\|_{L^2} \geq \delta \exp(-\delta^{-1}\rho^{m/(m-1)})\|f\|_{L^2}.$$

This is suggested by the possibility of expressing the fundamental solution for L_ζ in terms of $\psi_\zeta^{\pm}$ and their Wronskian, which should lead to good control of the inverse operator near $x_0(\rho)$. ∎

References

[C1]   M. Christ, *Some non-analytic-hypoelliptic sums of squares of vector fields*, Bulletin AMS 16 (1992), 137–140.

[C2]   ———, *Certain sums of squares of vector fields fail to be analytic hypoelliptic*, Comm. Partial Differential Equations 16 (1991), 1695–1707.

[CL]   E. Coddington and N. Levinson, *Theory of Ordinary Differential Equations*, McGraw–Hill, New York, 1955.

[HH]   N. Hanges and A. A. Himonas, *Singular solutions for sums of squares of vector fields*, Comm. Partial Differential Equations 16 (1991), 1503–1511.

[HN1]   B. Helffer and J. Nourrigat, *Caractérisation des opérateurs hypoelliptiques homogènes invariants à gauche sur un groupe de Lie nilpotent gradué*, Comm. Partial Differential Equations 4 (1979), 899–948.

[HN2]   ———, *Hypoellipticité Maximale Pour des Opérateurs Polynômes de Champs de Vecteurs*, Prog. Math. vol. 58, Birkhäuser, Boston, 1985.

[K] M. V. Keldysh, *On the completeness of the eigenfunctions of classes of non-selfadjoint linear operators,* Russian Math. Surveys 26 (1971), 15–44.

[M] D. Müller, *A new criterion for local non-solvability of homogeneous left invariant differential operators on nilpotent Lie groups,* J. Reine Angew. Math. 416 (1991), 207–219.

[N] J. Nourrigat, *Inégalités L^2 et représentations de groupes nilpotents,* J. Funct. Anal. 74 (1987), 300–327.

[PR] Pham The Lai and D. Robert, *Sur un probléme aux valeurs propres non linéaire,* Israel J. Math. 36 (1980), 169–186.

[RS] L. P. Rothschild and E. M. Stein, *Hypoelliptic differential operators and nilpotent groups* Acta Math. 137 (1976), 247–320.

5

Solvability of Second-Order PDO's on Nilpotent Groups—A Survey of Recent Results

Detlef Müller
Fulvio Ricci

5.1 Introduction

In these notes we shall present some recent results on solvability of second order left-invariant, homogeneous partial differential operators on two-step nilpotent Lie groups.

To put our results in a more general context, we have added some background material on doubly characteristic operators. At least for the case of transversally elliptic operators, part of this discussion is certainly well known to experts in PDE (compare e.g., [Ta]).

That analysis on nilpotent groups can be a powerful tool to understand regularity properties of general "subelliptic" partial differential operators, like for instance Hörmander's sum of squares operators, has been proven by the works of many authors (see e.g., [RS], [BeG]).

The question of solvability of a nonhypoelliptic partial differential operator is, however, of a different quality, and we do not have any hint if transference methods like those used in [RS] might be useful in one way or another in solvability problems. Still a general discussion should include our positive and negative results. So we believe that it is useful to indicate what invariants are significant in our context and what could be a possible interpretation of our results. Stein's result in [S] shows that an understanding of nonhomogeneous operators will require the introduction of at least one further invariant. In fact, very recently we succeeded in extending the results that will be described in this article to nonhomogeneous operators, and it would be interesting to see to which degree the quantities that rule solvability of these operators will play a role in the study of more general doubly characteristic differential operators.

For a full account of our results, we refer the reader to further forthcoming publications (in particular [MR3]).

5.2   Some historical background

Although a large part of the discussion in this section will apply to (classical) pseudodifferential operators as well, we shall restrict ourselves to the case of partial differential operators. So, let

$$P = p(x, D) = \sum_{|\alpha| \leq k} a_\alpha(x) D^\alpha \tag{1}$$

be a partial differential operator of order k on $\mathbb{R}^n$ with smooth coefficients a_α, where as usual $D^\alpha = \left(\frac{\partial}{i\partial x_1}\right)^{\alpha_1} \cdots \left(\frac{\partial}{i\partial x_n}\right)^{\alpha_n}$.

P is said to be *locally solvable* at $x_0 \in \mathbb{R}^n$, if there exists an open neighborhood $\mathcal{U}$ of x_0 such that the equation $Pu = f$ admits a distributional solution $u \in \mathcal{D}'(\mathcal{U})$ for every $f \in C_0^\infty(\mathcal{U})$ (for a slightly more general definition, see [H2]).

Around 1956, Malgrange and Ehrenpreis proved that every constant coefficient operator is locally solvable, and shortly later Lewy produced the following example of a nowhere solvable operator on $\mathbb{R}^3$:

$$Z = X - iY \quad \text{where} \quad X = \frac{\partial}{\partial x} - \frac{y}{2}\frac{\partial}{\partial u}, \quad Y = \frac{\partial}{\partial y} + \frac{x}{2}\frac{\partial}{\partial u}.$$

Not quite incidentally, Z is a left-invariant operator on a two-step nilpotent group, the Heisenberg group H_1 (for a detailed study of Z, based on this invariance property, see [GS]).

This example gave rise to an intensive study of so-called principal type operators, which eventually lead, most notably through the work of Hörmander, Maslov, Egorov, Nirenberg-Trèves, and Beals-Fefferman, to a complete solution of the problem of local solvability of such operators (see [H2]).

Let us recall some notation. The *complete symbol* of the operator (1) is the function $p(x, \xi) = \sum_{|\alpha| \leq k} a_\alpha(x)\xi^\alpha$ defined on $\mathbb{R}^n \times \mathbb{R}^n$. Its *principal symbol* is the homogeneous part of order k, i.e., $p_k(x, \xi) = \sum_{|\alpha|=k} a_\alpha(x)\xi^\alpha$. In contrast to the complete symbol, the principal symbol can be considered as an invariantly defined function on the cotangent bundle $T^*\mathbb{R}^n$ of $\mathbb{R}^n$, and because of the ξ-homogeneity of p_k, one usually regards p_k as a function on $\Omega = T^*\mathbb{R}^n \setminus 0 = \mathbb{R}^n \times (\mathbb{R}^n \setminus \{0\})$.

Let $\pi : \Omega \to \mathbb{R}^n$ denote the base projection. $T^*\mathbb{R}^n$ carries a canonical $1-$ form, in the usual coordinates given by $\Theta = \sum_{j=1}^n \xi_j dx_j$, so that Ω has a canonical symplectic structure, given by the $2-$ form $\omega = d\Theta = \sum_{j=1}^n d\xi_j \wedge dx_j$. In particular, for any smooth real function a on Ω, its corresponding Hamiltonian vector field H_a is well-defined, and explicitly given by $H_a = \sum \left(\frac{\partial a}{\partial \xi_j}\frac{\partial}{\partial x_j} - \frac{\partial a}{\partial x_j}\frac{\partial}{\partial \xi_j}\right)$.

If γ is an integral curve of H_a, i.e., if $\frac{d}{dt}\gamma(t) = H_a(\gamma(t))$, then a is constant along γ, and γ is called a *null bicharacteristic* of a, if a vanishes along γ. Finally, the Poisson bracket of two smooth functions a and b is given by $\{a, b\} = H_a b$.

Let $\Sigma = \{p_k = 0\} \subset \Omega$ denote the characteristic variety of P. P is said to be of *principal type*, if $\partial_\xi p_k$ does not vanish on Σ (or, more generally, if for every $\zeta \in \Sigma$ there is a complex number z such that $d(\Re e(zp_k))(\zeta)$ and $\Theta(\zeta)$ are nonproportional). The following condition $(\mathcal{P})$ of Nirenberg-Trèves rules the solvability of principal type operators:

$(\mathcal{P})$. The function $\Im(zp_k)$ does not take both positive and negative values along a null bicharacteristic $\gamma_z(t)$ of $\Re e(zp_k)$ for some $z \in \mathbb{C} \setminus \{0\}$.

In fact, P of principal type is locally solvable near x_0, if and only if, $(\mathcal{P})$ holds over some neighborhood of x_0. Notice that this is a condition based solely on the principal symbol of P.

The following examples of operators with double characteristics show that lower order terms will also become important in the presence of higher order characteristics.

Let $Z = X - iY$ be Lewy's operator, and set

$$L_1 = Z\bar{Z} = X^2 + Y^2 + iU,$$

$$L_2 = \frac{1}{2}(Z\bar{Z} + \bar{Z}Z) = X^2 + Y^2,$$

where $U = [X, Y] = \frac{\partial}{\partial u}$.

Then clearly L_1, like Z, is nowhere solvable, whereas L_2 is a sum of squares operator satisfying Hörmander's condition [H1] (in fact, L_2 is the Kohn-Laplacian on H_1, [K]), and hence hypoelliptic; together with the symmetry of L_2 this implies its local solvability.

5.3   Second-order PDO's on two-step nilpotent groups

Let G be a connected, simply connected Lie group whose Lie algebra $\mathfrak{g}$ decomposes into subspaces $\mathfrak{g} = \mathfrak{g}_1 \oplus \mathfrak{g}_2$, such that

$$[\mathfrak{g}, \mathfrak{g}] \subset \mathfrak{g}_2 \text{ and } [\mathfrak{g}, \mathfrak{g}_2] = 0.$$

Let $V_1, \ldots, V_m$ be a basis of $\mathfrak{g}_1$ and $U_1, \ldots, U_n$ be a basis of $\mathfrak{g}_2$. If we introduce exponential coordinates with respect to the corresponding basis of $\mathfrak{g}$, we may assume that $G = \mathfrak{g} = \mathbb{R}^m \times \mathbb{R}^n$ as manifolds, and that the group law in G is given by the Baker-Campbell-Hausdorff formula:

$$(x, u) \cdot (x', u') = \left(x + x', u + u' + \frac{1}{2}[x, x']\right), \qquad (2)$$

if $x, x' \in \mathfrak{g}_1, u, u' \in \mathfrak{g}_2$, where the Lie bracket is of the form

$$[x, x'] = \left(\sum_{i,j=1}^{m} J_{ij}^{k} x_i x_j' \right)_{k=1,\dots,n}, \tag{3}$$

with skew symmetric matrices $J^k = (J_{ij}^k)$.

If we identify as usual $X \in \mathfrak{g}$ with the left-invariant vector field

$$(Xf)(g) = \frac{d}{dt} f(g \exp t X)|_{t=0},$$

we get explicitly

$$V_j = \frac{\partial}{\partial x_j} + \frac{1}{2} \sum_{i,k} J_{ij}^{k} x_i \frac{\partial}{\partial u_k},$$

$$U_i = \frac{\partial}{\partial u_i}. \tag{4}$$

In the sequel we shall consider operators of the form

$$L = \sum_{j,k=1}^{m} a_{jk} V_j V_k + U, \tag{5}$$

where $A = (a_{jk})$ is a real, symmetric (and constant) matrix, and where $U \in \mathfrak{g}_2^{\mathbb{C}}$ is a complex, central vector field.

In addition to their invariance under left-translations on G, the operators (5) are also homogeneous of degree 2 with respect to the following automorphic dilations $\delta_r, r > 0$:

$$\delta_r(x, u) = (rx, r^2 u).$$

These properties of L imply that local solvability of L near any point of G is in fact equivalent to various forms of global solvability (compare [MR2]) on G, and we shall therefore simply speak of solvability of L in the sequel.

The operator (5) is said to be *transversally elliptic*, if the matrix A is definite. Examples of such operators are the operators L_1 and L_2 of Section 5.2, which are defined on the Heisenberg group $G = H_1$, which is $\mathbb{R}^2 \times \mathbb{R}$, with product given by (2), (3), where

$$J = J^1 = \begin{pmatrix} 0 & 1 \\ -1 & 0 \end{pmatrix}$$

is the "standard" symplectic matrix on $\mathbb{R}^2$.

Transversally elliptic operators have in fact been studied by many authors, in particular with respect to their regularity properties (see e.g., [G], [BGH], [RS], [Tr1], also for further references). With regard to solvability, some pioneering work had been done by Rothschild [R], and further results had been obtained, for instance, by Lion [Li] and Lévy-Bruhl [LB1], [LB2], [LB3].

If μ is a linear functional on $\mathfrak{g}_2$, let us define the skew form ω_μ on $\mathfrak{g}_1$ by

$$\omega_\mu(V, W) = \mu([V, W]). \tag{6}$$

Let J_μ be the associated matrix $(J_\mu)_{jk} = \omega_\mu(V_j, V_k)$, i.e., $J_\mu = \sum \mu_k J^k$, and set

$$S_\mu = -AJ_\mu.$$

Then one has

$${}^t S_\mu J_\mu + J_\mu S_\mu = 0,$$

i.e., S_μ is an element of the symplectic Lie algebra $\mathfrak{sp}(\omega_\mu)$ whenever ω_μ is nondegenerate. Only in this case the definition of S_μ will in fact be relevant.

We shall say that a property holds for *generic* μ, if it holds for all μ in some Zariski-open subset of $\mathfrak{g}_2^*$. We note that the following dichotomies arise:

(D1) Either ω_μ is degenerate for every $\mu \in \mathfrak{g}_2^*$, or ω_μ is nondegenerate for generic μ (in the latter case, we say that $\mathfrak{g}$ satisfies the Moore-Wolf condition, or that it is an *MW-algebra*);

(D2) either S_μ is semisimple for every μ, or S_μ is nonsemisimple for generic μ.

In the MW-case, let us say that $\mu \in \mathfrak{g}_2^*$ is *regular*, if

(i) ω_μ is nondegenerate;
(ii) the number of distinct eigenvalues of S_μ is maximal.

Here, *maximal* means maximal among all $\mu \in \mathfrak{g}_2^*$. The set $\mathcal{R}$ of all regular μ is Zariski-open and decomposes, by a theorem of Whitney, into a finite number of Euclidean connected components $\mathcal{R}_k$. Since the eigenvalues of S_μ are algebraic functions of μ in the sense that they satisfy certain algebraic equations with coefficients that are polynomials in μ, it is clear that these eigenvalues locally are analytic functions of μ on every component $\mathcal{R}_k$. However, they will in general not be globally defined as analytic functions on $\mathcal{R}_k$, but will ramify of finite order near singularities at the boundary of $\mathcal{R}_k$.

We also observe that, by passing to a suitable subgroup, if necessary, one may assume that the matrix $A = (a_{jk})$ is nondegenerate.

Let us write (5) as

$$L = \Delta_L + U_1 + iU_2,$$

where $\Delta_L = \sum a_{jk} V_j V_k$ and $U_1, U_2 \in \mathfrak{g}_2$, and let us denote by $\mathrm{spec}(S_\mu)$ the spectrum of S_μ.

THEOREM 1
Assume that A is nondegenerate.

(a) Each of the following properties implies that L is solvable:
   (i) ω_μ is degenerate for every $\mu \in \mathfrak{g}_2^*$, i.e. $\mathfrak{g}$ is not MW;
   (ii) $U_1 \neq 0$;
   (iii) $spec(S_\mu) \not\subseteq i\mathbb{R}$ for generic μ
   (iv) S_μ is nonsemisimple for generic μ.

(b) If $U_1 = 0$, $\mathfrak{g}$ is MW and S_μ is semisimple for generic μ, let us denote by $\mathcal{R}^3$ the union of all components $\mathcal{R}_k$ of $\mathcal{R}$ where $spec(S_\mu) \subset i\mathbb{R}$, and let $d = \frac{m}{2} \in \mathbb{N}$. Then, for each $\mu \in \mathcal{R}^3$ there exists a symplectic basis $X_1(\mu), \ldots, X_d(\mu), Y_1(\mu), \ldots, Y_d(\mu)$ of $\mathfrak{g}_1$ with respect to the symplectic form ω_μ, such that S_μ takes on normal form

$$
\begin{pmatrix}
0 & \cdots & 0 & \lambda_1(\mu) & & \\
\vdots & & \vdots & & \ddots & \\
0 & \cdots & 0 & & & \lambda_d(\mu) \\
-\lambda_1(\mu) & & & 0 & \cdots & 0 \\
& \ddots & & \vdots & & \vdots \\
& & -\lambda_d(\mu) & 0 & \cdots & 0
\end{pmatrix}
\tag{7}
$$

with respect to this basis. The "frequencies" $\lambda_j(\mu) \in \mathbb{R}$ are uniquely defined modulo permutations. In particular Δ_L is reduced to normal form

$$
\Delta_L = -\sum_{j=1}^{d} \lambda_j(\mu)(X_j(\mu)^2 + Y_j(\mu)^2).
\tag{8}
$$

For $m = (m_1, \ldots, m_d) \in \mathbb{N}^d$ let us set

$$
\Lambda(m, \mu) = \sum_{j=1}^{d} \lambda_j(\mu)(2m_j + 1) - \mu(U_2).
$$

Then the following conditions are equivalent:

(i) the operator $L = \Delta_L + iU_2$ is solvable;
(ii) there exist ε, $N > 0$, such that

$$
\int_{S \cap \mathcal{R}^3} \sum_{m \in \mathbb{N}^d} |\Lambda(m, \mu)|^{-\epsilon} (1 + |m|)^{-N} \, d\mu' < +\infty,
\tag{9}
$$

where $d\mu'$ denotes the surface measure of the unit sphere S in $\mathfrak{g}_2^*$;
(iii) there exists $N > 0$, such that for every nonempty, relatively compact ball B in $\mathcal{R}^3$

$$
\int_B |\Lambda(m, \mu)|^2 \, d\mu \geq C_B(1 + |m|)^{-N}, \quad m \in \mathbb{N}^d,
\tag{10}
$$

where C_B is independent of m.

The following remarks may be helpful:

(1) The function $\Lambda(m, \mu)$ is not really well defined, since it depends on the order in which the λ_js are listed in (7). But, any symmetric sum $\sum_{|m|=M} |\Lambda(m, \mu)|^a$, $a \in \mathbb{R}$, $M \in \mathbb{N}$, is well defined and independent of this order, so that condition (9) makes sense. On the other hand, on any ball $B \subset \mathcal{R}^{\mathfrak{I}}$ we may devise $\lambda_1, \ldots, \lambda_d$ as analytic functions of μ, and for any such choice $\Lambda(m, \mu)$ will be well defined and analytic on B. It is in this sense that condition (10) should be interpreted.

(2) One can give examples (see [MR3]) where S_μ has in fact a different type of spectrum on different components of $\mathcal{R}$, and in particular where $\mathcal{R}^{\mathfrak{I}} \neq \mathcal{R}$.

For transversally elliptic operators, our theorem implies the following result, which has been conjectured by Rothschild [R].

COROLLARY 1

Assume that $\pm A$ is positive definite, i.e., that L is transversally elliptic, and that $U_1 = 0$. Then, if $\mathfrak{g}$ is not MW, L is solvable. If $\mathfrak{g}$ is MW, L is solvable, if and only if $\Lambda(m, \cdot)$ does not vanish identically on any nonempty ball in $\mathcal{R} = \mathcal{R}^{\mathfrak{I}}$, for any multiindex $m \in \mathbb{N}^d$.

5.4   On the proof of Theorem 1

The most direct way to prove that a given operator is solvable consists in exhibiting an explicit fundamental solution.

The following lemma asserts however that it is sufficient to produce fundamental solutions "modulo solvable operators" (see Theorem 6.3 in [MR3] for a more general statement).

LEMMA 1

Let L be a left-invariant differential operator on a Lie group G, and assume that there exists a distribution K such that

$$LK = D\delta_0, \tag{11}$$

where D is a locally solvable left-invariant differential operator. Then also L is locally solvable.

In the cases when our operator L is solvable, such distributions K can be constructed in a rather explicit way. The tools at our disposal are

(1)   the formulas for the one-parameter groups of unitary operators generated by "twisted" differential operators,

(2)   the spectral analysis of the sub-Laplacian on the Heisenberg group.

Basically, (1) is needed in part (a) and (2) in part (b) of Theorem 1.

5.4.1   Twisted convolution and one-parameter groups generated by second-order differential operators

Let ω be a symplectic form on a vector space V of finite dimension $2d$. If dx denotes the volume form $\omega^{\wedge d}$, the *twisted convolution* of two functions f and g is given by

$$(f \times_\omega g)(x) = \int_V f(x - y)g(y)e^{-i\pi\omega(x.y)} \, dy.$$

Twisted convolution is associative, but not commutative. Left- and right-twisted translations are defined by twisted convolution on the left or on the right by Dirac delta's; correspondingly, given $v \in V$, the left twisted derivative in the v-direction is

$$(X_v f)(x) = \frac{d}{dt|_{t=0}}(f \times_\omega \delta_{tv})(x)$$

$$= \frac{\partial f}{\partial v}(x) + i\pi\omega(x, v) f(x). \tag{12}$$

It is convenient to introduce a symplectic basis $\{e_1, \ldots, e_d, f_1, \ldots, f_d\}$ of V, i.e., such that

$$\omega(e_i, e_j) = \omega(f_i, f_j) = 0$$

$$\omega(e_i, f_j) = \delta_{ij}.$$

In the corresponding coordinates, ω is represented by the matrix

$$J = \begin{pmatrix} 0 & I_d \\ -I_d & 0 \end{pmatrix}.$$

We write $X_1, \ldots X_{2d}$ instead of $X_{e_1}, \ldots, X_{e_d}, X_{f_1}, \ldots, X_{f_d}$.
Given a real symmetric matrix A, we form the operator

$$D = \sum_{i,j=1}^{2d} a_{ij} X_i X_j. \tag{13}$$

It can be proved that D is self-adjoint on an appropriate domain containing $S(V)$ [MR1].

It is natural to expect that the rules that govern calculus on the operators above reflect the symplectic structure on V. We call $\mathfrak{sp}(d, \mathbb{R})$ the *symplectic Lie algebra* formed by the $2d \times 2d$ real matrices S such that

$$SJ + J\,{}'S = 0,$$

i.e., such that SJ is symmetric, and denote by D_S the operator (13) with $A = SJ$. One easily checks that

$$[D_{S_1}, D_{S_2}] = -4\pi i D_{[S_1, S_2]}.$$

In [Hw] Howe proved that the map $S \longmapsto \left(\frac{i}{4\pi}\right) D_S$ can be exponentiated to a unitary representation of the metaplectic group $Mp(d, \mathbb{R})$, i.e., the two-fold covering of the symplectic group $Sp(d, \mathbb{R})$, which decomposes into countably many copies of the Shale-Weil representation. In [MR1] we give the precise form of the operators

$$T_{t,S} = e^{(i/4\pi)t D_S}.$$

Clearly enough, these are twisted convolution operators,

$$T_{t,S}\varphi = \varphi \times_\omega \gamma_{t,S}.$$

The next statement contains the basic estimates that we will need.

LEMMA 2
There is a Schwartz norm $\| \ \|_N$ on $S(V)$ such that

(a) for every $S \in \mathfrak{sp}(d, \mathbb{R})$, $t \in \mathbb{R}$, $\varphi \in S(V)$,

$$\left|\langle \gamma_{t,S}, \varphi\rangle\right| \le \|\varphi\|_N; \tag{14}$$

(b) If S has at least one eigenvalue $\lambda = \alpha + i\beta$ with $\alpha \neq 0$, then

$$\left|\langle \gamma_{t,S}, \varphi\rangle\right| \le \frac{1}{\cosh(\alpha t)} \|\varphi\|_N; \tag{15}$$

(c) if the nilpotent part of S has rank r, then

$$\left|\langle \gamma_{t,S}, \varphi\rangle\right| \le \frac{C_S}{(1 + |t|)^{r/2}} \|\varphi\|_N. \tag{16}$$

5.4.2 Construction of K in case (a) of Theorem 1

First of all, the subcase (i) can be reduced to (iv) by "adding new variables," i.e., by embedding $\mathfrak{g}$ into an appropriate larger MW-algebra.

We will assume therefore that $\mathfrak{g}$ is MW and that at least one of conditions (ii), (iii), or (iv) is satisfied.

By separation of variables, we apply L to functions of the form $f(x, u) = \varphi(x)e^{2\pi i \mu(u)}$, with $\mu \in \mathfrak{g}_2^*$. Then, if $\mu \in \mathcal{R}$,

$$L\big(\varphi(x)e^{2\pi i \mu(u)}\big) = \big(\Delta_L^\mu \varphi + 2\pi i \mu(U)\varphi\big)e^{2\pi i \mu(u)},$$

where

$$\Delta_L^\mu = \sum_{jk} V_j^\mu V_k^\mu$$

and the V_j^μ are the left-twisted derivatives (12) in the direction of the element $V_j \in \mathfrak{g}_1$, relative to the symplectic form ω_μ in (6).

Our aim is to solve (11) for some appropriate constant coefficient operator

$$D = P\left(\left(\frac{1}{2\pi}i\right)\partial_u\right)$$

in the u-variables (i.e., a central differential operator on G), which is solvable by the Malgrange-Ehrenpreis theorem.

If

$$K^\mu(x) = \int_{\mathfrak{g}_2} K(x, u)e^{-2\pi i\mu(u)}\, du,$$

then (11) forces the condition

$$\left(\Delta_L^\mu + 2\pi i\mu(U)\right)K^\mu = P(\mu)\delta_0.$$

When $\mu \in \mathcal{R}$, ω_μ is a symplectic form on $\mathfrak{g}_1$, we can try to construct K^μ from the distributions γ_t^μ such that

$$e^{it\Delta_L^\mu}\varphi = \varphi \times_{\omega_\mu} \gamma_t^\mu.$$

With respect to the results presented in Section 5.4.1, one has to notice that

(1)   the properties of the matrix S_μ vary with $\mu \in \mathfrak{g}_2^*$;
(2)   for each $\mu \in \mathcal{R}$, the estimates in Lemma 4 hold modulo a change of coordinates that reduces ω_μ to the canonical form.

To present the main ideas involved, we sketch the proof in the case where (iii) is satisfied for every $\mu \in \mathcal{R}$ and $U = iU_2$ is purely imaginary.

Formally

$$\varphi \times_{\omega_\mu} K^\mu = P(\mu)\left(\Delta_L^\mu - 2\pi\mu(U_2)\right)^{-1}\varphi$$

$$= iP(\mu)\int_0^{+\infty} e^{it\Delta_L^\mu}\varphi e^{-2\pi it\mu(U_2)}\, dt$$

$$= iP(\mu)\int_0^{+\infty} (\varphi \times_{\omega_\mu} \gamma_t^\mu)e^{-2\pi it\mu(U_2)}\, dt. \qquad (17)$$

We use Lemma 2(b) to give the necessary estimates that make this computation correct. Observe that (15) has to be modified by a factor that takes into account the change to symplectic coordinates. This factor will depend on μ and will be relevant in the next computations, which involve integration in μ.

It turns out that on each connected component Ω of $\mathcal{R}$ one can define transition matrices $R(\mu)$ reducing to symplectic coordinates in such a way that their entries are algebraic functions on Ω (to be more precise, they are functions defined on a finite covering of Ω, but this is a minor technical complication that can be dealt with without great difficulty).

It is a standard fact that nonzero algebraic functions can be estimated from above and from below by nonzero polynomials. We have therefore from (15)

$$|\langle \gamma_t^\mu, \varphi \rangle| \le \frac{(1 + |\mu|)^m}{|Q(\mu)| \cosh t\alpha(\mu)} \|\varphi\|_N,$$

where m is an integer and Q is a polynomial. As to $\alpha(\mu)$, we observe that the eigenvalues of S_μ, and their real parts, are also expressed by algebraic functions. Therefore, integrating in t and minorizing $\alpha(\mu)$ by another polynomial, we have

$$|\langle K^\mu, \varphi \rangle| \le |P(\mu)| \frac{(1 + |\mu|)^m}{|Q(\mu)|} \|\varphi\|_N.$$

We require now that $P(\mu)$ is divisible by $Q(\mu)$, and define K on G by

$$\langle K, f \rangle = \int_{\mathfrak{g}_2^*} \langle K^\mu, f^\mu \rangle \, d\mu, \tag{18}$$

where f^μ is the partial Fourier transform of f in u. If $f \in \mathcal{S}(G)$, the norms $\|f^\mu\|_N$ are rapidly decreasing in μ, and the integral in (18) converges absolutely.

This argument can be adapted to case (iv) of Theorem 1, exploiting the decay in t of $\langle \gamma_t^\mu, \varphi \rangle$ given by (16). If the rank of the nilpotent part of S_μ is either 1 or 2, this decay is not sufficient to assure the convergence of (18). The necessary modifications to the proof are presented in Section 7 of [MR3].

In case (ii), the decay in t is given by the presence of the factor $e^{2\pi t \mu(U_1)}$ in (17). It is then important to integrate over $t \in (-\infty, 0)$ when $\mu(U_1) > 0$ and over $t \in (0, +\infty)$ when $\mu(U_1) < 0$.

5.4.3   Construction of K in case (b) of Theorem 1

We assume here that $\mathcal{R}^3$ is nonempty, i.e., S_μ is semisimple and only has purely imaginary eigenvalues on some connected component of $\mathcal{R}$.

In contrast with the previous cases, if $\mu \in \mathcal{R}^3$, $\langle \gamma_t^\mu, \varphi \rangle$ has no decay in t that one can exploit in estimating the integral in (17). It is possible however to obtain the explicit spectral decomposition of Δ_L^μ and derive an expression for $\left(\Delta_L^\mu - 2\pi \mu(U_2)\right)^{-1}$ directly from that.

For simplicity, we will assume that $\mathcal{R}^3 = \mathcal{R}$.

LEMMA 3

Let (V, ω) be a symplectic vector space, and let $S \in \mathfrak{sp}(V, \omega)$ be semisimple with purely imaginary eigenvalues. Then there is a symplectic basis $\{e_1, \ldots, e_d,$

$f_1, \ldots, f_d\}$ *in which S takes the form*

$$
\begin{pmatrix}
 & & & & \lambda_1 & & \\
 & 0 & & & & \ddots & \\
 & & & & & & \lambda_d \\
 -\lambda_1 & & & & & & \\
 & \ddots & & & 0 & \\
 & & -\lambda_d & & & &
\end{pmatrix}
$$

with $\lambda_1, \ldots, \lambda_d \in \mathbb{R}$.

Correspondingly, in the notation of Section 5.4.1, D_S can be put in the form

$$
D_S = -\sum_{j=1}^{d} \lambda_j (X_j^2 + X_{j+d}^2).
$$

The analysis of each term $X_j^2 + X_{j+d}^2$ is well understood in terms of the Fourier analysis on the Heisenberg group (or, more precisely, of the Weyl correspondence for twisted convolution operators [F]), because the Schrödinger representation transforms this operator into the harmonic oscillator on the line.

If we denote by z the coordinates in V relative to the given symplectic basis, then

$$
X_j^2 + X_{j+d}^2 = -\sum_{m \in \mathbb{N}^d} (2m_j + 1)\pi_m;
$$

the orthogonal projection π_m is given by

$$
\pi_m f = f \times_\omega \left(e^{-\pi |z|^2 /2} L_{m_1}\left(\pi(z_1^2 + z_{d+1}^2)\right) \ldots L_{m_d}\left(\pi(z_d^2 + z_{2d}^2)\right) \right),
$$

where $L_m(t)$ is the Laguerre polynomial

$$
L_m(t) = \frac{1}{m!} e^t \frac{d^m}{dt^m}(e^{-t} t^m).
$$

Therefore

$$
D_S = \sum_{m \in \mathbb{N}^d} \left(\sum_{j=1}^{d} \lambda_j (2m_j + 1) \right) \pi_m.
$$

Going back to the situation where $\mu \in \mathcal{R}^3$ and $(V, \omega) = (\mathfrak{g}_1, \omega_\mu)$, we tentatively define

$$
\left(\Delta_L^\mu - 2\pi \mu(U_2) \right)^{-1} = \sum_{m \in \mathbb{N}^d} \Lambda(m, \mu)^{-1} \pi_m^\mu,
$$

where

$$
\Lambda(m, \mu) = \sum_{j=1}^{d} \lambda_j(\mu)(2m_j + 1) - \mu(U_2).
$$

One proves the following estimate.

LEMMA 4

For every $a > 0$ there exists a Schwartz norm $\| \ \|_N$ on $S(\mathfrak{g}_1)$, a polynomial Q on $\mathfrak{g}_2^*$, and an integer $M > 0$ such that

$$|\pi_m^\mu \varphi(0)| \leq \frac{1}{(|m|+1)^a} \frac{(|\mu|+1)^M}{|Q(\mu)|} \|\varphi\|_N$$

for all $\varphi \in S(\mathfrak{g}_1)$.

Proceeding as in Section 5.4.2, this approach leads to studying the convergence of

$$\int_{\mathfrak{g}_2^*} \sum_{m \in \mathbb{N}^d} |\Lambda(m,\mu)|^{-1} (|m|+1)^{-a} (|\mu|+1)^M \|f^\mu\|_N \, d\mu$$

for $f \in S(G)$. Because of the rapid decay in μ of $\|f^\mu\|_N$, this is the same as saying that

$$\int_{\mathfrak{g}_2^*} \sum_{m \in \mathbb{N}^d} |\Lambda(m,\mu)|^{-1} (|m|+1)^{-a} (|\mu|+1)^{-M} \, d\mu < \infty \qquad (19)$$

for some a and M.

It turns out that we can impose the weaker condition

$$\int_{\mathfrak{g}_2^*} \sum_{m \in \mathbb{N}^d} |\Lambda(m,\mu)|^{-\epsilon} (|m|+1)^{-a} (|\mu|+1)^{-M} \, d\mu < \infty \qquad (20)$$

for some $\epsilon > 0$. Formally this can be thought of as a solvability condition for $|L|^{-\epsilon}$.

The actual proof uses the fact that local solvability of L can be tested on functions on the group satisfying an appropriate covariance property under the action of a cocompact lattice in $\mathfrak{g}_2$. This has the effect of replacing the integration over $\mathfrak{g}_2^*$ in (19) and (20) by summation over a coset of the dual lattice.

It is then clear that the analogue of (20), with the integral replaced by a sum, implies the analogue of (19). It is also clear that (20) implies the existence of a coset of the given dual lattice for which the corresponding sum converges.

Integrating in polar coordinates, (20) reduces to

$$\int_S \sum_{m \in \mathbb{N}^d} |\Lambda(m,\mu)|^{-\epsilon} (|m|+1)^{-a} \, d\mu < \infty,$$

where S is the unit sphere in $\mathfrak{g}_2^*$.

So far we have sketched the proof of the implication (ii)⇒(i) in part (b) of Theorem 3. The implication (i)⇒(iii) depends on a criterion for local solvability that was proven independently by one of us in [Mü2] (see Lemma 9.2 in [MR3] for a formulation adapted to our situation).

Finally the implication (iii)⇒(ii) can be thought of as a form of "uniform reverse Hölder inequality" for the functions $\Lambda(m, \mu)$. To prove it, one first uses Hironaka's resolution of singularities to replace the algebraic functions $\Lambda(m, \mu)$ with real analytic functions on appropriate varieties. Next one uses the following lemma, based on a result in [Mü1].

LEMMA 5

Let $\eta_1, \ldots, \eta_k$ be real analytic functions on a closed ball $\overline{B}$. Then for every $p > 0$ and for sufficiently small ϵ there is a constant $C_{p,B,\epsilon}$ so that for all choices of real coefficients $\alpha_1, \ldots, \alpha_k$

$$\int_B |\alpha \cdot \eta(\mu)|^{-\epsilon} \, d\mu \leq C_{p,B,\epsilon} \left(\int_B |\alpha \cdot \eta(\mu)|^p \, d\mu \right)^{-\epsilon/p},$$

where $\alpha \cdot \eta = \sum_j \alpha_j \eta_j$.

5.5   Remarks on more general doubly characteristic operators and re-interpretation of the results in Section 5.3

5.5.1   A normal form

Let us assume here that the operator P in (1) is of second order. Adapting the notation of Section 5.2, we set

$$\Sigma_2 = \{(x, \xi) \in \Omega : \partial_\xi p_2(x, \xi) = 0\}.$$

Then, by Euler's identity, $\Sigma_2 \subset \Sigma$.

In the sequel, we shall always assume that p_2 is real and satisfies the following assumption.

Assumption 1

Σ_2 *is a submanifold of codimension $m < n$ in Ω, and* rank $D_\xi^2 p_2 = m$ *for $\xi \in \Sigma_2$, i.e., $D_\xi^2 p_2$ has maximal rank on Σ_2.*

PROPOSITION 1

Assume that p_2 is real. Then P satisfies Assumption 1 if, and only if, for any $x_0 \in \pi(\Sigma_2)$ there exist suitable linear coordinates near x_0 such that, in these coordinates, p_2 can be written

$$p_2(x, \xi) = \frac{1}{2}{}'(\xi' + E(x) \cdot \xi'') \cdot \tilde{A}(x) \cdot (\xi' + E(x) \cdot \xi'') \tag{21}$$

with respect to some splitting of coordinates $\xi = (\xi', \xi'')$, $\xi' \in \mathbb{R}^m$, $\xi'' \in \mathbb{R}^{n-m}$. Here $\tilde{A}(x)$ is a real, symmetric, nondegenerate $m \times m$ matrix, and $E(x)$ is a real $m \times (n - m)$ matrix, both depending smoothly on x.

In particular, $\Sigma_2 = \{\xi' + E(x) \cdot \xi'' = 0\}$ over a neighborhood of x_0.

PROOF We first remark that Assumption 1 is invariant under symplectic transformations of Ω induced by coordinate changes in x (however, not under arbitrary symplectic transformations).

Thus every operator whose principal symbol is given by (21) in suitable coordinates will satisfy Assumption 1.

Inversely, suppose that

$$p_2(x, \xi) = \sum_{j,k=1}^{n} a_{jk}(x)\xi_j\xi_k = \frac{1}{2}\,{}^t\xi \cdot A(x) \cdot \xi,$$

where $A(x) = (2a_{jk}(x))$ is real and, without restriction, symmetric, and that Assumption 1 is satisfied. Fix $x_0 \in \pi\,(\Sigma_2)$.

Since rank $A(x_0) = \text{rank } D_\xi^2 p_2(x_0, \xi) = m$, after a linear change of coordinates we may assume that $A(x_0) = \text{diag}(\lambda_j)$, with $\lambda_1, \ldots, \lambda_m \in \mathbb{R} \setminus 0$, $\lambda_{m+1} = \cdots = \lambda_n = 0$. So, if we split coordinates $\xi = (\xi', \xi'')$, $\xi' \in \mathbb{R}^m$, $\xi'' \in \mathbb{R}^{n-m}$, over a neighborhood $\mathcal{U}$ of x_0 we have

$$D_\xi^2 p_2(x, \xi) = A(x) = \begin{pmatrix} \tilde{A}(x) & B(x) \\ {}^t B(x) & D(x) \end{pmatrix},$$

where $\tilde{A}(x)$ is $m \times m$ and rank $\tilde{A}(x) = \text{rank } A(x) = m$. Consequently there exists a unique $(n - m) \times m$ matrix $Q(x)$ such that

$$Q(x) \cdot (\tilde{A}(x)|B(x)) = ({}^t B(x)|D(x)).$$

Thus

$${}^t B(x) = Q(x)\tilde{A}(x),$$

$$D(x) = Q(x)B(x),$$

so that $Q(x) = {}^t B(x)\tilde{A}(x)^{-1}$, and $D(x) = {}^t B(x)\tilde{A}(x)^{-1} B(x)$.

So one obtains

$${}^t\xi \cdot A(x) \cdot \xi = {}^t\xi' \cdot \tilde{A}(x) \cdot \xi' + 2\,{}^t\xi' \cdot B(x) \cdot \xi'' + {}^t\xi'' \cdot D(x) \cdot \xi''$$

$$= {}^t(\xi' + E(x) \cdot \xi'') \cdot \tilde{A}(x) \cdot (\xi' + E(x) \cdot \xi''),$$

where $E(x) = \tilde{A}^{-1} B(x)$. ∎

Proposition 1 shows that the operators under consideration are of the form

$$P = \sum_{j,k=1}^{m} \tilde{a}_{jk}(x) X_j X_k + \text{ lower order terms}, \tag{22}$$

with $X_1, \ldots, X_m$ independent smooth vector fields, and $(\tilde{a}_{jk}(x))_{jk}$ regular. In particular, we see that Assumption 1 implies that

$$\Sigma_2 = \{\zeta \in \Omega : dp_2(\zeta) = 0\}. \tag{23}$$

5.5.2 Invariants

Quantities associated with the symbol of a doubly characteristic partial differential operator P that rule its solvability should be invariant at least under symplectic transformations of Ω induced by coordinate changes.

The principal symbol is such a quantity, and also the restriction of the *subprincipal symbol*

$$p_{sub}(x, \xi) = \sum_{|\alpha|=1} a_\alpha(x)\xi^\alpha - \frac{1}{2i} \sum_{j=1}^{n} \frac{\partial^2}{\partial x_j \partial \xi_j} p_2(x, \xi)$$

to Σ_2 is invariantly defined (see e.g., [Tr2]).

Another invariant is the *Hamilton map* associated to p_2. If $Q_\zeta = D_\zeta^2 p_2(\zeta)$ denotes the Hessian of p_2 at $\zeta \in \Sigma_2$, regarded as a symmetric bilinear form on $T_\zeta \Omega$, then the Hamilton map $F_\zeta \in \mathfrak{sp}(\omega_\zeta, T_\zeta \Omega)$ is defined by

$$\omega_\zeta(u, F_\zeta v) = Q_\zeta(u, v), \quad u, v \in T_\zeta \Omega.$$

The invariance of F_ζ follows naturally from (23).

These two invariants play an important role in the solvability of our operators (5). In fact, they suffice to determine the solvability of this class of operators. However, for more general operators, at least one further invariant will be needed, which also takes into account the zero order terms of p, as the results in [S] as well as some recent work of ours indicate.

The following result applies in particular to symbols of the form (21) and sheds some more light on the meaning of the Hamilton map.

PROPOSITION 2

Assume that P is a pseudodifferential operator, with a principal symbol of the form

$$p_2 = \frac{1}{2} \sum_{j,k=1}^{m} a_{jk} q_j q_k,$$

where the $a_{jk} = a_{kj}$ and q_j are real, homogeneous symbols of order 0 and 1, respectively. Assume further that

$$\det(a_{jk}(x, \xi)) \neq 0, \tag{24}$$

$$dq_1(x, \xi), \ldots, dq_m(x, \xi) \text{ are linearly independent}, \tag{25}$$

the linear span of the $da_{jk}(x, \xi)$ *and the linear*

span of the $dq_j(x, \xi)$ *are transversal* (26)

for every (x, ξ). *For* $\zeta = (x, \xi) \in \Sigma_2$, *where here* Σ_2 *is defined by (23), let* F_ζ *denote the Hamilton map associated to* p_2, *let* A_ζ *be the matrix* $A_\zeta = (a_{jk}(\zeta))_{j,k}$, *and put*

$$J_\zeta = (\{q_j, q_k\}(\zeta))_{j,k=1,\dots,m},$$
$$S_\zeta = -J_\zeta A_\zeta,$$

where $\{\cdot, \cdot\}$ *denotes the Poisson bracket. Then*

(1) $\operatorname{spec} F_\zeta = \operatorname{spec} S_\zeta \cup \{0\}$, *with same multiplicities of the eigenvalues, except for 0.*
(2) *If* $\det J_\zeta = 0$, *then* F_ζ *is nonsemisimple.*
(3) *If* J_ζ *is nondegenerate, then* F_ζ *is conjugate to the matrix*

$$\begin{pmatrix} S_\zeta & 0 \\ 0 & 0 \end{pmatrix}.$$

In particular, F_ζ *is semisimple if and only if* S_ζ *is semisimple.*

PROOF Since

$$dp_2 = \frac{1}{2} \sum q_j q_k da_{jk} + \sum a_{jk} q_j dq_k, \tag{27}$$

we have $dp_2 = 0$ if and only if $q_j = 0$ for $j = 1, \dots, m$, and thus

$$\Sigma_2 = \{\zeta \in \Omega : q_1(\zeta) = \dots = q_m(\zeta) = 0\}. \tag{28}$$

And, because of (25), Σ_2 is a smooth, homogeneous submanifold of Ω.

If $\zeta \in \Sigma_2$, then $dq_1(\zeta), \dots, dq_m(\zeta)$ form a basis of $(T_\zeta \Sigma_2)^\perp \subset T_\zeta^* \mathbb{R}^n$. In a neighborhood of a given point, let us complement this basis by differentials $dq_{m+1}(\zeta), \dots, dq_{2n}(\zeta)$ to obtain a basis of $T_\zeta^* \mathbb{R}^n$. By (27) and (28) we obtain

$$D^2 p_2 = \sum_{j,k=1}^m a_{jk} dq_j \bigotimes dq_k \text{ on } \Sigma_2. \tag{29}$$

If we denote by $F_\zeta = J_n \cdot D^2 p_2(\zeta)$ the Hamilton map of p_2, where

$$J_n = \begin{pmatrix} 0 & I_n \\ -I_n & 0 \end{pmatrix},$$

and by $F_\zeta^* = -D^2 p_2(\zeta) \cdot J_n \in \operatorname{End}(T_\zeta^* \mathbb{R}^n)$ its adjoint, we find

$$F_\zeta^* = \sum_{j,k=1}^m a_{jk} dq_j \bigotimes H_{q_k} \text{ on } \Sigma_2. \tag{30}$$

Now, since $-dq_j(H_{q_k}) = \{q_j, q_k\}$ is the Poisson bracket of q_j and q_k, we find that F_ζ^* has the matrix representation

$$F_\zeta^* = \begin{pmatrix} S_\zeta^* & B_\zeta \\ 0 & 0 \end{pmatrix}, \quad \zeta \in \Sigma_2 \tag{31}$$

with respect to the blocks of coordinates $(dq_1, \ldots, dq_m)$ and $(dq_{m+1}, \ldots, dq_{2n})$, where $S_\zeta^* = AJ_\zeta$.

This proves (1).

To prove (2), assume that $\det J_\zeta = 0$, and that F_ζ were semisimple. Then, by (31), F_ζ^* would be conjugate to the matrix

$$\begin{pmatrix} \operatorname{diag}(\lambda_1, \ldots, \lambda_m) & 0 \\ 0 & 0 \end{pmatrix},$$

where $\{\lambda_1, \ldots, \lambda_m\} = \operatorname{spec} S_\zeta^*$. But then rank $F_\zeta^* < m$. On the other hand, by (24), (25) and (29), rank $F_\zeta^* = \operatorname{rank} A = m$, which leads to a contradiction.

Finally, if J_ζ is regular, the same is true of $S = S_\zeta^*$. Then, if (v, u) denote coordinates with respect to the blockform of (31), one has, with $B = B_\zeta$,

$$F_\zeta^* \begin{pmatrix} v \\ u \end{pmatrix} = \begin{pmatrix} Sv + Bu \\ 0 \end{pmatrix},$$

hence

$$F_\zeta^* \begin{pmatrix} v - S^{-1}Bu \\ u \end{pmatrix} = \begin{pmatrix} Sv \\ 0 \end{pmatrix},$$

which shows that F_ζ is conjugate to

$$\begin{pmatrix} S_\zeta & 0 \\ 0 & 0 \end{pmatrix}. \qquad \blacksquare$$

REMARK 1   One consequence of this proposition is that the Hamilton map of P given by (22) is entirely determined by the coefficient matrix $(\tilde{a}_{jk})$ and the *first-order commutators* $\{X_j, X_k\}$ of the vector fields X_j.

So, if one looks for a model situation in which first-order commutators do not necessarily vanish, but all higher order commutators do, *a posteriori* it is natural to consider left-invariant vector fields on a two-step nilpotent group. $\blacksquare$

5.5.3   Reinterpretation of the conditions in Theorem 1

Let us first observe that the principal symbol of the operator (5) is given by

$$p_2(x, u; \xi, \mu) = -{}^t\left(\xi - \frac{1}{2}J_\mu\right) A \left(\xi - \frac{1}{2}J_\mu\right) \tag{32}$$

and its subprincipal symbol by

$$p_{sub}(x, u; \xi, \mu) = i\mu(U). \tag{33}$$

Thus, if A is nondegenerate, by Proposition 1

$$\Sigma_2 = \left\{ \left(\xi - \frac{1}{2} J_\mu \right)_j = 0, \ j = 1, \ldots, m \right\}.$$

Moreover, direct calculation shows that for $(x, u; \xi, \mu) \in \Sigma_2$ the matrices $J_{(x,u;\xi,\mu)}$ and $S_{(x,u;\xi,\mu)}$ defined in Proposition 2 are given by

$$J_{(x,u;\xi,\mu)} = J_\mu, \quad S_{(x,u;\xi,\mu)} = 2J_\mu A = 2\,{}^t S_\mu \tag{34}$$

with J_μ and S_μ as in Section 4.3. Notice also that for J_μ nondegenerate, ${}^t S_\mu$ is conjugate to $-S_\mu$.

These observations in combination with Proposition 2 allow us to consider the conditions in Theorem 1 as conditions on the Hamilton map and the subprincipal symbol of L.

For example, conditions (a–i) and (a–iv) can both be subsumed under the condition that F_ζ be nonsemisimple for generic $\zeta \in \Sigma_2$, and the "frequencies" $\lambda_j(\mu)$ associated to the matrix S_μ that enter the definition of the function $\Lambda(m, \mu)$ can be considered as the corresponding frequencies of the matrix $\frac{1}{2} F_{(x,u;\xi,\mu)}$. Correspondingly, the integrals (8) could be interpreted as integrations over a portion of the intersection of the cosphere bundle of G with Σ_2. However, since for our operators the "frequencies" λ_j depend only on μ, but not on the remaining variables x, ξ, and u, there would be lot of ambiguity in such an interpretation, causing us to refrain from making a more precise statement.

References

[BeG]   Beals, R. and Greiner, P. *Calculus on Heisenberg manifolds,* Princeton University Press, Princeton, 1988.

[BGH]   Boutet de Monvel, L., Grigis, A. and Helffer, B. *Paramétrixes d'opérateurs pseudo-differentiels à caractèristiques multiples,* Astérisque 34–35 (1976), 93–121.

[F]   Folland, G. *Harmonic Analysis in Phase Space,* Princeton University Press, Princeton, 1989.

[G]   Grušin, V.V. *On a class of hypoelliptic operators,* Mat. Sbornik 83 (1970), 456–473; Math. USSR Sbornik 12 (1970), 458–476.

[GS]   Greiner, P.C. and Stein, E.M. *Estimates for the $\bar{\partial}$ complex and analysis on the Heisenberg group,* Math. Notes Princeton University Press, Princeton, 1976.

[H1]   Hörmander, L. *Hypoelliptic second-order differential equations,* Acta Math. 119 (1967), 147–171.

[H2]   —————— *The Analysis of Partial Differential Operators, IV,* Springer Verlag, New York, 1985.

[Hw]   Howe, R. *The oscillator semigroup,* Proc. Symp. Pure Appl. Math. 48 (1988), 61–132.

[K]   Kohn, J. *Harmonic integrals on strongly pseudo convex manifolds I,* Ann. of Math. 78 (1963), 112–147.

[LB1]   Levy-Bruhl, P. *Résolubilité d'opérateurs invariants du second ordre sur des groupes nilpotents,* Bull. Sci. Math. 104 (1980), 369–391.

[LB2]   —————— *Résolubilité locale d'opérateurs homogènes invariantes à gauche sur des groupes nilpotents d'ordre deux,* C.R.A.S. Paris 292 (1981), 197–200.

[LB3]   —————— *Application de la formule de Plancherel à la résolubilité d'opérateurs invariants à gauche sur des groupes nilpotents d'ordre deux,* Bull. Sci. Math. 106 (1982), 171–191.

[Li]   Lion, G. *Hypoelliplicité et résolubilité d'opérateurs differentiels sur les groupes de rang deux,* C.R.A.S. Paris 290 (1980), 271–274.

[Mü1]   Müller, D. *Twisted convolution with Calderón-Zygmund kernels,* J. Reine Angew. Math. 352 (1984), 133–150.

[Mü2]   —————— *A new criterion for local non-solvability of homogeneous left invariant differential operators on nilpotent Lie groups,* J. Reine Angew. Math. 416 (1991), 207–219.

[MR1]   Müller, D. and Ricci, F. *Analysis of second order differential operators on Heisenberg groups. I,* Inv. Math. 101 (1990), 545–582.

[MR2]   —————— *Analysis of second order differential operators on Heisenberg groups. II,* J. Funct. Anal. 108 (1992), 296–346.

[MR3]   —————— *Solvability for a class of doubly characteristic differential operators on two–step nilpotent groups,* to appear in Ann. of Math.

[R]   Rothschild, L.P. *Local solvability of second order differential operators on nilpotent Lie groups,* Arkiv för Math. 19 (1981), 145–175.

[RS]   Rothschild, L.P. and Stein, E.M. *Hypoelliptic differential operators and nilpotent groups,* Acta Math. 137 (1976), 247–320.

[S]   Stein, E.M. *An example on the Heisenberg group related to the Lewy operators,* Inv. Math. 69 (1982), 209–216.

[Ta]   Taylor, M.E. *Pseudodifferential Operators,* Princeton University Press, Princeton, 1981.

[Tr1] Trèves, F. *Analytic hypo-ellipticity of a class of pseudodifferential operators with double characteristics and applications to the $\bar{\partial}-$ Neumann problem,* Comm. Partial Diff. Eq. 3 (1978), 475–642.

[Tr2] ——— *Introduction to Pseudodifferential Operators and Fourier Integral Operators, I, II,* Plenum Press, New York, 1980.

6

Recent Work on Sharp Estimates in Second Order Elliptic Unique Continuation Problems

Thomas H. Wolff

6.1 Introduction and counterexamples

This is an expository article mainly concerned with the unique continuation properties of the differential inequality

$$\left| \sum_{ij} a_{ij} \frac{d^2 u}{dx_i dx_j} \right| \le A|u| + B|\nabla u|. \tag{1}$$

Here (a_{ij}) is elliptic on a domain $\Omega \subset \mathbb{R}^d$ with $d \ge 2$, that is, $a_{ij} : \Omega \to$ positive definite symmetric real $d \times d$ matrices. A and B are functions on Ω, possibly with singularities. We are interested in sharp results, i.e., in the minimal regularity conditions on (a_{ij}), A and B, which are needed for the classical results to hold.

First let us first clarify what we mean by a solution of (1) and by unique continuation. A solution of (1) means a function on Ω that belongs locally to the Sobolev space W^{22} and satisfies (1) pointwise a.e. In other words, we will consider strong solutions in W^{22}_{loc}. Most of the results are, however, valid somewhat more generally.

DEFINITION   *(1) has the unique continuation property (UCP) if every solution that vanishes on an open subset of Ω vanishes identically.*

   (1) has the strong unique continuation property (SUCP) if every solution that vanishes to infinite order at a point of Ω vanishes identically.

Here a function $u \in W^{22}_{\text{loc}}(\Omega)$ is said to vanish to infinite order at $p \in \Omega$ if $\lim_{r \to 0} r^{-n} \int_{D(p,r)} |u|^2 = 0$ for all $n \in \mathbb{Z}^+$, where $D(p,r) = \{x : |x - p| < r\}$. If u is smooth, then of course this is equivalent to all partial derivatives of u vanishing at p.

Let us now state some of the known results. There is extensive literature on unique continuation for (1) and we will not attempt to survey it (see also [K] in this connection). Instead we will just state a few classical results and then the results concerning optimal L^p conditions on A and B, which we will be discussing here. Other types of conditions on A and B have also been used (e.g., [CS], [Pa], [RV], [Sa], [SS], [St2]), but we will somewhat arbitrarily leave this aside.

The basic result is

(I)   *(1) has the SUCP if (a_{ij}) are locally Lipschitz and A and B are locally bounded.*

In $\mathbb{R}^2$ this is due to Carleman and in higher dimensions to Aronszajn and Cordes if (a_{ij}) are C^2 and by Hormander as stated. The Lipschitz regularity is needed according to the next two results.

(II)   *If $d \geq 3$, there is an inequality (1) with a_{ij} Holder of all orders less than one, A and B bounded, with a solution vanishing on an open set.*

This is due to Plis [P] who actually proved somewhat more. Subsequently Miller [M] was even able to show the following.

(III)   *If $d \geq 3$, there is an equation $div(A\nabla u) = 0$ with A Holder continuous and elliptic, with a smooth weak solution vanishing on an open set.*

So far as we have been able to determine, it is unknown whether such examples exist in $\mathbb{R}^2$. [*Note added in proof:* They do not exist, even if the Holder continuity is dropped. See [A] and references therein.]

Concerning optimal L^p conditions on A and B, the following conjecture is natural: (1) has the SUCP if (a_{ij}) are Lipschitz and $A \in L^{d/2}_{\text{loc}}$, $B \in L^d_{\text{loc}}$. To see why, consider, e.g., the case of the Laplace operator $(a_{ij} = \delta_{ij})$ and a function u, which vanishes to high finite order k at the origin. Then one would expect Δu to vanish to order $k - 2$, i.e., $\frac{\Delta u}{u}$ just fails being in $L^{d/2}_{\text{loc}}$, and similarly $\frac{\Delta u}{|\nabla u|}$ just fails being in L^d_{loc}. In fact it is very easy to see that the SUCP does not hold for $|\Delta u| \leq A|u|$, $\left(A \in L^p, p < \frac{d}{2}\right)$ or for $|\Delta u| \leq B|\nabla u|$, $(B \in L^p, p < d)$, take $u(x) = e^{-|x|^{-\varepsilon}}$, ε small [JK]. However, the situation turns out to be more complicated than the above argument suggests, and in particular the conjecture is wrong when $d \neq 3, 4$.

A clean result is known for the zero-order term, namely,

(IV)   *If $d \geq 3$ and $A \in L^{d/2}_{\text{loc}}$, then the inequality $|\Delta u| \leq A|u|$ has the SUCP.*

This is due to Jerison–Kenig [JK]. It has been extended to the case of (1) with (a_{ij}) smooth, $A \in L^{d/2}$, $B = 0$ by Sogge [So1] using a rather hard argument based on ψDO calculus – the case where (a_{ij}) is just Lipschitz is still open. Result (IV) is false when $d = 2$ [W2], although, as pointed out in [JK], it is true with the exponent $1 + \varepsilon$ instead of 1.

The gradient term B is more complicated. For the SUCP, the best known positive result is

(V) $|\Delta u| \leq B|\nabla u|$ *has the SUCP if* $B \in L_{\text{loc}}^p$ *with* $p = \max\left(d, \dfrac{3d-4}{2}\right)$.

This is described by the author [W1]. Earlier work [KRS], [J], and [Ki] should also be mentioned. In particular, in [Ki] the same result is proven with the exponent $\frac{3d-2}{2}$ instead of $\max\left(d, \frac{3d-4}{2}\right)$.

Note that (V) gives the conjectured sharp result when $d \leq 4$. This fails when $d \geq 5$:

(VI)   *If* $d \geq 5$, *then there is* $B \in L^d(\mathbb{R}^d)$ *such that* $|\Delta u| \leq B|\nabla u|$ *has a smooth solution vanishing to infinite order at the origin.*

Result (VI) is proven in [W6]. It is not known whether the SUCP holds for $|\Delta u| \leq B|\nabla u|$, $B \in L_{\text{loc}}^p$, $d \geq 5$, $p \in \left(d, \frac{3d-4}{2}\right)$. The situation for the UCP is different.

(VII)   *If* $d \geq 3$, $A \in L_{\text{loc}}^{d/2}$, $B \in L_{\text{loc}}^d$ *then* $|\Delta u| \leq A|u| + B|\nabla u|$ *has the UCP.*

This is proven in [W3] together with an analogous result for variable coefficients:

(VIII)   *If* $d \geq 5$, (a_{ij}) *Lipschitz,* $A \in L_{\text{loc}}^{d/2}$, $B \in L_{\text{loc}}^d$ *then (1) has the UCP.*

Results (VII) and (VIII) are *not* known to be sharp in the context of the UCP. In fact, the following question is open: If $A, B \in L^1$, does it follow that $|\Delta u| \leq A|u| + B|\nabla u|$ has the UCP? The analogous question is also open if $A \in L^p$, $B \in L^q$, for any $p \in \left[1, \frac{d}{2}\right)$ and $q \in [1, d)$. Below L^1 there are easy counterexamples obtained by starting from one-dimensional counterexamples and adjoining dummy variables; compare [Sa].

6.1.1   Carleman method

The proofs of all the positive results mentioned above use the classical Carleman method (originating in Carleman's work mentioned above), although in the case of (V), (VII), and (VIII), in somewhat modified form. We want to give an idea of how the method works and will first describe the version of it used by Jerison and Kenig to prove (IV).

The following *Carleman inequality* is the main result of [JK].

THEOREM 1
[JK] Suppose $d \geq 3$. *Let* p *satisfy* $\frac{1}{p} - \frac{1}{p'} = \frac{2}{d}$, *where* $p' \overset{\text{def}}{=} \frac{p}{p-1}$. *Then provided* $t > \frac{d-2}{2}$ *and* $t - \frac{d-2}{2} \notin \mathbb{Z}$,

$$\| \, |x|^{-t} u \|_{p'} \leq C \| \, |x|^{-t} \Delta u \|_p \tag{2}$$

for all functions $u \in C_0^\infty(\mathbb{R}^d \backslash \{0\})$, *where* C *depends only on* $\text{dist}\left(t - \frac{d-2}{2}, \mathbb{Z}\right)$.

REMARK 1   Note that if $t = 0$ Theorem 1 reduces to the Sobolev inequality. This explains the reason for the restriction $d \geq 3$. The restriction $t - \frac{d-2}{2} \notin \mathbb{Z}$ is also needed: otherwise one obtains counterexamples by considering a degree $t - \frac{d-2}{2}$ spherical harmonic truncated near 0 and ∞ [JK].   ∎

REMARK 2   Inequality (2) is called a Carleman inequality because the constant C is uniform in t for a sequence of $t \rightarrow +\infty$, namely, any sequence such that $t - \frac{d-2}{2}$ stays away from the integers. As $t \rightarrow \infty$, the weights $|x|^{-t}$ become highly concentrated near 0. This is the crucial point for the application to unique continuation.

We will comment on the proof of Theorem 1 in Section 6.2. For now, we assume Theorem 1 and explain the proof of (IV) following [JK]. We assume $u \in W_{\text{loc}}^{22}$ as above $\left(W_{\text{loc}}^{2p} \text{ with } \frac{1}{p} - \frac{1}{p'} = \frac{2}{d} \text{ would suffice} \right)$ and vanishes to infinite order at 0. The question is local so we may assume $A \in L^{d/2}$ (rather than $L_{\text{loc}}^{d/2}$). We fix ε small enough that $\|A\|_{L^{d/2}(D(a,2\varepsilon))} \leq \frac{1}{2C}$ for all a, where C is a constant for which (2) holds for a sequence of $t \rightarrow \infty$. Let $\phi \in C_0^{\infty}$ be 1 on $D(0, \varepsilon)$ and 0 on $\mathbb{R}^d \setminus D(0, 2\varepsilon)$. Inequality (2) was stated for functions in $C_0^{\infty}(\mathbb{R}^d \setminus \{0\})$, but a limiting argument using the infinite-order vanishing and the inequality $|\Delta u| \leq A|u|$ shows that it is also true for the function ϕu, see [JK]. So

$$\| |x|^{-t} \phi u \|_{p'} \leq C \| |x|^{-t} \Delta(\phi u) \|_p$$

$$\leq C \| |x|^{-t} \phi \Delta u \|_p + \| |x|^{-t} E \|_p.$$

Here E (for error) $= 2\Delta\phi \cdot \nabla u + \phi \Delta u$ and is an L^p function supported in $\{x : |x| \geq \varepsilon\}$. Using $|\Delta u| \leq A|u|$ and then Holder's inequality,

$$\| |x|^{-t} \phi u \|_{p'} \leq C \| |x|^{-t} \phi A u \|_p + \| |x|^{-t} E \|_p$$

$$\leq C \|A\|_{L^{d/2}(D(0,2\varepsilon))} \| |x|^{-t} \phi u \|_p + \| |x|^{-t} E \|_p.$$

By choice of ε the first term can be hidden and

$$\| |x|^{-t} \phi u \|_{p'} \leq C \| |x|^{-t} E \|_p.$$

Now the punchline: E is supported in $\{x : |x| \geq \varepsilon\}$, so

$$\| |x|^{-t} \phi u \|_{p'} \leq C \varepsilon^{-t} \| E \|_p$$

$$\left\| \left(\frac{\varepsilon}{|x|} \right)^t \phi u \right\|_{p'} \leq C \| E \|_p$$

and if we let $t \rightarrow \infty$, we conclude that ϕu vanishes on $D(0, \varepsilon)$. Hence u vanishes on $D(0, \varepsilon)$.

In other words, the set $Z = \{x : r \text{ vanishes to infinite order at } x\}$ is open, and in fact contains a disk of *fixed* radius ε centered at any one of its points. So Z must be everything, and the proof is complete.   ∎

In the preceding argument, it was not necessary to work with the functions $|x|^{-t}$: they could be replaced, e.g., by any functions $e^{t\psi}$ where t runs through a sequence tending to $+\infty$ and ψ is smooth on some set of the form $D(0, \varepsilon_0) \setminus \{0\}$ and satisfies (i) $\psi(x) < \liminf_{y\to 0} \psi(y)$ for all $x \neq 0$, and (ii) $\psi(x) \le \text{const.} \cdot \log \frac{1}{|x|}$.

The reason for requirement (i) is clear from the argument. (ii) is needed for the limiting argument, which we did not discuss, extending the inequality proven for C_0^∞ functions to the function ϕu (note that if (ii) fails, then infinite-order vanishing of u does not guarantee that $ue^{t\psi} \in L^1_{\text{loc}}$).

The difficulty with the gradient term $B|\nabla u|$ is that one would want to use an inequality

$$\|e^{t\phi}\Delta u\|_{p'} \le C\|e^{t\phi}\Delta u\|_p, \qquad \frac{1}{p} - \frac{1}{p'} = \frac{1}{d}$$

and no such inequality is true. In fact the following rather striking result is proven for a closely related result:

> Suppose $\Omega \subset \mathbb{R}^d$ is open, $\phi : \Omega \to \mathbb{R}$ a smooth function, and the inequality $\|e^{t\phi}\nabla u\|_q \le C\|e^{t\phi}\Delta u\|_p$ (some $p, q \in [1, \infty]$) holds for all $u \in C_0^\infty(\Omega)$, uniformly over a sequence of $t \to \infty$. Then either ϕ is constant, or else
>
> $$\frac{1}{p} - \frac{1}{q} \le \frac{2}{3d - 2}.$$

The number $\frac{2}{3d-2}$ is the correct one here [KRS], [J], and [Ki] which accounts for the result of [Ki] that the SUCP holds for $|\Delta u| \le B|\nabla u|$, $B \in L^{(3d-2)/2}_{\text{loc}}$. If one tries to sharpen this result, then the following issue of course arises: if the Carleman inequalities fail, then does unique continuation also fail, or is there some way of sharpening the Carleman method? As indicated by results (V), (VII), and (VIII), the latter can be the case in certain situations. We will explain one way of doing this in Section 6.3, but for the rest of Section 6.1 we will mainly discuss counterexamples.

One gains some insight by viewing unique continuation as a finite statement concerning the order of vanishing, rather than just a statement ruling out vanishing of infinite order. The following result appears to be the natural finite analogue of result IV.

PROPOSITION 1
Suppose $d \ge 3$, $n \in \mathbb{Z}^+ \cup 0$ and $u : \mathbb{R}^d \to \mathbb{R}$ is a (say) smooth function satisfying

(i) $u(x) = O(|x|^n)$ as $x \to \infty$.
(ii) $u(x) = O(|x|^{n+1})$ at $x \to 0$.

Then $\left\| \frac{\Delta u}{u} \right\|_{d/2} \ge C_0^{-1}$, where C_0 is a positive constant depending only on d.

In other words, the growth rate at ∞ is at least equal to the rate of vanishing at 0, for solutions of Schrodinger equations with $L^{d/2}$-small potentials. Results of this same general nature are proven, e.g., in [DF] and [GL].

PROOF   This is essentially the same as the proof of IV. We claim that inequality (2) is applicable to the function u, provided that $t \in \left(n + \frac{d}{2} - 1, n + \frac{d}{2}\right)$. If we accept this, then the proof of the proposition is as follows:

$$\| \, |x|^{-t}u\|_{p'} \leq C\| \, |x|^{-t}\Delta u\|_p$$

$$\leq C\| \, |x|^{-t}u\|_{p'} \left\| \frac{\Delta u}{u} \right\|_{d/2}$$

for (say) $t = n + \frac{d}{2} - \frac{1}{2}$, and therefore $\|\frac{\Delta u}{u}\|_{d/2} \geq C^{-1}$. (Note that (i) and (ii) guarantee that $\| \, |x|^{-t}u\|_{p'} < \infty$.)

To prove the claim, fix $\varepsilon > 0$ and let $\phi_\varepsilon \in C_0^\infty$ be a cutoff function such that

$$\phi_\varepsilon(x) = \begin{cases} 1, & \text{if } \varepsilon < |x| < \varepsilon^{-1} \\ 0, & \text{if } |x| < \frac{\varepsilon}{2} \text{ or } |x| > 2\varepsilon^{-1} \end{cases}$$

and satisfying the natural bounds, i.e., $|D^\alpha\phi_\varepsilon| \leq C_\alpha\varepsilon^{-|\alpha|}$ when $\frac{\varepsilon}{2} < |x| < \varepsilon$ and $|D^\alpha\phi_\varepsilon| \leq C_\alpha\varepsilon^{|\alpha|}$ where $\varepsilon^{-1} < |x| < 2\varepsilon^{-1}$. Certainly (2) is applicable to $\phi_\varepsilon u$. Thus by the product rule,

$$\| \, |x|^{-t}\phi_\varepsilon u\|_{p'}$$
$$\leq C(\| \, |x|^{-t}\phi_\varepsilon \Delta u\|_p + 2\| \, |x|^{-t}\nabla\phi_\varepsilon \cdot \nabla u\|_p + \| \, |x|^{-t}u\Delta\phi_\varepsilon\|_p) \quad (3)$$

Let $A_\varepsilon = \left\{x : \frac{\varepsilon}{2} < |x| < \varepsilon\right\}$, $\tilde{A}_\varepsilon = \left\{x : \frac{\varepsilon}{4} < |x| < 2\varepsilon\right\}$, $B_\varepsilon = \left\{x : \varepsilon^{-1} < |x| < 2\varepsilon^{-1}\right\}$, $\tilde{B}_\varepsilon = \left\{x : \frac{1}{2}\varepsilon^{-1} < |x| < 4\varepsilon^{-1}\right\}$. Then the last two terms in (3) live on $A_\varepsilon \cup B_\varepsilon$. We have

$$\| \, |x|^{-t}\nabla\phi_\varepsilon \cdot \nabla u\|_{L^p(A_\varepsilon)} \leq C_t\varepsilon^{-(t+1)}\|\nabla u\|_{L^p(A_\varepsilon)}$$

$$\leq C_t\varepsilon^{-(t+1)}(\|u\|_{L^p(\tilde{A}_\varepsilon)}\|\Delta u\|_{L^p(\tilde{A}_\varepsilon)})^{1/2}$$

by a standard interpolation inequality. Assuming as we may that $\|\frac{\Delta u}{u}\|_{d/2} < \infty$, we obtain

$$\| \, |x|^{-t}\nabla\phi_\varepsilon \cdot \nabla u\|_{L^p(A_\varepsilon)} \leq C_{tu}\varepsilon^{-(t+1)}(\|u\|_{L^p(A_\varepsilon)}\|u\|_{L^{p'}(A_\varepsilon)})^{1/2}$$

uniformly in ε. Hypothesis (ii) then implies that

$$\| \, |x|^{-t}\nabla\phi_\varepsilon \cdot \nabla u\|_{L^p(A_\varepsilon)} \leq C_{tu}\varepsilon^{-(t+1)}\varepsilon^{n+1}(\varepsilon^{d/p}\varepsilon^{d/p'})^{1/2}$$

$$= C_{tu}\varepsilon^{n+(d/2)-t}.$$

One also has

$$\| \, |x|^{-t} u \Delta \phi_\varepsilon \|_{L^p(A_\varepsilon)} \leq C_t \varepsilon^{-(t+2)} \|u\|_{L^p(A_\varepsilon)}$$
$$\leq C_{tu} \varepsilon^{n+1-t-2+(d/p)}$$
$$= C_{tu} \varepsilon^{n+(d/2)-t}.$$

Similar estimates using (ii) instead of (i) show that

$$\| \, |x|^{-t} \nabla \phi_\varepsilon \cdot \nabla u \|_{L^p(A_\varepsilon)} + \| \, |x|^{-t} u \Delta \phi_\varepsilon \|_{L^p(A_\varepsilon)} \leq C_{tu} \varepsilon^{-(n+(d/2)-t-1)}.$$

Accordingly, by (3),

$$\| \, |x|^{-t} \phi_\varepsilon u \|_{p'} \leq C \| \, |x|^{-t} \phi_\varepsilon \Delta u \|_p + C_{tu} \left(\varepsilon^{n+(d/2)-t} + \varepsilon^{-(n+(d/2)-t-1)} \right).$$

Both powers of ε appearing here are positive, so by letting $\varepsilon \to 0$ one obtains the result. ∎

We now explain why result (IV) fails in the two-dimensional case [W2]. First of all, it is very easy to give a counterexample to the two-dimensional analogue of Proposition 1. Let $v_n(x) = \log \frac{1}{|x-e_1|} - p_n(x)$ where e_1 is (say) the first standard basis vector and p_n is the degree n Taylor polynomial of $\log \frac{1}{|x-e_1|}$ at 0. Define u_n by smoothing off v_n slightly, keeping u_n equal to v_n outside a small disc $D(e_1, \varepsilon)$. If ε is small, then this can be done with $u_n \geq \text{const} \cdot \log \frac{1}{\varepsilon}$ and $|\Delta u_n| \leq \text{const} \cdot \varepsilon^{-2}$ on $D(e_1, \varepsilon)$. Since u_n is harmonic outside $D(e_1, \varepsilon)$ we have $\| \frac{\Delta u_n}{u_n} \|_1 \lesssim C \left(\log \frac{1}{\varepsilon} \right)^{-1}$, and clearly, $u_n = O(|x|^n)$ at ∞, $u_n = O(|x|^{n+1})$ at 0. We conclude: if $n \in \mathbb{Z}^+$, $\delta > 0$, then there is a smooth function $u_n : \mathbb{R}^2 \to \mathbb{R}$ such that $u_n(x) = O(|x|^n)$ at ∞, $u_n(x) = O(|x|^{n+1})$ at 0 and $\| \frac{\Delta u_n}{u_n} \|_1 < \delta$.

Note that the behavior of u_n at 0 will be the same as the behavior of $-u_{n+1}$ at ∞, i.e.,

$$u_n(x) = T_n(x) + O(|x|^{n+2}) \qquad \text{at } 0$$

$$-u_{n+1}(x) = T_n(x) + O(|x|^n) \qquad \text{at } \infty$$

where T_n is the degree $n+1$ term in the expansion of $\log \frac{1}{|x-e_1|}$ at 0. Because of this fact, it is possible to piece together dilations of the functions $(-1)^n u_n$ to obtain a function u vanishing to infinite order at 0, and with

$$\left\| \frac{\Delta u}{u} \right\|_1 = \sum_n \left\| \frac{\Delta u_n}{u_n} \right\|_1 + \text{small error}.$$

This is finite if $\delta = \delta_n$ are small enough. Details are in [W2].

We now consider the gradient term and result VI. This is different in that even though the inequality $\| \, |x|^{-t} \nabla u \|_{p'} \leq C \| \, |x|^{-t} \Delta u \|_p$ must fail when $\frac{1}{p} - \frac{1}{p'} = \frac{1}{d}$, an analogous inequality still holds with a constant depending on t. Namely, the following result is a special case of Lemma 3.1 in [W1] (it is however closely related to [JK] and could perhaps be derived from their proof).

PROPOSITION 2

Let p satisfy $\frac{1}{p} - \frac{1}{p'} = \frac{1}{d}$. Then provided t is large and $t - \frac{d-1}{2} \notin \mathbb{Z}$,

$$\| |x|^{-t} \nabla u \|_{p'} \leq C t^{(1/2)-(1/d)} \| |x|^{-t} \Delta u \|_p, \qquad u \in C_0^\infty(\mathbb{R}^d \setminus \{0\})$$

As in Theorem 1, the constant C depends only on the distance from $t - \frac{d-1}{2}$ to the integers. By repeating the proof of Proposition 1 one therefore obtains Proposition 3.

PROPOSITION 3

Suppose $u : \mathbb{R}^d \to \mathbb{R}$ is a smooth function satisfying

   (i) $\nabla u(x) = O(|x|^n)$ at ∞

   (ii) $\nabla u(x) = O(|x|^{n+1})$ at 0.

where n is large. Then

$$\left\| \frac{\Delta u}{|\nabla u|} \right\|_d \geq C^{-1} n^{-[(1/2)-(1/d)]}.$$

In [W6] we show first that this is sharp when $d \geq 4$; if $d \geq 4$, then for every sufficiently large n there is a smooth function $u_n : \mathbb{R}^d \to \mathbb{R}$ satisfying (i) and (ii) of Proposition 3 and with $\left\| \frac{\Delta u_n}{|\nabla u_n|} \right\|_d \leq C n^{-[(1/2)-(1/d)]}$. It is then possible to piece together the u_n's as above to obtain a function u vanishing to infinite order at the origin and with

$$\left\| \frac{\Delta u}{|\nabla u|} \right\|_d^d \leq \sum_n \left\| \frac{\Delta u_n}{|\nabla u_n|} \right\|_d^d + \text{ small error}$$

$$\lesssim \sum_n n^{-[(d/2)-1]},$$

which is finite if $d \geq 5$. This proves result VI.

Finally we will make some remarks about inequalities like (2) in the case of variable coefficients. We will base the discussion on a result from [W5]; however, the paper of Miller [M] should also be mentioned here. The result from [W5] is the following.

PROPOSITION 4

Suppose $d \geq 3$, Ω is a domain in $\mathbb{R}^d$ containing 0, and v_1 and v_2 are harmonic functions on Ω whose critical points in $\mathbb{R}^d \setminus 0$ are isolated and with $v_1(0) = v_2(0)$. Then there are positive numbers R_1 and R_2, $R_2 > R_1$, $\overline{D(0, R_2)} \subset \Omega$ and a smooth function $u : \Omega \to \mathbb{R}$ such that

   (i) $u = v_1$ on $D(0, R_1)$

   (ii) $u = v_2$ on $\Omega \setminus \overline{D(0, R_2)}$

   (iii) *u satisfies an elliptic equation $\frac{d}{dx_i} a_{ij} \frac{d}{dx_j} u = 0$ where (a_{ij}) are smooth and $a_{ij} = \delta_{ij}$ on $\overline{D(0, R_1)} \cup (\Omega \setminus D(0, R_2))$.*

An immediate consequence is that for any $N < \infty$, there is a smooth function which is $O(|x|)$ at ∞, $O(|x|^N)$ at 0 and satisfies a smooth divergence form elliptic equation coinciding with the Laplacian near 0 and ∞. By the proof of Proposition 1 we may conclude the following.

PROPOSITION 5

If $d \geq 3$, then for any $n < \infty$ there is a smooth elliptic operator $\operatorname{div}A\nabla$ that coincides with the Laplacian near 0 and ∞, such that no inequality

$$\| \, |x|^{-t}u\|_{p'} \leq C_t \| \, |x|^{-t}\operatorname{div}A\nabla u\|_p, \qquad u \in C_0^\infty(\mathbb{R}^d \setminus \{0\}) \tag{4}$$

holds for any $t \in \left(\frac{d}{2}, n\right)$.

PROOF   Take u to be $O(|x|)$ at ∞ and $O(|x|^N)$, $N = N(n)$ large, at 0. Define A by the above remarks. By the proof of Proposition 1, any inequality (4) with $\frac{d}{2} < t < n$ would be applicable to u, which is a contradiction since the right-hand side is zero.   ∎

It is easy to see that Proposition 4 fails in $\mathbb{R}^2$ with (say) $v_1(x) = \operatorname{re}((x_1 + ix_2)^m)$, $v_2(x) = \operatorname{re}((x_1 + ix_2)^n)$, $m > n$. This is because in $\mathbb{R}^2$ the function $d(r) = \deg_{\partial D(0,r)} \nabla u$ must be monotone nonincreasing by the Poincare-Hopf theorem if u (is Morse and) satisfies the maximum principle, since the index at a saddle is odd.

6.2   Carleman inequalities

By a Carleman inequality we mean here an inequality of the form

$$\|e^{t\phi}u\|_q \leq C\|e^{t\phi}\Delta u\|_p, \qquad u \in C_0^\infty(\Omega) \tag{5}$$

or

$$\|e^{t\phi}\nabla u\|_q \leq C\|e^{t\phi}\Delta u\|_p, \qquad u \in C_0^\infty(\Omega) \tag{6}$$

where Ω is a domain in $\mathbb{R}^d$, $p, q \in (0, \infty]$ and C is a constant, which is uniform in t as $t \to +\infty$.

More generally one can look for inequalities like (5) and (6) where the constant C is allowed to depend on t but in a controlled way (as in Proposition 3 and (16) below), and needless to say, one can study differential operators other than the Laplacian, but we will not do this here except for briefly mentioning the d, d^* system at the end of the section.

Before discussing the known positive results on (5) and (6) let us make the following simple observations concerning necessary conditions.

PROPOSITION 6

Suppose that inequality (5) (or (6)) is valid for some p and q and a sequence of $t \to +\infty$. Then there is no point $a \in \Omega$ such that the Hessian $H_\phi(a)$ is negative definite.

PROOF Assume (5) holds. If $H_\phi(a)$ is negative definite then there is a linear function $k \cdot x$ such that $\phi(x) - k \cdot x$ has a strict local maximum at a. Let $\ell \in \mathbb{R}^d$ satisfy $\ell \perp k$, $|\ell| = |k|$, and let χ be supported in a small neighborhood $D(a, \varepsilon_0)$ and equal to 1 on $D(a, \delta_0)$ for some $\delta_0 > 0$. Let $u_t(x) = e^{-t\langle k+i\ell, x \rangle} \chi(x)$. Then u_t is harmonic on $D(a, \delta_0)$. We have

$$\|e^{t\phi} u_t\|_q \geq \|e^{t(\phi(x)-k\cdot x)}\|_{L^q(D(a,\delta_0))} \tag{7}$$

$$\|e^{t\phi} \Delta u_t\|_p \leq \|e^{t(\phi(x)-k\cdot x)}(\Delta\chi - 2t\langle k + i\ell, \nabla\chi \rangle)\|_p. \tag{8}$$

Let $M = \inf_{D(a,\delta_0/2)}(\phi(x) - k \cdot x)$, $m = \sup_{D(a,\varepsilon_0)\setminus D(a,\delta_0)}(\phi(x) - k \cdot x)$. Then $M > m$ provided ε_0 and δ_0 have been chosen appropriately. Also,

$$\|e^{t\phi} u_t\|_q \geq C^{-1} e^{tM}$$

$$\|e^{t\phi} \Delta u_t\|_p \leq C e^{tm}$$

with C independent of t. Letting $t \to \infty$, we find a contradiction. The argument for (6) is essentially the same. If $k \neq 0$ one can use the same function u_t; if $k = 0$, use instead $\tilde{u}_t(x) = x_1 u_t(x)$. ∎

Proposition 6 can be extended somewhat. Let us say that a symmetric $d \times d$ matrix A satisfies (*) if for every $\lambda \in \mathbb{R}$ the numbers λ and $-\lambda$ have the same multiplicity as eigenvalues of A. Thus every matrix satisfying (*) has trace zero, and the converse is true if $d = 2$. Note also that if d is odd then every matrix satisfying (*) has a null space.

PROPOSITION 7

Assume there is a point $a \in \Omega$ and a matrix A satisfying (*) such that $H_\phi(a) - A$ is negative definite. If d is odd, assume also the following: there is a vector e with $Ae = 0$ and $\nabla\phi(a) \perp e$. Then neither (5) nor (6) can hold for any p and q and sequence of $t \to +\infty$.

PROOF We can assume $a = 0$, and also $A \neq 0$ since the case $A = 0$ has already been done. Let $k = \nabla\phi(0)$. The assumptions on A and k imply that

$$A = \sum_{j=1}^{[d/2]} \lambda_j \left(e_j \otimes e_j - \tilde{e}_j \otimes \tilde{e}_j \right), \qquad \lambda_j \in \mathbb{R}$$

for a certain orthonormal set $\{e_j, \tilde{e}_j\}$ with $k \in \mathrm{sp}(\{e_j, \tilde{e}_j\})$. Here by $e \otimes f$ we mean the operator $e \otimes f(x) = \langle f, x \rangle e$. Express k as $\sum_j k_j e_j + \tilde{k}_j \tilde{e}_j$, $k_j, \tilde{k}_j \in \mathbb{R}$ and define

$$B = \sum_j \lambda_j \left(e_j \otimes \tilde{e}_j + \tilde{e}_j \otimes e_j \right)$$

$$\ell = \sum_j k_j \tilde{e}_j - \tilde{k}_j e_j$$

$$Q(x) = \frac{1}{2} \langle (A + iB)x, x \rangle + \sum \langle k + i\ell, x \rangle.$$

Then $|\ell| = |k|$, $\ell \perp k$, $\langle Ax, k \rangle = \langle Bx, \ell \rangle$, $\langle Ax, \ell \rangle = -\langle Bx, k \rangle$, $|Ax| = |Bx|$ and $Ax \perp Bx$ for all $x \in \mathbb{R}^d$. These relations say that $\nabla \phi$ is self-perpendicular for the bilinear form $\sum z_j w_j$ on $\mathbb{C}^d$.

Since A and B here have trace zero, it follows that e^{tQ} is harmonic for all t. Now choose χ with sup $\chi \subset D(0, \varepsilon_0)$, $\chi = 1$ on $D(0, \delta_0)$, for suitable $\varepsilon_0, \delta_0 > 0$. Let $u_t = \chi e^{-tQ}$ and

$$M = \inf_{x \in D(0, \delta_0/2)} (\phi(x) - \mathrm{re}\, Q(x)),$$

$$m = \sup_{x \in D(0, \varepsilon_0) \setminus D(0, \delta_0)} (\phi(x) - \mathrm{re}\, Q(x)).$$

Here $\mathrm{re}\, Q(x) = \langle k, x \rangle + \frac{1}{2} \langle Ax, x \rangle$, so the assumption implies $M > m$, provided ε_0 and δ_0 have been chosen appropriately. Also,

$$\|e^{t\phi} u_t\|_q \geq \|e^{t(\phi - \mathrm{re}\, Q)}\|_{L^q(D(0, \delta_0))} \geq C^{-1} e^{tM},$$

$$\|e^{t\phi} \nabla u_t\|_q \geq t \|e^{t(\phi - \mathrm{re}\, Q)}(k + i\ell - \nabla Q)\|_{L^q(D(0, \delta_0))}$$

$$\geq C_\varepsilon^{-1} t e^{t(M - \varepsilon)}$$

for any $\varepsilon > 0$, since the assumption $A \neq 0$ implies of course that $k + i\ell - \nabla Q$ does not vanish on an open set. Likewise

$$\|e^{t\phi} \Delta u_t\|_p = \|e^{t\phi} \Delta u_t\|_{L^p(D(0, \varepsilon_0) \setminus D(0, \delta_0))}$$

$$\leq C e^{tm},$$

and it follows that neither (5) nor (6) can hold. $\blacksquare$

REMARK 3 If $d = 2$ then the hypothesis of Proposition 7 reduces to ϕ being nonsubharmonic. Conversely if ϕ is subharmonic and $d = 2$, then (5) and (6) are valid for any bounded domain Ω and $p = q = 2$ ([J2]; see [C] for a related argument). $\blacksquare$

What we want to do now is to discuss some of the techniques that have been used to prove positive results on (5) and (6). Roughly speaking, many results are known when ϕ is convex or satisfies related conditions. The relevance of convexity goes back at least to [H1], although we will only discuss a few recent results.

We will start by sketching a proof of the Jerison–Kenig result, Theorem 1. Many proofs are known, all based on more or less the same ideas. We will follow a line of argument from [W1], which can easily be generalized as is done there to give, e.g., the inequalities needed for result V. It is quite simple in principle but requires a fair amount of calculation, which we will mainly omit. We believe that the reader who is interested and is familiar with $L^p \to L^{p'}$ oscillatory integral estimates, e.g., [So2], [St1] will be able to fill in the details. A less calculational approach is in [J] (and presented also in [So2]).

The proof of the usual Sobolev inequality

$$\|u\|_q \leq C\|u\|_p, \qquad \frac{1}{p} - \frac{1}{q} = \frac{2}{d}, \qquad u \in C_0^\infty$$

may be thought of as taking place in two steps: construction of a fundamental solution

$$u(x) = c_d \int |x - y|^{2-d} \Delta u(y) \, dy$$

and then estimation of the fundamental solution via $L^p \to L^q$ estimates for integral operators.

The proof of (2) may be accomplished in these same two steps. However, now the fundamental solution involves an oscillatory factor, so the $L^p \to L^q$ estimates require one or another oscillatory integral technique. We will use the variable coefficient Plancherel theorem of Hormander [H2] and [St1]. The relevant fundamental solution is as follows.

LEMMA 1

If $\alpha \in \mathbb{R}$, $n \in \mathbb{Z}^+$, then for $u \in C_0^\infty(\mathbb{R}^d \setminus \{0\})$,

$$|x|^{-(n+(d-2/2)-\alpha)} u(x) = \int K_{n,\alpha}(x, y)|y|^{-(n+(d-2/2)-\alpha)} \Delta u(y) \, dy \qquad (9)$$

where $K_{n,\alpha}$ obeys the following estimates:

(i) $|K_{n,\alpha}(x, y)| \leq C|x - y|^{2-d}$ $\left(|x - y| \leq \dfrac{1}{2n}|y|\right)$

(ii) $|K_{n,\alpha}(x, y)| \leq C \, n^{d-3} \min\left(n, \left|1 - \dfrac{|x|}{|y|}\right|^{-1}\right) |x|^{-((d/2)-1-\alpha)},$

     $\cdot |y|^{-((d/2)-1+\alpha)}\left(|x - y| \geq \dfrac{1}{2n}|y|\right)$

(iii)   *Let $\theta = \angle x 0 y \in [0, \pi]$ be the angle subtended by x and y at the origin $\left(\text{equivalently, } \theta = \cos^{-1}\frac{(x,y)}{|x||y|}\right)$. Then*

$$K_{n,\alpha}(x, y) = |x|^{-((d/2)-1-\alpha)}|y|^{-((d/2)-1+\alpha)}\mathrm{re}\left(a_n\left(\dfrac{|x|}{|y|}, \theta\right)e^{in\theta}\right)$$

where a_n is a complex valued function satisfying estimates

$$\left|\dfrac{d^j}{d\theta^j}a_n(r, \theta)\right| \lesssim n^{(d/2)-2}\left(|\sin\theta| + |1 - r|\right)^{-1}|\sin\theta|^{-((d/2)-1+j)})$$

when $|\sin\theta| > \frac{1}{n}$.

PROOF   See [W1], Proposition 1.1. We note that $K_{n,\alpha}$ is equal to $|x|^{-((d/2)-1-\alpha)}$ $\cdot |y|^{-((d/2)-1+\alpha)} \cdot I_n$ where I_n is the kernel in [W1].

The key point for making the estimates is that the function $\theta : S^{d-1} \times S^{d-1} \to \mathbb{R}$ defined by $\theta(e, f) = \angle e 0 f$ has the property that its mixed Hessian $\nabla_e\nabla_f\theta$ (we regard this as a linear operator from $T_e S^{d-1}$ to $T_f S^{d-1}$) has rank $d - 2$ and takes the form

$$\begin{pmatrix} \lambda I & 0 \\ 0 & 0 \end{pmatrix}$$

with $|\lambda| \approx (\sin\theta)^{-1}$ in appropriate smooth coordinate systems at $e, f \in S^{d-1}$. (Namely: the last coordinate in each system should be $\angle e 0 f$; the others should be orthogonal to this one at e and f respectively.) We introduce a partition of unity $\psi_j|_{j=0}^{\log_2 n}$ on $[0, 1]$ with sup $\psi_0 \subset \left\{t : |t| \leq \frac{2}{n}\right\}$ and sup $\psi_j \subset \left\{t : \frac{1}{2}\frac{2^j}{n} \leq t \leq 2 \cdot \frac{2^j}{n}\right\}$ for $j \geq 1$, and define $K_{n,\alpha}^{j,s,t} : S^{d-1} \times S^{d-1} \to \mathbb{R}$ by

$$K_{n,\alpha}^{j,s,t}(e, f) = \psi_j(\sin\theta)K_{n,\alpha}(se, tf).$$

Because of the preceding property of the mixed Hessian of θ and the bounds in Lemma 1, standard techniques based on the variable coefficient Plancherel theorem lead to the estimate $\left(\text{with } \frac{1}{p} - \frac{1}{p'} = \frac{2}{d}\right)$.

$$\| K^{j,s,t}_{n,\alpha} \|_{L^p(S^{d-1}) \to L^{p'}(S^{d-1})}$$

$$\leq C s^{-(\frac{d}{2}-1-\alpha)} t^{-(\frac{d}{2}-1+\alpha)} \left(\frac{2^j}{n} + |1 - \frac{s}{t}| \right)^{-1} n^{-2/d} \qquad (10)$$

provided $j \geq 1$ or $|s - t| > \frac{|t|}{n}$.

We now record the following version of the "product space Young's inequality" (see [W1], Lemmas 2.1 and 2.2).

LEMMA 2
Suppose $1 \leq p \leq 2$, $X \times Z$, $d\mu \times d\nu$ is a product measure space and K : $(X \times Z) \times (X \times Z) \to \mathbb{R}$. For $z, w \in Z$ let $n(z, w)$ be the $L^p(X, \mu) \to L^{p'}(X, \mu)$ norm of the integral operator with kernel $K((\cdot, z), (\cdot, w))$. Let u be any positive function on Z and define

$$A = \sup_{w \in Z} \left(\int \left[(u(z)u(w))^{-1/p'} n(z, w) \right]^q u(z) d\nu(z) \right)^{1/q}$$

$$B = \sup_{z \in Z} \left(\int \left[(u(z)u(w))^{-1/p'} n(z, w) \right]^q u(w) d\nu(w) \right)^{1/q}$$

where q satisfies $\frac{1}{p} + \frac{1}{q} = \frac{1}{p'} + 1$. Then the norm of K as an integral operator from $L^p(X \times Z, \mu \times \nu)$ to $L^{p'}(X \times Z, \mu \times \nu)$ is $\leq (AB)^{1/2}$.

To prove (2) let $\phi : \mathbb{R}^d \times \mathbb{R}^d \to \mathbb{R}$ be a cutoff function that is 1 when $|x - y| < \frac{|y|}{2n}$ and 0 when $|x - y| > \frac{|y|}{n}$ and write $K_{n,\alpha} = \widetilde{K}_{n,\alpha} + \widetilde{\widetilde{K}}_{n,\alpha}$, $\widetilde{\widetilde{K}}_{n,\alpha} = \phi K_{n,\alpha}$. By Lemma 1, $|\widetilde{\widetilde{K}}_{n,\alpha}| \lesssim |x - y|^{2-d}$ and therefore maps L^p to $L^{p'}$. To estimate $\widetilde{K}_{n,\alpha}$ we first consider $K^j_{n,\alpha}(x, y) = \psi_j(\sin \theta) \widetilde{K}_{n,\alpha}(x, y)$. Regarding $\mathbb{R}^d$ as $S^{d-1} \times \mathbb{R}^+$ with measure $d\sigma \times r^{d-1} dr$ and applying Lemma 1 with $u(r) = r^{-d}$, and using (10) to estimate A and B we obtain (with $\frac{1}{p} - \frac{1}{p'} = \frac{2}{d}$, $\frac{1}{p} + \frac{1}{q} = \frac{1}{p'} + 1$),

$$\| K^j_{n,\alpha} \|_{p \to p'} \leq (AB)^{1/2}$$

$$A \lesssim n^{-2/d} \sup_t \left(\int \left[\left(\frac{s}{t} \right)^\alpha \left(\frac{2^j}{n} + |1 - \frac{s}{t}| \right)^{-1} \right]^q \frac{ds}{s} \right)^{1/q}$$

$$B \lesssim n^{-2/d} \sup_s \left(\int \left[\left(\frac{s}{t} \right)^\alpha \left(\frac{2^j}{n} + |1 - \frac{s}{t}| \right)^{-1} \right]^q \frac{dt}{t} \right)^{1/q}.$$

If $0 < \alpha < 1$, then one checks that $A \lesssim 2^{-2j/d}$, $B \lesssim 2^{-2j/d}$, so $\sum_j \| K^{(j)}_{n,\alpha} \|_{L^p \to L^{p'}} \leq$ const. This completes the proof. ∎

We now discuss an osculation technique from [W1] (Hormander's paper [H1] should however be mentioned here) that can be used, e.g., to pass from (2) to related inequalities with respect to other weights. We first give a simple example.

PROPOSITION 8
(Kenig–Ruiz–Sogge [KRS]) *If $d \geq 3$ and $\frac{1}{p} - \frac{1}{p'} = \frac{2}{d}$, then*

$$\|e^{k \cdot x} u\|_{p'} \leq C \|e^{k \cdot x} \Delta u\|_p$$

uniformly in $k \in \mathbb{R}^d$ and $u \in C_0^\infty$.

PROOF   It should be mentioned that the original proof of Kenig–Ruiz–Sogge is technically quite simple, simpler than the proof of (2). This is due to the fact that the operator taking $e^{k \cdot x} u$ to $e^{k \cdot x} \Delta u$ commutes with translation, so the problem reduces to a multiplier problem. However, it may also be of interest that Proposition 8 is a formal consequence of (2). Namely, write down (2) with the origin at the point $\frac{tk}{|k|^2}$, which is outside the support of u when t is large. Thus

$$\left\| \left| x - \frac{tk}{|k|^2} \right|^{-t} u \right\|_{p'} \leq C \left\| \left| x - \frac{tk}{|k|^2} \right|^{-t} \Delta u \right\|_p$$

for a sequence of $t \to \infty$. Equivalently,

$$\left\| \left(\frac{t^2}{|k|^2} - 2t \frac{k}{|k|^2} \cdot x + |x|^2 \right)^{-t/2} u \right\|_{p'} \leq C \left\| \left(\frac{t^2}{|k|^2} - 2t \frac{k}{|k|^2} \cdot x + |x|^2 \right)^{-t/2} \Delta u \right\|_p ,$$

i.e.,

$$\left\| \left(1 - \frac{2}{t} k \cdot x + \frac{|k|^2 |x|^2}{t^2} \right)^{-t/2} u \right\|_{p'} \leq C \left\| \left(1 - \frac{2}{t} k \cdot x + \frac{|k|^2 |x|^2}{t^2} \right)^{-t/2} \Delta u \right\|_p .$$

Now let $t \to \infty$. The expression $\left(1 - \frac{2}{t} k \cdot x + \frac{|k|^2 |x|^2}{t^2} \right)^{-t/2}$ converges to $e^{k \cdot x}$ uniformly on sup a, and the proof is complete.   ∎

In [W1] we used an osculation argument to prove (6) and related inequalities for appropriate functions ϕ, which roughly speaking are strictly convex functions of $\log \frac{1}{|x|}$. We will give a sketch of this argument in the case of inequality (6). Thus Proposition 9 below generalizes the result of Kim [Ki].

Let us first explain what happens when one attempts to adapt the calculation in the above proof of (2) to the case of the gradient inequality (6). A fundamental solution $L_{n.\alpha}$,

$$|x|^{-(n+[(d-1)/2]-\alpha)} \nabla u(x) = \int L_{n.\alpha}(x, y) |y|^{-(n+[(d-1)/2]-\alpha)} \Delta u(y) \, dy \qquad (11)$$

may be obtained as follows. Replace n by $n + 1$ in (2.4) and differentiate for x:

$$|x|^{-(n+(d/2)-\alpha)}\left(\nabla u(x) - \left(n + \frac{d}{2} - \alpha\right)u(x)\frac{x}{|x|^2}\right)$$

$$= \int \nabla_x K_{n+1,\alpha}(x, y)|y|^{-(n+(d/2)-\alpha)}\Delta u(y)\,dy.$$

Now substitute in (8) (with n replaced by $n + 1$) for $u(x)$ on the left side:

$$|x|^{-(n+(d/2)-\alpha)}\nabla u(x) = \int \left(\nabla_x K_{n+1,\alpha}(x, y)\right.$$

$$\left. + \left(n + \frac{d}{2} - \alpha\right)K_{n+1,\alpha}(x, y)\frac{x}{|x|^2}\right)|y|^{-(n+(d/2)-\alpha)}\Delta u(y)\,dy.$$

One may therefore take

$$L_{n,\alpha}(x, y) = \left(\frac{|x|}{|y|}\right)^{1/2}\left(\nabla_x K_{n+1,\alpha}(x, y) + \left(n + \frac{d}{2} - \alpha\right)K_{n+1,\alpha}(x, y)\frac{x}{|x|^2}\right).$$

One has the analogue of the estimates in Lemma 1, namely,

$$|L_{n,\alpha}(x, y)| \le C|x - y|^{-(d-1)} \qquad \left(|x - y| < \frac{1}{2n}|y|\right)$$

$$|L_{n,\alpha}(x, y)| \le Cn^{d-2}\min\left(1, \left|1 - \frac{|x|}{|y|}\right|^{-1}\right)|x|^{-((d-1/2)-\alpha)}|y|^{-((d-1/2)+\alpha)}$$

$$\left(|x - y| > \frac{1}{2n}|y|\right)$$

$$L_{n,\alpha}(x, y) = \mathrm{Re}\left(a\left(\frac{|x|}{|y|}, \theta\right)e^{in\theta}\right)|x|^{-((d-1/2)-\alpha)}|y|^{-((d-1/2)+\alpha)}$$

when $|\sin\theta| > \frac{1}{n}$, with a $\mathbb{C}^d$ valued function $a(r, \theta)$ satisfying

$$\left|\frac{d^j}{d\theta^j}a\right| \le n^{(d/2)-1}(|\sin\theta| + |1 - r|)^{-1}|\sin\theta|^{-((d/2)-1+j)}.$$

However, repeating the $L^p \to L^{p'}$ estimates in the proof of (2) leads only to an inequality of the form $\| |x|^{-t}\nabla u(x)\|_{p'} \le C(t)\| |x|^{-t}\Delta u(x)\|_p$, $u \in C_0^\infty(D(0, 1)\setminus \{0\})$ where $C(t)$ blows up as $t \to \infty$ if $p < 2$ (e.g., Proposition 2 is proven this way). This is necessarily the case in view of counterexamples in [KRS, J]. On the other hand, we have Proposition 9.

PROPOSITION 9

Suppose $\phi : \mathbb{R}^+ \to \mathbb{R}$ satisfies

(i)   *ϕ is increasing and convex*

(ii)   *$0 < \phi' \le$ const., $\phi'' \le$ const., and for each $\delta > 0$, there is $C_\delta > 0$ such that*

$$\frac{\phi''(t)}{\phi'(t)} \ge C_\delta^{-1} e^{-\delta t} \qquad \text{for all } t \in \mathbb{R}^+ \tag{12}$$

Then for $u \in C_0^\infty(D(0, 1)\backslash\{0\})$, $v \in \mathbb{R}^+$, v large,

$$\|e^{v\phi(\log(1/|x|))}\nabla u(x)\|_{L^{p'}(dx)} \le C\|e^{v\phi(\log(1/|x|))}\Delta u(x)\|_{L^p(dx)} \tag{13}$$

if $\frac{1}{p} - \frac{1}{p'} = \frac{2}{3d-2}$.

REMARK 4   As mentioned above this is due to Kim [Ki] for certain special ϕ. It clearly implies the SUCP for $|\Delta u| \le B|\nabla u|$, $B \in L^{(3d-2)/2}$. Proposition 9 as stated is a special case of [W1], Lemma 3.2.   ∎

REMARK 5   On the other hand, in [W1] the SUCP is proven for $|\Delta u| \le B|\nabla u|$ when $B \in L^r$, $r = \max\left(d, \frac{3d-4}{2}\right)$. When $d \ge 5$, this involves using an inequality like (13) with $\frac{1}{p} - \frac{1}{p'} = \frac{2}{3d-4}$, but with C allowed to depend on v in a controlled way as $v \to \infty$, together with a modification of the Carleman method to be discussed in Section 6.3.

We now sketch the proof of Proposition 9. The idea is to obtain a fundamental solution P_ϕ^v

$$e^{v\phi(\log(1/|x|))}\nabla u(x) = \int P_\phi^v(x, y)e^{v\phi(\log(1/|y|))}\Delta u(y)\,dy$$

by starting from $L_{n,\alpha}$ and osculating the convex function ϕ by linear ones. It will be convenient to denote $L_{n,\alpha}, 0 \le \alpha < 1$ by L_t where $t = n + \frac{d-1}{2} - \alpha$.

If x and y are points of $D(0, 1)$ then we will denote $\log\frac{1}{|x|}$ by σ, and $\log\frac{1}{|y|}$ by τ. Then

$$e^{v\phi(\sigma)}\nabla u(x) = e^{v(\phi(\sigma)-\phi'(\sigma)\sigma)}e^{v\phi'(\sigma)\sigma}\nabla u(x)$$

$$= e^{v(\phi(\sigma)-\phi'(\sigma)\sigma)}\int L_{v\phi'(\sigma)}(x, y)e^{v\phi'(\sigma)\tau}\Delta u(y)\,dy$$

$$= \int e^{v(\phi(\sigma)-\phi(\tau)-\phi'(\sigma)(\sigma-\tau))}L_{v\phi'(\sigma)}(x, y)e^{v\phi(\tau)}\Delta u(y)\,dy,$$

i.e., we can take

$$P_\phi^{(\nu)'}(x, y) = e^{-\nu(\phi(\tau) - \phi(\sigma) - \phi'(\sigma)(\tau - \sigma))} L_{\nu\phi'(\sigma)}(x, y).$$

The expression $\phi(\tau) - \phi(\sigma) - \phi'(\sigma)(\tau - \sigma)$ is positive by convexity of ϕ, and under the "strict convexity" hypothesis (10), $e^{-\nu(\phi(\tau) - \phi(\sigma) - \phi'(\sigma)(\tau - \sigma))}$ will decay rapidly when (roughly speaking) $|\tau - \sigma| \gg \nu^{-1/2}$. It follows that $P_\phi^{(\nu)}$ behaves roughly like L_ν when $|\tau - \sigma| \lesssim \nu^{-1/2}$ but is negligibly small when $|\tau - \sigma| \gg \nu^{-1/2}$. If one goes through the calculation proving (2) with this in mind, one obtains Proposition 9 ([W1], Section 3).   ∎

We will now discuss recent work of Evasius [E]. His main result is the following.

PROPOSITION 10
Let p and q satisfy $p = \frac{d}{2}$, $q < \frac{2d}{d-3}$ if $d \geq 3$, and $p > 1$, $q < \infty$ if $d = 2$. Then if Ω is a bounded convex domain in $\mathbb{R}^d$, there is a constant $C = C(p, q, \Omega)$ such that

$$\|e^\phi u\|_q \leq C \|e^\phi \Delta u\|_p$$

for all convex functions $\phi : \Omega \to \mathbb{R}$ and all $u \in C_0^\infty(\Omega)$.

In other words, if ϕ is convex then inequality (5) holds for suitable p and q with a constant that is independent not only of t but also of ϕ. If ϕ is kept smooth and strictly convex, then this was proven (with $p = q = 2$) in [H1]. Note that the sets $S(k)$ below are equal size discs in that case. We will give a sketch of Evasius' proof. For $k \in \mathbb{R}^d$ define

$$a(k) = \min_{y \in \Omega} \phi(y) - k \cdot y,$$

$$S(k) = \{x \in \Omega : \phi(x) - k \cdot x \leq a(k) + 1\}.$$

Thus $S(k)$ is the set where ϕ is well approximated by an affine function of the form $k \cdot x + \text{const}$. The $S(k)$'s form a covering of Ω by convex sets, which may have arbitrary directions and eccentricities. The main step is to obtain a covering lemma in this context.

LEMMA 3
There is a subcollection $\{S_{k_j}\}$ such that

(i) $\cup_j S_{k_j} = \Omega$

(ii) $\sum_j |S_{k_j}| \leq \text{const} |\Omega|$.

This is nontrivial, and we do not discuss it here; see [E]. However, once Lemma 3 is known, Proposition 10 is proven as follows [E]. Assume $d \geq 3$ for simplicity.

Start with the inequality

$$\|e^{k \cdot x} u\|_q \le C \|e^{k \cdot x} \Delta u\|_r, \qquad \frac{1}{r} - \frac{1}{q} = \frac{2}{d}, \quad \frac{2d}{d-1} < q < \frac{2d}{d-3}. \qquad (14)$$

This inequality is an improvement of Proposition 8 and like Proposition 8 is due to Kenig–Ruiz–Sogge [KRS]. Clearly it suffices to prove Proposition 10 when $\frac{2d}{d-1} < q < \frac{2d}{d-3}$. With r corresponding to q in (14), and $\{k_j\}$ as in Lemma 3,

$$\begin{aligned}
\|e^{\phi} u\|_q^q &= \sum_j \|e^{\phi} u\|_{L^q(S_{k_j})}^q \\
&\le C \sum_j \|e^{k_j \cdot x + a_{k_j}} u\|_{L^q(S_{k_j})}^q \\
&\le C \sum_j \|e^{k_j \cdot x + a_{k_j}} \Delta u\|_r^q \\
&\le C \sum_j \|e^{-(\phi - k_j \cdot x - a_{k_j})}\|_q^q \|e^{\phi} \Delta u\|_{d/2}^q.
\end{aligned}$$

Evasius then shows, using elementary properties of convex functions, that $\|e^{-(\phi - k \cdot x - a_k)}\|_q^q \le C |S_k|$ for any k. So

$$\|e^{\phi} u\|_q \le C \|e^{\phi} \Delta u\|_{d/2} \left(\sum_j |S_{k_j}| \right)^{1/q},$$

and the proof is complete because of (ii) in the lemma.

As was previously discussed, it is less understood to what extent (5) is true when the Hessian of ϕ is neither positive nor negative definitively. As was pointed out to me by Jerison [J2], this type of question also arises in connection with the following long-open problem.

Bers' problem Assume u is a harmonic function on the unit ball in $\mathbb{R}^d$, which is smooth up to the boundary, and ∇u vanishes on a boundary set of positive measure. Does it follow that u is constant?

If $d = 2$ or ∇u vanishes on an open subset of the boundary, then the answer is yes by the F. and M. Riesz theorem or Cauchy–Kowalewski theorem, respectively. If u is just $C^{1+\varepsilon}$ up to the boundary for small ε, then the answer is no [W4].

PROPOSITION 11

[J2] *Assume that for some* $p \in (1, \infty)$, $q \in (0, \infty)$ *and* $C > 0$ *the inequality*

$$\|e^{\phi} \omega\|_q \le C(\|e^{\phi} d\omega\|_p + \|e^{\phi} d^* \omega\|_p) \qquad (15)$$

is valid for all 1-forms $\omega \in C_0^{\infty}(D(0, 1))$ *and all subharmonic (or just all harmonic)* $\phi : D \to \mathbb{R}^+$. *Then there is no nonconstant harmonic function* $v : D \to \mathbb{R}$,

$C^{1+\alpha}$ *up to the boundary with* $\alpha = \frac{1}{p'}$, *and such that* ∇v *vanishes on a boundary set of positive measure.*

REMARK 6   The argument was shown to me with $p = q = 2$, v smooth up to the boundary but is the same in the general case. I have worked through it to obtain the dependence of α on p.   ∎

REMARK 7   Observe that by the result of [W4], (13) must fail if p is close to 1.   ∎

REMARK 8   Inequality (15) is of course not exactly of the form (5) or (6) but is very closely related, and it is also not known whether (5) holds (on a disc) for harmonic ϕ and suitable p and q.   ∎

PROOF OF LEMMA 3   [J2] In addition to the Carleman argument this uses another classical argument, the Privalov tent construction. Let $E \subset \partial D$ be a compact set with positive measure on which ∇v vanishes. For each $x \in E$, let

$$\Gamma_2(x) = \{z \in D : |z - x| \le 2(1 - |z|)\}$$
$$\Gamma_4(x) = \{z \in D : |z - x| \le 4(1 - |z|)\}$$

and

$$\Gamma_2(E) = \bigcup_{x \in E} \Gamma_2(x),$$
$$\Gamma_4(E) = \bigcup_{x \in E} \Gamma_4(x).$$

It is standard that if $z \in \Gamma_2(E)$, then dist $(z, \mathbb{R}^d \backslash \Gamma_4(E)) \ge C^{-1}(1 - |z|)$. We may therefore choose a smooth cutoff function $\phi : D \to \mathbb{R}$ with $\phi = 1$ on $\Gamma_2(E)$, $\phi = 0$ on $D \backslash \Gamma_4(E)$ and $|\nabla \phi(z)| \le C(1 - |z|)^{-1}$.

Next let h_E be the harmonic measure of E, i.e., h_E is harmonic on D with boundary values 1 on E, and 0 on $\partial D \backslash E$. Then

$$\|e^{t h_E} \omega\|_q \le C(\|e^{t h_E} d\omega\|_p + \|e^{t h_E} d^* \omega\|_p) \tag{16}$$

for $C_0^\infty(D)$ 1-forms ω, uniformly in $t > 0$.

Suppose $\varepsilon > 0$, $\eta > 0$. Then there are $\delta > 0$ and a smooth function $\psi : \mathbb{R} \to [0, 1]$ such that $\psi(s) = 1$ if $s > \varepsilon$, $\psi(s) = 0$ if $s < \delta$ and $(\int |\psi'(s)|^p s^{p/p'} ds)^{1/p} < \eta$. (One can take ψ of the form max $\left(1, \frac{\log_+ t/\delta}{\log \varepsilon/\delta}\right)$ and then smooth off slightly.) Inequality (16) applies to $f\phi\, dv$ where $f(z) = \psi(1 - |z|)$. Thus

$$\|e^{t h_E} f\phi\, dv\|_q \le C(\|e^{t h_E} d(f\phi\, dv)\|_p + \|e^{t h_E} d^*(f\phi\, dv)\|_p$$

$$\leq C(\|e^{th_E}|df||\phi \, dv|\|_p + \|e^{th_E}|f||d\phi||dv|\|_p)$$

since $d \, dv = d^* \, dv = 0$.

The first term here is $\leq C_t (\int |\psi'(s)|^p s^{p/p'} ds)^{1/p}$, since $|dv| \leq C(1-|z|)^\alpha = C(1-|z|)^{1/p'}$ when $z \in \Gamma_4(E)$ and $\phi = 0$ on $D \backslash \Gamma_4(E)$. Since $|f| \leq 1$ everywhere and $f = 1$ when $1 - |z| > \varepsilon$, we obtain

$$\|e^{th_E}\phi \, dv\|_{L^q(\{z : 1-|z| > \varepsilon\})} \leq C_t \eta + C\|e^{th_E}|d\phi||dv|\|_p.$$

Letting $\eta \to 0$, then $\varepsilon \to 0$,

$$\|e^{th_E}\phi \, dv\|_q \leq C\|e^{th_E}|d\phi||dv|\|_p.$$

We now use another fact about the sawtooth domain, namely, that $h_E \leq \beta$ for some $\beta < 1$ on $D \backslash \Gamma_2(E)$. Since $d\phi = 0$ on $\Gamma_2(E)$,

$$\|e^{t(h_E - \beta)}\phi \, dv\|_q \leq C\||d\phi||dv|\|_p.$$

The right side is finite and independent of t. Letting $t \to \infty$, we conclude the following:

$$\text{if } h_E(z) > \beta, \text{ then } \phi(z)\nabla v(z) = 0.$$

On the other hand, $\phi = 1$ on $\Gamma_2(E)$, and $\Gamma_2(E)$ contains points with $h_E(z) > \beta$ since h_E has radial limit 1 almost everywhere on E. Such points of course form an open set, so ∇v vanishes on an open set and therefore everywhere. ∎

REMARK 9   The inequality analogous to (15) for 0-forms (and subharmonic ϕ), $p = q = 2$ is easily proved by integration by parts. Furthermore as we have already mentioned, when $d = 2$ inequality (15) is true for subharmonic ϕ, $p = q = 2$ [J2]. When $d \geq 3$, it is of course quite unclear whether (15) is true or not. Proposition 11 is included with Jerison's permission—to my mind, it is of some interest whether or not (15) is true as stated. ∎

6.3   A modification of the Carleman method

In this section we will explain the proof of result VII; a similar idea is used to prove results V and VIII.

We will use the Kenig-Ruiz-Sogge inequality

$$\|e^{k \cdot x} u\|_{p'} \leq C\|e^{k \cdot x} \Delta u\|_p, \qquad \frac{1}{p} - \frac{1}{p'} = \frac{2}{d}. \tag{17}$$

The analogous inequalities for the gradient term are false as discussed in Section 6.1. However, it is possible to show that (with the same p)

$$\|e^{k \cdot x} \nabla u\|_{L^2(E)} \leq C(|k|^d |E|)^{1/d} \|e^{k \cdot x} \Delta u\|_p \tag{18}$$

for any set E with $|E| \geq |k|^{-d}$. Note that $\frac{1}{p} - \frac{1}{2} = \frac{1}{d}$. The proof of (18) is similar to the proof of (17); see ([W3], Lemma 6.2). The challenge is determining how to use (18); it is not a true Carleman inequality since the coefficient $(|k|^d |E|)^{1/d}$ blows up as $|k| \to \infty$ with E held fixed. We will base the argument on a certain covering lemma, Theorem 2 below.

To state it let us recall some facts about convex sets. A *convex body* is a compact convex set with interior. If E is a convex body then we let b_E be its barycenter, $b_E = \frac{\int_E x \, dx}{\int_E dx}$. We write CE to mean the dilation of E by a factor C around its barycenter, $CE = \{x \in \mathbb{R}^d : b_E + C^{-1}(x - b_E) \in E\}$. If $b_E = 0$, then we let $\tilde{E}$ be the dual body, $\tilde{E} = \{\xi \in \mathbb{R}^d : \langle \xi, x \rangle \leq 1 \text{ for all } x \in E\}$. In what follows, we will be working with a family of convex bodies E_k depending on a parameter k. In such a situation, we define

$$E_k^* = \widetilde{(E_k - b_{E_k})} + k.$$

Thus E_k^* is defined by translating E_k so its barycenter is at the origin, passing to the dual convex body and then translating by k.

THEOREM 2
Suppose μ is a positive measure in $\mathbb{R}^d$ with faster than exponential decay, i.e.,

$$\lim_{T \to \infty} T^{-1} \log \mu(\{x : |x| \geq T\}) = -\infty. \tag{19}$$

Define μ_k by $d\mu_k(x) = e^{k \cdot x} d\mu(x)$. Suppose C is a convex body. Then there is a sequence $\{k_j\} \subset C$ and, for each j, a convex body E_{k_j} with

$$\mu_{k_j}(\mathbb{R}^d \setminus (1 + T) E_{k_j}) \leq \frac{1}{2} e^{-T} \|\mu_{k_j}\| \tag{20}$$

and such that

$$\{E_{k_j}\} \quad \text{are pairwise disjoint} \tag{21}$$

$$|\cup_j E_{k_j}^* \cap C| \geq C^{-1} |C| \tag{22}$$

where C depends on d only.

REMARK 10   We note again that $E_{k_j}^* = \widetilde{(E_{k_j} - b_{E_{k_j}})} + k_j$.   ∎

REMARK 11   Theorem 2 was formulated slightly differently in [W3, Lemma 1]. In particular, instead of (20), we stated that $\sum |E_{k_j}|^{-1} \geq C^{-1} |C|$. This follows from (22) since $|E_{k_j}^*| \approx |E_{k_j}|^{-1}$ (this is a general fact about dual convex bodies). *A priori* (22) is stronger. However, the proof in [W3] also proves (22). We will justify this in a remark at the end of the section. ∎

In the first part of this section we will discuss Theorem 2, and then we will use it to prove result VII.

A simple example illustrating the meaning of Theorem 2 is a Gaussian. Take $d\mu(x) = e^{-|x|^2/2}\,dx$. Then μ_k is another Gaussian, $d\mu_k(x) = C_k e^{-|x-k|^2/2}\,dx$ with $C_k = e^{|k|^2/2}$. So if b is a large fixed constant, then $\mu_k(\mathbb{R}^d \setminus (1+T)D(k,b)) \leq \frac{1}{2}e^{-T}\|\mu_k\|$ for all k. A maximal $2b$-separated subset $\{k_j\} \subset C$ has cardinality $\gtrsim |C|$, and since $E_{k_j}^* = D(k_j, b^{-1})$, one easily checks (22). Another natural example is surface measure on the unit sphere, $C = D(0, N)$, N large. Then one can take $\{k_j\} = $ maximal $CN^{1/2}$-separated subset of $\partial D(0, N)$, $E_{k_j} = $ rectangle centered at $\frac{k_j}{|k_j|}$ with side CN^{-1} in $\frac{k_j}{|k_j|}$-direction, $CN^{-1/2}$ in perpendicular directions.

We will give an essentially complete proof of Theorem 2 in the one-dimensional case and will then indicate how the argument extends to higher dimensions. The one-dimensional argument is constructed with this in mind and is not likely to be the simplest possible.

PROOF OF THEOREM 2 WHEN $d = 1$ We may assume μ is not a single point mass.

We define an interval E_k for each $k \in \mathbb{R}$, as follows. Let $z_k = \dfrac{\int x\,d\mu_k(x)}{\int d\mu_k(x)}$ be the barycenter of μ_k. Then it is easy to see that there is a unique $\rho_k^+ > 0$ such that

$$\mu_k((z_k + (1+t)\rho_k^+, \infty)) \leq \frac{1}{4}e^{-2t}\|\mu_k\| \tag{23}$$

for all $t \geq 0$, and

$$\mu_k([z_k + (1+t)\rho_k^+, \infty)) \geq \frac{1}{4}e^{-2t}\|\mu_k\| \tag{24}$$

for some $t \geq 0$.

We need only take ρ_k^+ to be the infimum of all ρ such that (23) holds. ρ_k^+ is then finite by (19) and nonzero since μ is not a point mass, and (23) and (24) are easily checked. See [W3], Lemma 2.1 for the details. Likewise there is a unique ρ_k^- such that

$$\mu_k((-\infty, z_k - (1+s)\rho_k^-)) \leq \frac{1}{4}e^{-2s}\|\mu_k\| \tag{25}$$

for all $s \geq 0$

$$\mu_k((-\infty, z_k - (1+s)\rho_k^-]) \geq \frac{1}{4}e^{-2s}\|\mu_k\| \tag{26}$$

for some $s \geq 0$.

We let $a_k = z_k - \rho_k^-$, $b_k = z_k + \rho_k^+$, and $E_k = [a_k, b_k]$. It is then clear that $(1 + T)E_k$ contains $[z_k - (1 + \frac{1}{2}T)\rho_k^-, z_k + (1 + \frac{1}{2}T)\rho_k^+]$ (the factor of $\frac{1}{2}$ is needed since z_k is not necessarily the midpoint of E_k), and therefore $\mu_k(\mathbb{R}\setminus(1 + T)E_k) \leq \frac{1}{2}e^{-T}\|\mu_k\|$, i.e., (20) holds. Now fix an interval $C \subset \mathbb{R}$. We must find a sequence $\{k_j\} \subset C$ for which we can prove (21) and (22). The possibility of extracting such a sequence is due to Lemma 4.

LEMMA 4
$|E_j \cap E_k| \leq \frac{C}{|j-k|}$.

PROOF   We may assume $j > k$, and also $b_k > a_j$ since otherwise $E_j \cap E_k = \emptyset$. By (22) and (24) there are $s, t \geq 0$ such that

$$\mu_j((-\infty, a_j - s\rho_j^-]) \geq \frac{1}{4}e^{-2s}\|\mu_j\|$$

$$\mu_k([b_k + t\rho_k^+, \infty)) \geq \frac{1}{4}e^{-2t}\|\mu_k\|.$$

We will assume for simplicity that $s = t = 0$—the general case is similar and is done in detail in [W3], Lemma 4.4. Assuming $s = t = 0$, let $I = (-\infty, a_j]$ and $J = [b_k, \infty)$. Then

$$\frac{\mu_j(I)}{\mu_k(I)} \geq \frac{\mu_j(I)}{\|\mu_k\|} \geq \frac{1}{4}e^{-2s}\frac{\|\mu_j\|}{\|\mu_k\|} \geq \frac{1}{16}e^{-2(s+t)}\frac{\mu_j(J)}{\mu_k(J)}.$$

On the other hand, $x \in I$ and $y \in J$ imply $x - y \leq a_j - b_k$. Therefore,

$$\frac{\mu_j(I)}{\mu_k(I)} = \frac{\int_I e^{(j-k)x}d\mu_k(x)}{\mu_k(I)} \leq e^{(j-k)(a_j-b_k)}\frac{\int_J e^{(j-k)y}d\mu_k(y)}{\mu_k(J)} = e^{(j-k)(a_j-b_k)}\frac{\mu_j(J)}{\mu_k(J)}.$$

Combining this inequality and the preceding one, we have $(j - k)(a_j - b_k) \leq \log 16$, so $|E_j \cap E_k| \leq b_k - a_j \leq \frac{\log 16}{j-k}$, and we're done.   ∎

We now finish the proof of the one-dimensional case of Theorem 2. Let ϵ_0 be a fixed small positive constant. We may assume that $|E_k| \geq |C|^{-1}$, since otherwise the result is obvious ($\{k_j\}$ = singleton). Choose $k_1 \in C$ such that $|E_{k_1}| \leq 2\inf_{k \in C}|E_k|$. Recursively, if $k_1, \ldots, k_{n-1}$ have been chosen; then let $C_n = C\setminus\cup_{i<n} (k_i - \epsilon_0^{-1}|E_{k_i}|^{-1}, k_i + \epsilon_0^{-1}|E_{k_i}|^{-1})$. If $C_n \neq \emptyset$, then choose $k_n \in C_n$ with $|E_{k_n}| \leq 2\inf_{k \in C_n}|E_k|$. If $C_n = \emptyset$, then cease the construction.

This results in a sequence $\{k_n\}$, which in principle may be finite or infinite. We show next that $\cap_n C_n = \emptyset$. Namely, if this fails, then fix $k \in \cap_n C_n$. By construction, $\{k_n\}$ is an infinite sequence, and $|E_{k_n}| \leq 2|E_k|$ for all n. Thus the k_n form an infinite discrete set, since $|k_j - k_i| \geq (2\epsilon_0)^{-1}|E_k|^{-1}$. But $k_n \in C$ for all n and C is compact, so we have a contradiction.

Equivalently, we have shown that

$$C \subset \cup_j 2\epsilon_0^{-1} E_{k_j}^*. \tag{27}$$

Now fix i and j with $i < j$. Then

$$|E_{k_i} \cap E_{k_j}| \leq \frac{C}{|k_i - k_j|} \leq C\epsilon_0 |E_{k_i}|.$$

The first inequality follows by Lemma 2 and the second since $k_j \in C_j$. On the other hand $|E_{k_i}| \leq 2|E_{k_j}|$ by the selection rule so

$$|E_{k_i} \cap E_{k_j}| \leq 2C\epsilon_0 \min(|E_{k_i}|, |E_{k_j}|).$$

Provided $\epsilon_0 < (4C)^{-1}$, this implies that no E_{k_i} is contained in the union of two others. Consequently $\{E_{k_i}\}$ is the union of two families of pairwise disjoint intervals. In view of (27), one of these sequences must satisfy (22), so we are finished with the one-dimensional case of Theorem 2.

We now discuss the higher dimensional case. To keep things simple we will sketch only the proof of a slightly weaker result: instead of (19) we assume Supp $\mu \subset D(0, 1)$, and instead of (22) we will obtain only

$$\sum |E_{k_i}|^{-1} \geq C_\epsilon^{-1} |C|^{1-\epsilon} \tag{28}$$

for an arbitrary fixed $\epsilon > 0$. Theorem 2 corresponds roughly to the endpoint $\epsilon = 0$; this requires an additional combinatorial argument given in Section 5 of [W3].

We first indicate how the sets E_k are constructed. We may assume μ is not supported on a hyperplane. Let C_0 be a suitable constant and σ a small positive constant. Constants C below are independent of the choice of σ. Let $z_k = \dfrac{\int x \, d\mu_k}{\int d\mu_k}$. For each unit vector $e \in \mathbb{R}^d$, let $\rho_k(e)$ be the unique positive number with

$$\mu_k(\{x : \ll x - z_k, e \gg (1 + t)\rho_k(e)\}) \leq \sigma e^{-C_0 t} \|\mu_k\|$$

for all $t \geq 0$, and

$$\mu_k(\{x : \langle x - z_k, e \rangle \geq (1 + t)\rho_k(e)\}) \geq \sigma e^{-C_0 t} \|\mu_k\|$$

for some $t \geq 0$. Let $D_k = \cap_k \{x : \langle x - z_k, e \rangle \leq \rho_k(e)\}$ and let $E_k = (1 + \sigma^{1/d})D_k$, where we dilate around the barycenter of D_k as previously indicated. Using a little bit of convexity theory one can show that (20) holds, provided σ is small. See [W3], Section 2. What remains is to find a sequence $\{k_j\}$ satisfying (21) and (28).

The appropriate analogue of Lemma 4 is the following Lemma 5. If A is a positive constant then let us write $j \overset{A}{\underset{\sim}{\frown}} k$ if both $E_j \subset AE_k$ and $E_k^* \subset AE_j^*$. If $j \overset{A}{\underset{\sim}{\frown}} k$ and $k \overset{A}{\underset{\sim}{\frown}} j$ then we will write $j \overset{A}{\approx} k$. We let γ be an appropriate positive constant. (Any γ less than $\frac{1}{d}$ will work.)

LEMMA 5

If $E_j \cap E_k \neq \emptyset$ then either

(i) $\quad j \overset{C_1\sigma^{-1}}{\approx} k$ *or*

(ii) $\quad \exists \ell \in \overline{jk}, \ell \overset{C_1\log(1/\sigma)}{\lesssim} k, |E_\ell| \leq C_1\sigma^\gamma |E_k|$ *or*

(iii) *Same as (ii) reversing the roles of j and k.*

This lemma says that if E_j and E_k intersect, then either they are essentially the same, even when regarded as sets in phase space via the identification $E_j \to E_j \times E_j^*$, or else there is a much smaller E_ℓ situated close to one of them. Since ℓ belongs to the line segment connecting j and k, it will of course belong to any convex set containing j and k. We will not discuss the proof of Lemma 5—see Section 4 of [W3].

To finish the proof of Theorem 2 (in the weaker form mentioned above), fix C. We may assume that $|E_k| \geq |C|^{-1}$ for all $k \in C$. Define k to be *minimal* if there is no $\ell \in C$ such that $\ell \overset{C_1\log(1/\sigma)}{\lesssim} k$ and $|E_\ell| \leq C_1\sigma^\gamma|E_k|$, with C_1 as in Lemma 5. An immediate consequence of Lemma 5 is that if j and k are *minimal* and $E_j \cap E_k \neq \emptyset$ then $j \overset{C_1\sigma^{-1}}{\approx} k$.

On the other hand, if $j \in C$, then we may form a chain $j = j_1, j_2, \ldots$ with $j_{i+1} \overset{C_1\log(1/\sigma)}{\lesssim} j_i$ and $|E_{j_{i+1}}| \leq C_1\sigma^\gamma|E_{j_i}|$, stopping when we reach a minimal E_{j_n}. If C_2 is a suitable constant, then there can be no such chain of length $n > C_2\frac{\log|C|}{\log(1/\sigma)}$: this would give the contradiction

$$|E_{j_n}| \leq (C\sigma^\gamma)^n|E_j| < |C|^{-1}$$

since $|E_j| \lesssim 1$ by the assumption that μ is supported in the unit disc. It follows that for any $j \in C$ there is a minimal k such that $k \overset{(C\log(1/\sigma))^{n_0}}{\lesssim} j$, where $n_0 = C_2\frac{\log|C|}{\log(1/\sigma)}$. Thus the sets

$$\left(C\log\frac{1}{\sigma}\right)^{C\frac{\log|C|}{\log(1/\sigma)}} \quad \text{with } k \text{ minimal}$$

cover C.

Now choose (using Zorn's lemma) a sequence $\{k_j\}$ such that (i) each k_j is minimal (ii) no two E_{k_j}'s intersect (iii) every E_k with k minimal intersects some E_{k_j}. Then every minimal k satisfies $E_k^* \subset C\sigma^{-1}E_{k_i}^*$ for some i and therefore the sets

$$C\sigma^{-1}\left(C\log\frac{1}{\sigma}\right)^{\frac{C\log|C|}{\log(1/\sigma)}}E_{k_i}^*$$

cover C. If σ is small, then the coefficient $C\sigma^{-1}\left(C\log\frac{1}{\sigma}\right)^{C\frac{\log|C|}{\log(1/\sigma)}}$ is $\leq C_\sigma|C|^\epsilon$, so $\sum_i|E_{k_i}^*|\gtrsim|C|^{1-\epsilon}$, completing the proof.

We now discuss the proof of result VII. We will only show that if $u\in W^{2p}$, $\frac{1}{p}-\frac{1}{p'}=\frac{2}{d}$ satisfies the inequality $|\Delta u|\leq A|u|+B|\nabla u|$, $A\in L^{d/2}$, $B\in L^d$ and has *compact support,* then u vanishes identically. The general case where u is only assumed to vanish on an open set may be performed using the Kelvin transform and then a variant of the same argument, see [W3], Section 7.

Let K be the convex hull of supp u, and let χ be a cutoff function equal to 1 on a neighborhood of ∂K and such that $||A||_{L^{d/2}(\text{Supp }\chi)}+||B||_{L^d(\text{Supp }\chi)}<\alpha$, where α is a small positive number to be chosen. Let $v=\chi u$. Then $|\Delta v|\leq A|v|+B|\nabla v|+E$, where $E=|2\nabla\chi\cdot\nabla u+u\Delta\chi|$ is supported on a compact subset of int K and is in L^p. We may assume that $A\geq1$ on K, and then since a linear function on a compact convex set reaches its maximum at an extreme point, it is easy to show that

$$\|e^{k\cdot x}E\|_p\leq\|e^{k\cdot x}(A|v|+B|\nabla v|)\|_p\tag{29}$$

for any k with $|k|$ sufficiently large. Now define $d\mu(x)=(A|v|+B|\nabla v|)^p\,dx$ and apply Theorem 2 with $C=D(a,N)$, N large, $|a|=2N$. Let $\{k_j\}$ and $\{E_{k_j}\}$ be as there and $F_j=E_{k_j}\cap\text{Supp }\chi$, $\ell_j=\frac{k_j}{p}$. Then we have

$$\|e^{\ell_j\cdot x}(A|v|+B|\nabla v|)\|_p$$
$$\leq C(\|A\|_{L^{d/2}(\text{Supp }\chi)}\|e^{\ell_j\cdot x}v\|_{p'}+\|B\|_{L^d(F_j)}\|e^{\ell_j\cdot x}\nabla v\|_2)$$
$$\leq C(\|A\|_{L^{d/2}(\text{Supp }\chi)}+(|\ell_j|^d|E_{k_j}|)^{1/d})\|e^{\ell_j\cdot x}(A|v|+B|\nabla v|+E)\|_p$$

by Holder's inequality and (17) and (18). We may drop E here by (29) and then cancel the last factor on the right side against the left side. Since $|\ell_j|\approx N$ we obtain

$$\|A\|_{L^{d/2}(\text{Supp }\chi)}+\|B\|_{L^d(F_j)}(N^d|E_{k_j}|)^{1/d}\geq\text{ const},$$

and therefore $\|B\|_{L^d(F_j)}^d\geq\text{ const }N^{-d}|E_{k_j}|^{-1}$, since $\|A\|_{L^{d/2}(\text{Supp }\chi)}$ is small. Then

$$\|B\|_{L^d(\text{ sup }\chi)}^d\gtrsim N^{-d}\sum_j|E_{k_j}|^{-1}\gtrsim1$$

by (22). This is a contradiction for small α. ∎

REMARK 12 Results V and VIII are proven by variants of the preceding argument. To get V one applies the one-dimensional version of Theorem 2 in the radial variable, see [W1]. For VIII, see Section 8 of [W3]. We will give a brief sketch. It is possible to show that if C in Theorem 2 is a disc of radius N and μ has compact support, then the $\{E_{k_j}\}$ may be taken with diameter $\lesssim N^{-1/2}$. On the other hand there is a Carleman inequality for the Hessian as follows:

$$||e^{k \cdot x} H_u||_{L^2(D(a,R))} \leq C|k|R||e^{k \cdot x} \Delta u||_2.$$

If L has Lipschitz coefficients and $|k| \approx N$, then on a disc of radius $N^{-1/2}$ Lu is $L_0 u + E$ where L_0 is obtained by freezing coefficients and $|E| \lesssim N^{-1/2}|H_u|$. It follows by the preceding Carleman inequality that on any such disc D,

$$||e^{k \cdot x} E||_{L^2(D)} \lesssim ||e^{k \cdot x} L_0 u||_2.$$

Since we need to make our estimates only on the sets E_{k_j}, which are contained in such discs, this permits the argument to go through as before. ∎

REMARK 13   In [W3], Theorem 2 was stated with the *a priori* weaker conclusion that

$$\sum |E_{k_j}|^{-1} \geq C^{-1}|C| \tag{30}$$

instead of (22).

The argument in [W3] however proves (22), and we would like to take this opportunity to indicate why. Let us first state a somewhat stronger form of Theorem 2. ∎

THEOREM 3
Suppose that μ is a measure satisfying (19) and C_0 is a convex body that contains the double of a rectangle C with sides $N_1, \ldots, N_d$. Let R be the rectangle centered at the origin with the same axes as C and sides $\frac{1}{N_1}, \ldots, \frac{1}{N_d}$. Then there is a sequence $\{k_j\} \subset C$ and for each j, a convex body E_{k_j} containing a translate of $C^{-1} R$, such that (20), (21) and (30) hold and

$$\text{The sets } F_{k_j}^{C^{-1}} \stackrel{\text{def}}{=} \{\xi : \underset{x \in E_{k_j}}{\mathrm{Supp}} \langle \xi - k_j, x - b_{E_{k_j}} \rangle \leq C^{-1}\}$$

are pairwise disjoint and contained in C_0. $\qquad (31)$

If in addition Supp μ is contained in a translate of rC for a certain $r \geq \max_j N_j^{-2}$, then the sets E_{k_j} may be taken so that diam $E_{k_j} \leq Cr^{1/2}$.

Every convex body contains a rectangle of comparable volume. So to derive Theorem 16 (for the convex body C_0) from Theorem 3, simply note that (30) and (31) imply (22), since $|F_{k_j}^{C^{-1}}| \approx |E_{k_j}|^{-1}$.

PROOF Modulo the remark at the bottom of page 261 of [W3], Theorem 3 differs from Lemma 1' of [W3] only in that (31) is asserted to hold. However, disjointness of the sets $F_{k_j}^{C^{-1}}$ for sufficiently large fixed C is stated and proved on page 261, line 13 ff. of [W3]. Furthermore it is stated and proved on page 261, line 8 of [W3] that $E_{k_j}^*$ (called F_{k_j} in [W3]) will be contained in a fixed multiple $(1 + C_2)C$. Since $k_j \in C$, it follows that $F_{k_j}^{C^{-1}} \subset 2C \subset C_0$ for large fixed C. This completes the proof. ∎

References

[A]   G. Allesandrini, to appear.

[C]   M. Christ, *Embedding three-dimensional compact CR manifolds of finite type in* $\mathbb{C}^n$, Ann. Math. 129 (1989), 195–213.

[CS]   S. Chanillo and E. Sawyer, *Unique continuation for* $\Delta + V$ *and the C. Fefferman-Phong class,* Trans. Amer. Math. Soc. 318 (1990), 275–300.

[DF]   H. Donnelly and C. Fefferman, *Nodal sets of eigenfunctions on Riemannian manifolds,* Inv. Math. 93 (1988), 161–183.

[E]   D. Evasius, *Carleman inequalities with convex weights,* Caltech thesis, 1992, and to appear.

[GL]   N. Garofalo and F.H. Lin, *Monotonicity properties of variational integrals, A_p weights and unique continuation,* Indiana Univ. Math. J. 35 (1986), 245–267.

[H1]   L. Hormander, *On the uniqueness of the Cauchy problem,* Math. Scand. 6 (1958), 213–225.

[H2]   L. Hormander, *Oscillatory integrals and multipliers of FL^p,* Ark. Mat. 11 (1971), 1–11.

[J1]   D. Jerison, *Carleman inequalities for the Dirac and Laplace operators and unique continuation,* Adv. Math. 62 (1986), 118–134.

[J2]   D. Jerison, private communication, 1987.

[JK]   D. Jerison and C. Kenig, *Unique continuation and absence of positive eigenvalues for Schrodinger operators,* Ann. Math. 121 (1985), 463–488.

[K]   C. Kenig, in the conference proceedings, Harmonic Analysis, El Escorial, 1988.

[KRS]   C. Kenig, A. Ruiz, and C. Sogge, *Sobolev inequalities and unique continuation for second order constant coefficient elliptic operators,* Duke Math. J. 55 (1987), 329–347.

[Ki]   Y. M. Kim, MIT thesis, 1989.

[M]   K. Miller, *Non-unique continuation for certain ODE's in Hilbert space and for uniformly parabolic and elliptic equations in self-adjoint divergence form,* pp. 85–101 in "Symposium on non-well-posed problems and logarithmic convexity," ed. R. J. Knops, Springer Lecture Notes in Mathematics, vol. 316, 1973.

[Pa]   Y. F. Pan, *Unique continuation for Schrodinger operators with singular potentials,* Comm. PDE, to appear.

[P]   A. Plis, *On non-uniqueness in the Cauchy problem for an elliptic second order differential equation,* Bull. Acad. Sci. Polon., Ser. Sci. Math. Astro. Phys. 11 (1963), 95–100.

[RV]   A. Ruiz and L. Vega, *Unique continuation for the solutions of the Laplacian plus a drift,* Ann. Inst. Fourier (Grenoble) 41 (1991), 651–663.

[Sa]   E. Sawyer, *Unique continuation for Schrodinger operators in dimension three or less,* Ann. Inst. Fourier (Grenoble) 33 (1984), 189–200.

[SS]   M. Schechter and B. Simon, *Unique continuation for Schrodinger operators with unbounded potential,* J. Math. Anal. Appl. 77 (1980), 482–492.

[So1]   C. Sogge, *Strong uniqueness theorems for second order elliptic differential equations,* Amer. J. Math 112 (1990), 943–984.

[So2]   C. Sogge, "Fourier integrals in classical analysis," Cambridge University Press, 1993.

[St1]   E. M. Stein, *Oscillatory integrals in Fourier analysis,* pp. 307-357 in "Beijing lectures in harmonic analysis," ed. E. M. Stein, Princeton University Press, 1986.

[St2]   E. M. Stein, editor's note, Ann. Math. 121 (1985), 489–494.

[W1]   T. Wolff, *Unique continuation for $|\Delta u| \leq V|\nabla u|$ and related problems,* Revista Math. Iberoamericana 6 (1990), 155–200.

[W2]   T. Wolff, *Note on counterexamples in strong unique continuation problems,* Proc. Amer. Math. Soc. 114 (1992), 351–356.

[W3]   T. Wolff, *A property of measures in $\mathbb{R}^n$ and an application to unique continuation,* Geometric and Functional Analysis 2 (1992), 225–284.

[W4]   T. Wolff, *Counterexamples with harmonic gradients in $\mathbb{R}^3$,* to appear, conference proceedings in honor of E. M. Stein.

[W5]   T. Wolff, *Some constructions with solutions of variable coefficient elliptic equations,* to appear, Journal of Geometric Analysis.

[W6]   T. Wolff, *A counterexample in a unique continuation problem,* to appear, Communications in Analysis and Geometry.

Part III

Contribution Articles

7

Weighted Lipschitz Spaces Defined by a Banach Space

Oscar Blasco[1]

ABSTRACT   We consider general weighted Lipschitz spaces defined by a Banach space. Under certain assumptions on the weight and the space, we find a Littlewood-Paley type formulation for such spaces. This allows us to give a formulation for the predual space as a generalized Besov space. We also prove that operators acting on certain weighted Besov spaces correspond to vector valued functions in a natural way.

7.1   Introduction

It is well known that the use of Calderón's reproducing formula allows for a Littlewood-Paley formulation of many functions spaces (see [FJW]).

The aim of this paper is to show that this method can also be used in the general setting of weighted Lipschitz spaces and then to give some applications to duality results and to characterize the bounded operators acting on certain weighted Besov classes.

In [J1], Janson considered the spaces $Lip(\rho, E)$ defined by distributions whose moduli of continuity in the Banach space E are dominated by a function ρ, which grows arbitrarily slowly. He also considered spaces $B(\rho, E)$ defined using convolution with test functions ψ such that $\int \psi \neq 0$. These spaces give a unified approach to certain Besov and Lipschitz classes. We shall define some closely related Banach spaces, but for a more general weight functions and under slightly different assumptions on the space E. The main difference from the spaces in [J1] comes from the fact that our test functions will be of mean zero.

Let us start by recalling some definitions on the weights and the Banach space that will be the setting for our results.

[1] This research has been partially supported by the Spanish DGICYT. Proyectos PB89-0106 and PB92-069.

DEFINITION 1   A weight ω will be a measurable function $\omega : \mathbb{R}^+ \to \mathbb{R}^+$ such that $\omega > 0$ a.e. We shall say that ω satisfies Dini condition if

$$\int_0^s \frac{\omega(t)}{t} \, dt \leq C\omega(s)$$

and ω is called a b_1-weight if

$$\int_s^\infty \frac{\omega(t)}{t^2} \, dt \leq C\frac{\omega(s)}{s}.$$

REMARK 1   This type of weights has been used by different authors (see [J1], [B1], [BS], and [BlS]) and they turned out to be the natural setting for certain weighted versions of classical results.

The main examples are $\omega(t) = t^\alpha(1 + |\log t|)^\beta$ for $0 < \alpha < 1$.   ∎

As usual S denotes the Schwartz class of test functions on $\mathbb{R}^n$ and S' the space of tempered distributions. We denote by S_0 the set of functions in S with mean zero, and S'_0 its topologic dual.

We prefer working with measurable functions rather than distributions. Our Banach spaces will be formed by measurable functions satisfying

$$\int_{\mathbb{R}^n} \frac{|f(x)|}{(1 + |x|)^{n+1}} \, dx < \infty.$$

DEFINITION 2   Let $(E, \|.\|_E)$ be a Banach space included in $L^1\left(\mathbb{R}^n, \frac{dx}{(1+|x|)^{n+1}}\right)$. E is said to be bounded under translations space if

(i) $S \subset E \subset S'$ (with continuity)

(ii)   If $\tau_x(f)(y) = f(x + y)$ for $x \in \mathbb{R}^n$, then $x \to \tau_x(f)$ is an E-valued bounded measurable function, that is $\sup_{x \in \mathbb{R}^n} \|\tau_x(f)\|_E \leq C_0\|f\|_E$.

REMARK 2   A bounded under translations space E is a Banach L^1-module, that is

if $f \in L^1, g \in E$ then $f * g \in E$ and $\|f * g\|_E \leq C_0\|f\|_1\|g\|_E$.

Indeed, given $f \in L^1$ and $g \in E$ then $y \to f(-y)\tau_y(g) \in L^1(\mathbb{R}^n, E)$. Therefore, since $f * g = \int_{\mathbb{R}^n} f(-y)\tau_y(g) \, dy$, we have

$$\|f * g\|_E \leq C_0\|f\|_1\|g\|_E. \tag{1}$$

∎

DEFINITION 3 *Given a weight* ω, *a function* $\phi \in S_0$ *and a bounded under translations Banach space* E *we define*

$$\Lambda_\omega^E = \left\{ f \in L^1\left(\mathbb{R}^n, \frac{dx}{(1+|x|)^{n+1}}\right) : \|\Delta_x f\|_E \leq C\omega(|x|), x \in \mathbb{R}^n \right\}$$

$$B_{w,\phi}^{E,\infty} = \left\{ f \in L^1\left(\mathbb{R}^n, \frac{dx}{(1+|x|)^{n+1}}\right) : \|\phi_t * f\|_E \leq C\omega(t), t > 0 \right\}$$

$$B_{\omega,\phi}^{E,1} = \left\{ f \in L^1\left(\mathbb{R}^n, \frac{dx}{(1+|x|)^{n+1}}\right) : \phi_t * f \in L^1\left(\omega(t)\frac{dt}{t}, E\right) \right\}$$

where $\Delta_x f(y) = f(x+y) - f(y)$ *and* $\phi_t(x) = \frac{1}{t^n}\phi(\frac{x}{t})$.

They are complete spaces under the following seminorms

$$\|f\|_{\Lambda_\omega^E} = inf\{C > 0 : \|\Delta_x f\|_E \leq C\omega(|x|)\}$$

$$\|f\|_{B_{w,\phi}^{E,\infty}} = inf\{C > 0 : \|\phi_t * f\|_E \leq C\omega(t)\}$$

$$\|f\|_{B_{\omega,\phi}^{E,1}} = \int_0^\infty \|\phi_t * f\|_E \omega(t)\frac{dt}{t}.$$

REMARK 3 Note that the assumption $\int_{\mathbb{R}^n} \frac{|f(x)|}{(1+|x|)^{n+1}} dx < \infty$ gives sense to the convolution $\phi_t * f$.

Observe also that $\int_0^\infty \omega(t)\frac{dt}{t} < \infty$ implies that E embeds into $B_{\omega,\phi}^{E,1}$.

Note that $\int_{\mathbb{R}^n} \phi(x)\, dx = 0$ implies that constant functions have seminorm equal zero in $B_{w,\phi}^{E,\infty}$ and $B_{\omega,\phi}^{E,1}$.

To avoid more notation we use the same formulation for the Banach spaces consisting of equivalence classes coming from the kernel of the seminorms. ∎

Let us now recall Calderón's formula and formulate the version to be used later on.

Let $\phi \in S_0$ be a real and radial function with $\int_0^\infty (\hat{\phi}(t\xi))^2 \frac{dt}{t} = 1$ for $\xi \in \mathbb{R}^n \setminus \{0\}$. If $\psi \in S$, then for $\xi \in \mathbb{R}^n \setminus \{0\}$,

$$\hat{\psi}(\xi) = \int_0^\infty (\phi_t * \phi_t * \psi)\hat{}(\xi)\frac{dt}{t}.$$

This shows that $\psi_{\varepsilon,\delta} = \int_\varepsilon^\delta \phi_t * \phi_t * \psi \frac{dt}{t}$ converges to ψ in S.

LEMMA 1
(see [FJW]) Let $\phi \in S_0$ *be a real and radial function with* $\int_0^\infty (\hat{\phi}(t\xi))^2 \frac{dt}{t} = 1$

for $\xi \in \mathbb{R}^n \setminus \{0\}$. *Let* f *be a measurable function with* $\int_{\mathbb{R}^n} \frac{|f(x)|}{(1+|x|)^{n+1}} dx < \infty$, $0 < \varepsilon < \delta$ *and write*

$$f_{\varepsilon,\delta} = \int_\varepsilon^\delta \phi_t * \phi_t * f \frac{dt}{t}$$

then $\lim_{\varepsilon \to 0, \delta \to \infty} f_{\varepsilon,\delta} = f$ *in* S'_0.

7.2 The theorems and their proofs

Let us mention some properties for test functions that are used later on.
    If $\psi \in S$, then for any $\varepsilon > 0$, taking

$$C_1 = \max \left(\max\{|\psi(y)| : |y| \le 1\}, \ \max\{|y|^{n+\varepsilon}|\psi(y)| : |y| \ge 1\} \right)$$

one has

$$|\psi_t(x)| \le C_1 \min \left(\frac{1}{t^n}, \frac{t^\varepsilon}{|x|^{n+\varepsilon}} \right). \tag{2}$$

Since $\|\Delta_x(\psi_t)\|_1 = \|\Delta_{\frac{x}{t}} \psi\|_1$, then taking

$$C_2 = \max \left(2\|\psi\|_1, \int_{\mathbb{R}^n} \max_{|z-y|\le 1} |\nabla \psi(z)| \, dy \right),$$

one has

$$\|\Delta_x(\psi_t)\|_1 \le C_2 \min \left(1, \frac{|x|}{t} \right). \tag{3}$$

LEMMA 2
Let $\phi \in S_0$. *Then*

$$\|\phi_t * \phi_s\|_1 \le C \min \left(\frac{s}{t}, \frac{t}{s} \right). \tag{4}$$

PROOF   Let us assume $s \le t$. From the mean zero assumption we write

$$\phi_t * \phi_s(y) = \int_{\mathbb{R}^n} \phi_s(-x) \Delta_x(\phi_t)(y) \, dx.$$

From Fubini, (2) for $\varepsilon = 2$ and (3)

$$\|\phi_t * \phi_s\|_1 \le \int_{\mathbb{R}^n} |\phi_s(-x)| \, \|\Delta_x(\phi_t)\|_1 \, dx$$

$$\le C \int_{\mathbb{R}^n} \min \left(\frac{1}{s^n}, \frac{s^2}{|x|^{n+2}} \right)$$

$$\leq C\left(\int_{|x|<s} \frac{|x|}{ts^n}\,dx + \int_{s\leq|x|\leq t} \frac{s^2}{t|x|^{n+1}}\,dx + \int_{|x|>t} \frac{s^2}{|x|^{n+2}}\,dx\right)$$

$$\leq C\left(\frac{1}{ts^n}\int_0^s u^n du + \frac{s^2}{t}\int_s^t \frac{du}{u^2} + s^2\int_t^\infty \frac{du}{u^3}\right)$$

$$\leq C\left(\frac{s}{t} + \frac{s^2}{t}\left(\frac{1}{s} - \frac{1}{t}\right) + \frac{1}{2}\left(\frac{s}{t}\right)^2\right) \leq C\frac{s}{t}. \qquad \blacksquare$$

THEOREM 1

Let E be a Banach space bounded under translations, ω a b_1-weight satisfying Dini condition, and $\phi \in S_0$ a real radial function with $\int_0^\infty (\hat{\phi}(t\xi))^2 \frac{dt}{t} = 1$, $(\xi \neq 0)$, then

$$\Lambda_\omega^E = B_{w,\phi}^{E,\infty} \text{ (with equivalent norms).}$$

PROOF   Let us take $f \in \Lambda_\omega^E$, since $\int_{\mathbf{R}^n} \phi(x)dx = 0$ then

$$\phi_t * f = \int_{\mathbf{R}^n} \phi_t(-x)\Delta_x f\,dx.$$

Using (2) for $\varepsilon = 1$, we have

$$\int_{\mathbf{R}^n} |\phi_t(-x)|\,\|\Delta_x f\|_E\,dx \leq C\left(\frac{1}{t^n}\int_{|x|<t} \omega(|x|)\,dx + t\int_{|x|>t} \omega(|x|)\frac{dx}{|x|^{n+1}}\right)$$

$$\leq C\left(\int_0^t \left(\frac{s}{t}\right)^n \omega(s)\frac{ds}{s} + t\int_t^\infty \omega(s)\frac{ds}{s^2}\right) \leq C\omega(t).$$

This shows that $F_t(x) = \phi_t(-x)\Delta_x f$ belongs to $L^1(\mathbf{R}^n, E)$ and moreover

$$\|\phi_t * f\|_E \leq C\omega(t).$$

Conversely, let us take $f \in B_{w,\phi}^{E,\infty}$. Given $0 < \varepsilon < \delta$ we have

$$\Delta_x(f_{\varepsilon,\delta}) = \int_\varepsilon^\delta (\Delta_x \phi_t) * \phi_t * f \frac{dt}{t}.$$

For each $x \in \mathbf{R}^n$, let us denote F_x the E-valued function defined by

$$F_x(t) = (\Delta_x \phi_t) * \phi_t * f.$$

Using (1) and (3) we have

$$\|F_x(t)\|_E \leq C\|\Delta_x \phi_t\|_1 \|\phi_t * f\|_E \leq C\min\left(1, \frac{|x|}{t}\right)\omega(t).$$

Note that

$$\int_0^\infty \min\left(1, \frac{|x|}{t}\right) \omega(t) \frac{dt}{t} \le C\left(\int_0^{|x|} \omega(t) \frac{dt}{t} + |x| \int_{|x|}^\infty \frac{\omega(t)}{t^2}\, dt\right) \le C\omega(|x|).$$

This allows us to say that $F_x \in L^1((0,\infty), \frac{dt}{t}, E)$. This implies $\Delta_x(f_{\varepsilon,\delta})$ is a Cauchy net in E and hence convergent in E. On the other hand, from Lemma 1, $f_{\varepsilon,\delta} \to f$ in S'_0 what implies $\Delta_x(f_{\varepsilon,\delta}) \to \Delta_x(f)$ in S'. Combining both facts we have $\Delta_x(f_{\varepsilon,\delta}) \to \Delta_x(f)$ in E.

Finally take limit as $\varepsilon \to 0$ and $\delta \to \infty$ in

$$\|\Delta_x(f_{\varepsilon,\delta})\|_E \le C \int_0^\infty \min\left(1, \frac{|x|}{t}\right) \omega(t) \frac{dt}{t} \le C\omega(|x|)$$

to show that $f \in \Lambda_\omega^E$. ∎

REMARK 4 This produces a Littlewood-Paley formulation of the Lipschitz classes Λ_α and Λ_α^p for $0 < \alpha < 1$ (see [FJW]). ∎

LEMMA 3

Let $\phi \in S_0$ and ω a b_1-weight such that $\omega_1(t) = t\omega(t)$ satisfies Dini condition.
If $t > 0$ and $f \in E$ then $\phi_t * f \in B_{\omega,\phi}^{E,1}$ and

$$\|\phi_t * f\|_{B_{\omega,\phi}^{E,1}} \le C\omega(t)\|f\|_E. \tag{5}$$

PROOF

$$
\begin{aligned}
\|\phi_t * f\|_{B_{\omega,\phi}^{E,1}} &= \int_0^\infty \|\phi_t * \phi_s * f\|_E \omega(s) \frac{ds}{s} \\
&\le C \int_0^\infty \|\phi_t * \phi_s\|_1 \|f\|_E \omega(s) \frac{ds}{s} \\
&\le C \int_0^\infty \min\left(\frac{t}{s}, \frac{s}{t}\right) \omega(s) \frac{ds}{s} \\
&\le C\|f\|_E \left(\frac{1}{t} \int_0^t \omega_1(s) \frac{ds}{s} + t \int_t^\infty \frac{\omega(s)}{s^2} ds\right) \le C\|f\|_E \omega(t). \quad\blacksquare
\end{aligned}
$$

THEOREM 2

Let E, E^* be Banach spaces bounded under translations such that S is dense in E. Let ω be a b_1–weight such that $\omega_1(t) = t\omega(t)$ satisfies Dini condition and $\omega \in L^1(\frac{dt}{t})$. Let $\phi \in S_0$ be a real radial function with $\int_0^\infty \hat{\phi}^2(t\xi) \frac{dt}{t} = 1$, $(\xi \ne 0)$. Then the dual space of $B_{\omega,\phi}^{E,1}$ is isomorphic to $\Lambda_\omega^{E^*}$.

PROOF   Let us take $f \in \Lambda_\omega^{E^*}$ and $g \in S$.
From Calderón's formula

$$\int_{\mathbf{R}^n} f(x)g(x)\,dx = \int_{\mathbf{R}^n} \left(\int_0^\infty \phi_t * \phi_t * g(x) \frac{dt}{t} \right) f(x)\,dx$$

$$= \int_0^\infty \int_{\mathbf{R}^n} \phi_t * f(x)\phi_t * g(x)\,dx \frac{dt}{t}.$$

This implies

$$\left| \int_{\mathbf{R}^n} f(x)g(x)\,dx \right| \leq \int_0^\infty \|\phi_t * g\|_E \|\phi_t * f\|_{E^*} \frac{dt}{t}$$

$$\leq \|f\|_{\Lambda_\omega^{E^*}} \int_0^\infty \|\phi_t * g\|_E \omega(t) \frac{dt}{t}.$$

Denoting by $\Phi(g) = \int_{\mathbf{R}^n} f(x)g(x)\,dx$, we obtain

$$|\Phi(g)| \leq \|f\|_{\Lambda_\omega^{E^*}} \|g\|_{B_{\omega,\phi}^{E,1}}.$$

Hence Φ extends to an operator on $B_{\omega,\phi}^{E,1}$.

Conversely, let us take $\Phi \in \left(B_{\omega,\phi}^{E,1} \right)^*$. It is immediately clear that $|\Phi(g)| \leq C\|g\|_E$ for any $g \in S$. Hence Φ can be extended to a functional in E^*.
Therefore there exists $f \in E^*$ so that

$$\Phi(g) = \int_{\mathbf{R}^n} f(x)g(x)\,dx, \qquad (g \in B_{\omega,\phi}^{E,1}).$$

Let us show that $f \in \Lambda_\omega^{E^*}$:

$$\|\phi_t * f\|_{E^*} = \sup \left\{ \left| \int_{\mathbf{R}^n} (\phi_t * f)(x)g(x)\,dx \right| : \|g\|_E \leq 1 \right\}$$

$$= \sup \left\{ \left| \int_{\mathbf{R}^n} (\phi_t * g)(x)f(x)\,dx \right| : \|g\|_E \leq 1 \right\}$$

$$= \sup \left\{ |\Phi(\phi_t * g)| : \|g\|_E \leq 1 \right\}.$$

Applying the boundedness of Φ and Lemma 3 one has

$$\|\phi_t * f\|_{E^*} \leq C\omega(t). \qquad \blacksquare$$

REMARK 5   See [J1] for a proof of a similar result for ϕ with $\int \phi \neq 0$. Duality results for $E = L^p$ where achieved in [F1], [F2], [T1], [T2], and [T3]. The reader is referred to [BS] for a proof under the same conditions on the weight but for spaces defined on the torus $\mathbb{R}$ and convolution with the derivative of the Poisson kernel. $\blacksquare$

THEOREM 3

Let ω be a b_1-weight such that $\omega_1(t) = t\omega(t)$ satisfies Dini condition and $\omega \in L^1(\frac{dt}{t})$. Let $\phi \in S_0$ be a real radial function with $\int_0^\infty \hat{\phi}^2(t\xi)\frac{dt}{t} = 1$, ($\xi \neq 0$). Given a Banach space $(X, \|.\|_X)$ and a linear map $T : S \to X$, define $F : \mathbb{R}_+^{n+1} \to X$ by

$$F(x,t) = T(\tau_x(\phi_t)).$$

The following are equivalent:

(i) $T : B_{\omega,\phi}^{L^1,1} \to X$ *is bounded.*

(ii) $\sup_{x\in\mathbf{R}^n} \|F(x,t)\|_X \leq C\omega(t).$

PROOF   Assume that T is bounded. From (4) it follows that

$$\|\tau_x(\phi_t)\|_{B_{\omega,\phi}^{L^1,1}} \leq C\int_0^\infty \|\phi_t * \phi_s\|_1 \omega(s)\frac{ds}{s} \leq C\omega(t).$$

Hence

$$\|F(x,t)\|_X \leq \|T\| \, \|\tau_x(\phi_t)\|_{B_{\omega,\phi}^{L^1,1}} \leq C\omega(t).$$

Conversely let us assume $\sup_{x\in\mathbf{R}^n} \|F(x,t)\|_X \leq C\omega(t)$ and take $g \in B_{\omega,\phi}^{L^1,1}$. Let $0 < \varepsilon < \varepsilon' < \delta' < \delta$.
Since

$$\|g_{\varepsilon,\delta} - g_{\varepsilon',\delta'}\|_{B_{\omega,\phi}^{L^1,1}}$$

$$= \int_0^\infty \left\| \int_\varepsilon^{\varepsilon'} \phi_s * \phi_t * \phi_t * g\frac{dt}{t} + \int_{\delta'}^\delta \phi_s * \phi_t * \phi_t * g\frac{dt}{t} \right\|_1 \omega(s)\frac{ds}{s}$$

$$\leq C\int_0^\infty \left(\int_\varepsilon^{\varepsilon'} + \int_{\delta'}^\delta \right) \|\phi_s * \phi_t * \phi_t * g\|_1 \frac{dt}{t}\omega(s)\frac{ds}{s}$$

$$\leq C\left(\int_\varepsilon^{\varepsilon'} + \int_{\delta'}^\delta \right) \|\phi_t * g\|_1 \left(\int_0^\infty \|\phi_s * \phi_t\|_1 \omega(s)\frac{ds}{s} \right)\frac{dt}{t}$$

$$\leq C\left(\int_\varepsilon^{\varepsilon'} + \int_{\delta'}^\delta \right) \|\phi_t * g\|_1 \left(\int_0^\infty min(\frac{s}{t}, \frac{t}{s})\omega(s)\frac{ds}{s} \right)\frac{dt}{t}$$

$$\leq C\left(\int_\varepsilon^{\varepsilon'} + \int_{\delta'}^\delta \right) \|\phi_t * g\|_1 \omega(t)\frac{dt}{t}.$$

This implies that

$$\lim_{\varepsilon \to 0, \delta \to \infty} \int_{\varepsilon}^{\delta} \int_{\mathbf{R}^n} \phi_t(x - y)\phi_t * g(y)\, dy \frac{dt}{t} = g \qquad \left(\text{in } B_{\omega,\phi}^{L^1,1}\right);$$

which shows that

$$\int_0^\infty \int_{\mathbf{R}^n} F(x, t)\phi_t * g(y)\, dy \frac{dt}{t} = T(g).$$

Hence

$$\|T(g)\|_X \leq \int_0^\infty \int_{\mathbf{R}^n} \|F(x, t)\|_X |\phi_t * g(y)|\, dy \frac{dt}{t}$$

$$\leq C \int_0^\infty \|\phi_t * g\|_1 \omega(t) \frac{dt}{t} \leq C\|g\|_{B_{\omega,\phi}^{E,1}}. \qquad \blacksquare$$

REMARK 6 In [B1] a similar and more general result is established and there are several applications of it to Carleson measures, multipliers, and composition operators. Applications of a similar nature, with the analogue formulation in $\mathbf{R}^n$, can be obtained from Theorem 3. $\blacksquare$

References

[AB] J. L. Ansorena and O. Blasco, *Characterization of weighted Besov spaces,* Math. Nachr. 170 (1994).

[B1] O. Blasco, *Operators on weighted Bergman spaces and applications,* Duke Math. J. 66 (1992) 443–467.

[BS] O. Blasco and G. S. de Souza, *Spaces of analytic functions on the disc where the growth of $M_p(F, r)$ depends on a weight,* J. Math. Anal. and Appl. 147 (1990), 580–598.

[BlS] S. Bloom and G. S. de Souza, *Atomic decomposition of generalized Lipschitz spaces,* Illinois J. Math. 33 (1989), 181–189.

[FJW] M. Frazier, B. Jawerth and G. Weiss, *Littlewood-Paley characterization in functions spaces,* Amer. Math. Soc., 1991.

[F1] T. M. Flett, *Temperatures, Bessel potentials and Lipschitz spaces,* Proc. London Math. Soc. 20 (1970), 749–768.

[F2] T. M. Flett, *Lipschitz spaces of functions on the circle and the disc,* J. Math. Anal. and Appl. 39 (1972), 125–158.

[J1]   S. Janson, *Generalization on Lipschitz spaces and applications to Hardy spaces and bounded mean oscillation,* Duke Math. J. 47 (1980), 959–982.

[T1]   M. Taibleson, *On the theory of Lipschitz spaces of distributions on Euclidean n-space. I,* J. Math. Mech. 13 (1964), 407–480.

[T2]   M. Taibleson, *On the theory of Lipschitz spaces of distributions on Euclidean n-space. II,* ibid. 14 (1965), 821–840.

[T3]   M. Taibleson, *On the theory of Lipschitz spaces of distributions on Euclidean n-space. III,* ibid. 15 (1966), 973–981.

8

A Note on Monotone Functions

María J. Carro

Javier Soria[1]

ABSTRACT   *In this note, we obtain a representation formula for integral operators acting on radial decreasing functions. As a consequence, a simple proof for the almost everywhere convergence of the Fourier integral means is also given.*

8.1   Introduction

We say that a nonnegative function is a weight, if it is locally integrable. The weighted Lorentz space $\Lambda_\mu^p(w)$ is defined as those measurable functions in $\mathbb{R}^n$ or $\mathbb{R}^+$ such that $\|f\|_{\Lambda_\mu^p(w)} = \left(\int_0^\infty (f_\mu^*(x))^p w(x)\, dx \right)^{1/p} < +\infty$, where f_μ^* denotes the decreasing rearrangement function with respect to the measure μ (see [BS]). These spaces have been widely studied in multiple contexts. In general, they are not Banach spaces, however they are quasi-Banach spaces, if and only if, $W(x) = \int_0^x w(t)\, dt$ satisfies the Δ_2-condition (see [CS1]).

The chapter is organized as follows. In Section 8.2, we use a distribution formula for decreasing functions to obtain some simple consequences in several contexts. This formula was proven for particular choices of w in [So] and then in [CS1] for a general weight w. We mention two applications already given in [CS1] and [CS2] in the setting of weighted Lorentz spaces and integral operators, respectively, and also present a third application in the setting of integral operators, but now defined on $\mathbb{R}^n$ and acting on radial decreasing functions. Section 8.3 is devoted to the same type of results as in Section 8.2 but for increasing functions.

[1] Both authors have been partially supported by the DGICYT PB91-0259.

1980 *Mathematics Subject Classification* (1985 *Revision*). 42B25.

8.2   Distribution formula

Let $(\mathcal{M}, \mu)$ be a measure space. We call $\lambda_f^\mu(y) = \mu\{x : |f(x)| > y\}$, the distribution function of f with respect to the measure $d\mu$. When μ equals the Lebesgue measure, we simply write λ_f.

THEOREM 1

Let w be a locally integrable function in $\mathbb{R}^+$. Then, for every positive, decreasing simple function f, we obtain

$$\int_0^\infty f(t)w(t)\,dt = \int_0^\infty \left(\int_0^{\lambda_f(y)} w(t)\,dt\right) dy. \tag{1}$$

We observe that if w is a weight then (1) holds for every positive, decreasing measurable function f. And, in general, (1) holds for every measurable function f such that f is positive, decreasing, bounded and with compact support.

A first application of this result was given in [CS1] in the setting of weighted Lorentz spaces. In that paper, we made use of (1) to get a distribution formula for the "norm" in $\Lambda_u^p(w)$ and studied the natural generalization of the well-known embeddings for the classical Lorentz spaces, $L^{p,1} \subset \cdots \subset L^p \subset \cdots \subset L^{p,q} \subset \cdots \subset L^{p,\infty}$, for $p \leq q$. We also study conditions on the weights u_j and w_j for the embedding $\Lambda_{u_0}^{p_0}(w_0) \subset \Lambda_{u_1}^{p_1}(w_1)$ to hold. The characterization of the weights w_j for this embedding, when $u_0 = u_1 = 1$, was done by Sawyer in [Sa2]. His proof can be applied to the case $u_0 = u_1$. We study the boundedness of the Hardy operator $Sf(x) = \int_0^x f(t)\,dt$ from $\Lambda_{u_0}^{p_0}(w_0)$ into $\Lambda_{\mu_1}^{p_1}(w_1)$. The idea is to obtain a unified version of the boundedness of S in both $L^p(u)$ (as in [Br], [Ma], [Mu], [Sa1]) and $\Lambda^p(w)$, (as in [AM], [Sa2]).

A second application was given in [CS2]. In that paper, we deal with integral operators of the form

$$T_k f(x) = \int_0^\infty k(x,t)f(t)\,dt,$$

with k a positive function.

The boundedness of this operator

$$T_k : L_{\text{dec}}^{p_0}(w_0) \longrightarrow L^{p_1}(d\mu),$$

where w_0 is a weight in $\mathbb{R}^+$, $d\mu$ is a measure on $\mathcal{M}$, and $L_{\text{dec}}^{p_0}(w_0)$ is the space of decreasing functions in $L^{p_0}(w_0)$, has been widely studied for particular choices of the kernel k (see [An], [AM], [N], [Sa2], and [Sh]).

In particular, if $k(x, t) = x^{-1}a(tx^{-1})$, with a an arbitrary function, the weak boundedness of

$$T_k : L^{p_0}(w_0) \longrightarrow L^{p_1,\infty}(w_1),$$

with w_1 a weight in $\mathbb{R}^+$, has been completely solved by Andersen in [An]. In [Sa2], Sawyer solved question (1), for $1 < p_0, p_1$, via the study of T_k^*, the adjoint operator, whenever this operator can be easily identified and its boundedness easily studied. His argument is based upon a duality type result for nonincreasing functions. Results about particular cases of operators T_k have many other proofs (see [AM], [N], ...).

In [CS2], the boundedness of the operator $T_k : L_{dec}^{p_0}(w_0) \longrightarrow L^{p_1}(w_1)$ is completely solved in the range $0 < p_0 \leq 1$, $p_0 \leq p_1$.

In this note we present a third application of (1), also in the setting of integral operators, but now the kernel $K(x, y)$ is not supposed to be positive. Let $\mathcal{N}_0$ be the set of measurable functions in $\mathbb{R}^+$ such that f is positive, decreasing, bounded, and with compact support, and let

$$\mathcal{N} = \{ f : \mathbb{R}^n \longrightarrow \mathbb{R} \text{ measurable; } f_0 \in \mathcal{N}_0 \},$$

where $f_0(|x|) = f(x)$. We have the following representation formula for $f \in \mathcal{N}$.

PROPOSITION 1
If $K : \mathcal{M} \times \mathbb{R}^n \longrightarrow \mathbb{R}$ is a locally integrable function with respect to the second variable and

$$(T_K f)(x) = \int_{\mathbb{R}^n} K(x, y) f(y) \, dy, \qquad (2)$$

then, for every $f \in \mathcal{N}$,

$$T_K f(x) = \int_0^\infty T_K \left(\chi_{B(0, \lambda_{f_0}(y))} \right)(x) \, dy.$$

PROOF We use (1) and polar coordinates to obtain

$$T_K f(x) = \int_0^\infty f_0(r) r^{n-1} \int_{S_{n-1}} K(x, r\xi) \, d\xi \, dr$$

$$= \int_0^\infty \int_0^{\lambda_{f_0}(y)} r^{n-1} \int_{S_{n-1}} K(x, r\xi) \, d\xi \, dr \, dy$$

$$= \int_0^\infty T_K \left(\chi_{B(0, \lambda_{f_0}(y))} \right)(x) \, dy. \qquad (3)$$

∎

This formula gives us a necessary and sufficient condition for any integral operator of the form (2) to be bounded from $L_{dec}^p(w)$ with $p \leq 1$, into any Banach space F.

THEOREM 2

Let T_K be defined as in (2), $p \leq 1$ and let F be any Banach space. Let w be a weight in $\mathbb{R}^n$ or $\mathbb{R}^+$, then $T_K : L^p(w) \cap \mathcal{N} \longrightarrow F$ is bounded if and only if there exists a positive constant C such that, for every $r > 0$,

$$\| T_K \left(X_{B(0,r)} \right) \|_F \leq C \left(\int_{B(0,r)} w(x)\, dx \right)^{1/p}.$$

PROOF By (3), if $f \in \mathcal{N}$, we deduce that $\| T_K f \|_F \leq \int_0^\infty \| T_K(X_{B(0,\lambda_{f_0}(y))}) \|_F\, dy$ and, hence, by hypothesis, $\| T_K f \|_F \leq \int_0^\infty \left(\int_{B(0,\lambda_{f_0}(y))} w(x)\, dx \right)^{1/p} dy$. Now, since λ_{f_0} is a decreasing function, one can estimate the previous integral by summing between 2^k and 2^{k+1} and, hence, one can easily see that

$$\| T_K f \|_F \leq C \left(\int_0^\infty y^{p-1} \int_{B(0,\lambda_{f_0}(y))} w(x)\, dx\, dy \right)^{1/p}.$$

Finally, it remains to observe that, by (1), the right side of the previous formula coincides with $\| f \|_{L^p(w)}$. ∎

Using (3) with $K(x, y) = e^{ixy}$ we deduce that, for every function in $\mathcal{N}$,

$$\widehat{f}(\xi) = \int_0^\infty \widehat{X}_{B(0,\lambda_{f_0}(y))}(\xi)\, dy,$$

and, hence, if we want to characterize the weights w_0, w_1 such that the Fourier transform is bounded from $L^{p_0}(w_0) \cap \mathcal{N}$ into $L^{p_1}(w_1)$ with $p_0 \leq 1$ and $p_1 \geq 1$, it is enough to deal with characteristic functions of balls (see [CH]).

Now if we write $S_R f(x) = \int_{B(0,R)} \widehat{f}(\xi) e^{-2\pi ix\xi}\, d\xi$, then simple computations show that, for every $f \in \mathcal{N}$,

$$S_R f(x) = \int_0^\infty S_{R\lambda_{f_0}(y)} \left(X_{B(0,1)} \right) \left(\frac{x}{\lambda_{f_0}(y)} \right) dy.$$

From this formula and the well known fact (see [V]) that

$$S_R(X_{B(0,1)})(x) \rightarrow X_{B(0,1)}(x), \tag{4}$$

a.e. x, one can easily get the almost everywhere convergence for $f \in \mathcal{N}$. The convergence is known for radial functions (see [P]) but the point here is that such a convergence can be trivially transferred from that of X_B for such a particular subclass of radial functions. Let us see how.

If we denote by N the set of measure zero such that $S_R(X_{B(0,1)})(x)$ does not converge to $X_{B(0,1)}(x)$ and by

$$M = \left\{ (x, y);\ \frac{x}{\lambda_{f_0}(y)} \in N \right\},$$

we deduce that, since λ_{f_0} is a decreasing function, M also has measure zero. On the other hand, since $h = \sup_{R>0} |S_R(X_{B(0,1)})|$ is in $L^1 + L^\infty$, we can write $h = g_0 + g_1$ where $g_0 \in L^1$ and $g_1 \in L^\infty$. We now know that

$$\int_{\mathbb{R}^n} \int_0^\infty \left| g_0\left(\frac{x}{\lambda_{f_0}(y)}\right) \right| dy\, dx = \int_0^\infty \lambda_{f_0}(y)^n \int_{\mathbb{R}^n} |g_0(x)|\, dx\, dy < +\infty,$$

and, hence,

$$\int_0^\infty \left| g_0\left(\frac{x}{\lambda_{f_0}(y)}\right) \right| dy < +\infty,$$

a.e. x. Now since $g_1 \in L^\infty$ and λ_{f_0} is monotone, one can easily see that the same condition holds for g_1 and therefore, for h. Using (4) and dominated convergence we see that

$$\begin{aligned}
\lim_{R\to\infty} S_R f(x) &= \int_0^\infty X_{B(0,1)}\left(\frac{x}{\lambda_{f_0}(y)}\right) dy \\
&= \int_0^\infty X_{\{y;\ |x|<\lambda_{f_0}(y)\}}\, dy \\
&= |\{y;\ |x| < \lambda_{f_0}(y)\}| \\
&= f_0(|x|) \\
&= f(x).
\end{aligned}$$

Finally, the localization principle (see [Cas]) shows that the result can be extended to any radial function such that f_0 is decreasing.

8.3 Increasing functions

Applying Theorem 1 to the function $g(u) = f\left(\frac{1}{u}\right)$, we can obtain similar formulae to (1) and Proposition 1 for increasing functions. The results of this section can be applied to obtain similar results to those in [Sh].

PROPOSITION 2

(a) *Let w be a weight in $\mathbb{R}^+$. Then, for every positive, increasing function f, we get*

$$\int_0^\infty f(t)w(t)\, dt = \int_0^\infty \left(\int_{1/\lambda_g(y)}^\infty w(t)\, dt \right) dy. \tag{5}$$

(b) *For a general w, (5) holds for every positive, increasing function f such that f is bounded and there exists $a > 0$ such that $f(x) = 0$ for every $x \in [0, a)$.*

Now, if we define I_0 to be the set of measurable functions such that $g(u) = f\left(\frac{1}{u}\right)$ is in $\mathcal{N}_0$ and $I = \{f : \mathbb{R}^n \longrightarrow \mathbb{R} \text{ measurable}; \ f_0 \in I_0\}$, we obtain the following result.

PROPOSITION 3

Let T_K defined as in (2) and let us assume that $K(x, \cdot)$ is integrable at infinity for almost every x. Let us write $C(r, \infty) = \{x \in \mathbb{R}^n; \ r \leq |x|\}$. Then, for every $f \in I$,

$$T_K f(x) = \int_0^\infty T_K\left(\chi_{C(\lambda_{1/g_0}(y), \infty)}\right)(x)\, dy,$$

where $g_0(u) = f_0\left(\frac{1}{u}\right)$.

THEOREM 3

Let T_K be defined as in (2) and such that $K(x, \cdot)$ is integrable at infinity for almost every $x \in \mathbb{R}^n$, $p \leq 1$ and let F be any Banach space. Then the operator $T_K : L^p(w) \cap I \longrightarrow F$ is bounded if and only if there exists a positive constant C such that, for every $r > 0$,

$$\|T_K\left(\chi_{C(r, \infty)}\right)\|_F \leq C\left(\int_{C(r, \infty)} w(x)\, dx\right)^{1/p}.$$

Finally, if φ is a measurable function such that $\varphi(x) \neq 0$ for every x and we define

$$\mathcal{N}_\varphi = \{f : \mathbb{R}^+ \to \mathbb{R}^+; \ f\varphi \in \mathcal{N}_0\},$$

one can easily extend the results in this note and determine a distribution formula for $\mathcal{N}_\varphi$ together with a representation formula for integral operators, similarly for radial functions f such that $f_0 \in \mathcal{N}_\varphi$.

References

[An]   K. Andersen, *Weighted generalized Hardy inequalities for nonincreasing functions,* Canad. J. Math. 43 (1991), 1121–1135.

[AM]   M. Ariño and B. Muckenhoupt, *Maximal functions on classical Lorentz spaces and Hardy's inequality with weights for nonincreasing function,* Trans. Amer. Math. Soc. 320 (1990), 727–735.

[BS]   C. Bennett and R. Sharpley, *Interpolation of operator,* Academic Press, 1988.

[Br]   J. Bradley, *Hardy inequalities with mixed norms,* Canad. Math. Bull. 21 (1978), 405–408.

[Cas]   A. Carbery and F. Soria, *Almost-everywhere convergence of Fourier integrals for functions in Sobolev Spaces, and an L^2-localisation principle*, Revista Mat. Iber. 4 (1988), 319–337.

[CS1]   M. J. Carro and J. Soria, *Weighted Lorentz spaces and the Hardy operator*, Jour. Funct. Anal. 112 (1993), 480–494.

[CS2]   M. J. Carro and J. Soria, *Boundedness of some integral operators*, Canad. J. Math., 45 (1993), 1155–1166.

[CH]   C. Carton-Lebrun and H. P. Heinig, *Weighted Fourier transform inequalities for radially decreasing functions*, Siam J. Math. Anal. 23 (1992), 785–798.

[Ma]   V. Maz'ya, *Sobolev Spaces*, Springer series in Soviet Mathematics, Springer Verlag, (1985).

[Mu]   B. Muckenhoupt, *Hardy's inequality with weights*, Studia Math. 34 (1972), 31–38.

[N]   C. J. Neugebauer, *Weighted norm inequalities for general operators of monotone functions*, Publi. Mat. 35 (1991), 429–447.

[P]   E. Prestini, *Almost everywhere convergence of the spherical partial sums for radial functions*, Monatsh. Math. 105 (1988), 207–216.

[Sa1]   E. Sawyer, *Weighted Lebesgue and Lorentz norm inequalities for the Hardy operator*, Trans. Am. Math. Soc. 281 (1984), 329–337.

[Sa2]   E. Sawyer, *Boundedness of classical operators on classical Lorentz spaces*, Studia Math. 96 (1990), 145–158.

[Sh]   L. Shanzhong, *Weighted norm inequalities for general operators on monotone functions*, Trans. A.M.S., to appear.

[So]   J. Soria, *Weighted tent spaces*, Math. Nachr. 155 (1992), 231–256.

[V]   N. J. Vilenkin, *Special functions and the theory of group representations*, Translations of Math. Monographs A.M.S., 1968.

9

Hilbert Transforms in Weighted Distribution Spaces

C. Carton-Lebrun

ABSTRACT   Let $w(x) = (1 + x^2)^{1/2}$, $k \in \mathbb{N}$ and $V_k' = w^{k+1} \mathcal{D}'_{L^1} = \{f \in S' : w^{-k-1} f \in \mathcal{D}'_{L^1}\}$. Suppose $\xi \in \mathcal{D}$, $0 \leq \xi(x) \leq 1$, $\xi(x) = 1$ in a neighborhood of the origin and $\eta = 1 - \xi$. For $f \in V_k'$, let $S_k f = H(\xi f) + x^k H\left(\frac{\eta f}{t^k}\right)$, where $H = S_0$ denotes the Hilbert transform defined by Schwartz on $V_0' = w \mathcal{D}'_{L^1}$. We prove that S_k is a continuous mapping from V_k' into V_{k+1}' and $S_k = -T_k'$, where $T_k \psi = \xi H \psi + \left(\frac{\eta}{x^k}\right) H(t^k \psi)$ for $\psi \in V_k$. The inversion formula $S_{k+1} S_k f = -f - p(x; \eta, k, f)$ is next established, where the last term is a polynomial of degree inferior or equal to k, the coefficients of which are explicitly determined in terms of η, k and f. Additional results are obtained for distributions in the subspace $\{f \in V_k' : S_k f \in V_k'\}$. In case $k = 0$ for instance, $H(Hf) = -f$ whenever f and Hf both belong to V_0'. The associativity of the product $S_{k+1} S_{k+1} S_k$ is also proven.

9.1   Introduction

We consider distribution spaces of the form $V_k' = w^{k+1} \mathcal{D}'_{L^1} = \{f \in S' : w^{-k-1} f \in \mathcal{D}'_{L^1}\}$, where $k \in \mathbb{N}$, $w(x) = (1 + x^2)^{1/2}$ and $\mathcal{D}'_{L^1}$ denotes the space of integrable distributions. Recall that $\mathcal{D}'_{L^1}$ is the dual of $\mathcal{B}_0$, where $\mathcal{B}_0$ denotes the set of C^∞ functions that combined with all their derivatives, vanish at infinity. The topology on $\mathcal{B}_0$ is defined by the family of norms $s_k(\psi) = \sup\{\|D^\alpha \psi\|_\infty : 0 \leq \alpha \leq k\}$. $\mathcal{D}'_{L^1}$ can also be considered as the dual of the space $\mathcal{B}_c$ of C^∞ functions that are bounded on $\mathbb{R}$ altogether with all their derivatives. Here, $\mathcal{B}_c$ is endowed with the following topology: $\psi_j \to 0$ in $\mathcal{B}_c$ if $\psi_j \to 0$ in $\mathcal{E}$ and if, for each $\alpha \in \mathbb{N}$, $\|D^\alpha \psi_j\|_\infty \leq C_\alpha$ with C_α independent of j (see [S2], for instance).

According to a result of Schwartz [S1], the Hilbert transform Hf of $f \in V'_0$ can be defined by

$$_{S'}\langle Hf, \ \psi \rangle_S = -_{\mathcal{D}'_{L^1}}\langle w^{-1}f, \ wH\psi \rangle_{\mathcal{B}_c} \tag{1}$$

for all $\psi \in S$, where $H\psi$ denotes the classical Hilbert transform of ψ. The mapping $H : w\mathcal{D}'_{L^1} \to S'$ is continuous and coincides with the usual Hilbert transform on the subspaces $\mathcal{E}'$ and $\mathcal{D}'_{L^p}(1 \le p < \infty)$ of $w\mathcal{D}'_{L^1}$ [S1].

The aim of this work is to study more general Hilbert transforms for tempered distributions. For each $k \in \mathbb{N}$, we define the Hilbert transform $S_{\eta,k}$, of index k on V'_k, by

$$S_{\eta,k}f = H(\xi f) + x^k H\left(\frac{\eta f}{t^k}\right), \qquad f \in V'_k, \tag{2}$$

where $\xi \in \mathcal{D}, 0 \le \xi(x) \le 1, \ \xi(x) = 1$ in a neighborhood of the origin and $\eta = 1 - \xi$. In the above equality, H denotes the Hilbert transform defined by (1) on V'_0. Obviously, $S_{\eta,0} = H$ on V'_0 and $S_{\eta,0}$ is thus independent of the auxiliary function η.

For each $f \in V'_k$, Hilbert transforms $S_{\eta,m}$, of higher index $m \ge k + 1$, will also be used. They are defined in a similar way, by

$$S_{\eta,m}f = H(\xi f) + x^m H\left(\frac{\eta f}{t^m}\right), \qquad f \in V'_k, \tag{3}$$

and satisfy some useful algebraic properties. These are given in Section 9.2, with additional notation and lemmas. In Section 9.3, continuity properties of the above transforms are proven for appropriate topologies. In Section 9.4, we establish inversion formulas of the form $S_{\eta,m}S_{\eta,k}f = -f - p(x; \eta, k, m, f)$, where the last term is a polynomial of degree inferior or equal to $m - 1$, the coefficients of which are explicitly determined in terms of η, k, m, and f. In Section 9.5, additional results are obtained for distributions in subspaces of the form $\{f \in V'_k : S_{\eta,k} \in V'_k\}$. The associative ability of the product $S_{\eta,k+1}S_{\eta,k+1}S_{\eta,k}$ is proven too.

Let us mention that the results of this chapter are concerned with concrete elements of S' and not with equivalence classes in quotient spaces of the form $\frac{S'}{P}$, where P is some set of polynomials. In addition, the operators $S_{\eta,m}$, considered here, are real, in the sense that, for real valued test functions, $S_{\eta,m}f$ is real valued whenever f is. For all functions in the Hardy class $H^1(\mathbb{R})$, $S_0 f$ coincides with the classical Hilbert transform. For $b \in BMO$, $S_1 b$ coincides with the Hilbert transform introduced by Fefferman ([F], [CL2]). In the case of arbitrary distributions, continuity and inversion problems have been discussed in [CL3].

9.2 Notation and preliminary lemmas

Functions under consideration here are real valued functions of one variable. We utilize the usual notation with regards to spaces of test functions and distributions. We refer, for instance to [H] and [S2] for the definitions of $\mathcal{D}$, $\mathcal{D}'$, $\mathcal{E}$, $\mathcal{E}'$, S, S', $\mathcal{D}_{L^q}$, and $\mathcal{D}'_{L^p}$. Additional definitions related to $\mathcal{B}_0$ and $\mathcal{B}_c$ have been mentioned in the introduction. We also adhere to the convention of writing $\mathcal{B}$ instead of $\mathcal{B}_c$ when no topology is needed on this set. The integral of $f \in \mathcal{D}'_{L^1}$ is $\langle f, 1 \rangle$. Its "normalized moment" is noted

$$M(f) = \left(\frac{1}{\pi}\right)\langle f, 1 \rangle. \tag{4}$$

The following notation and definitions are needed too:

(a) A sequence ξ_j is an approximate unit, for short $\xi_j \in AU$, if $\xi_j \in \mathcal{D}$ for each j, $\xi_j \to 1$ in $\mathcal{B}_c$ and if, for every compact $K \subset \mathbb{R}$, there exists $J \in \mathbb{N}$, such that $\xi_j(x) = 1$ for all $x \in K$, $j \geq J$. A distribution f belongs to $\mathcal{D}'_{L^1}$, if and only if, for any $\xi_j \in AU$, $\lim_{j \to \infty} \langle f, \xi_j \rangle$ exists. In the latter case, $\lim_{j \to \infty} \langle f, \xi_j \rangle = \langle f, 1 \rangle$.

(b) Let $w(x) = (1 + x^2)^{1/2}$. For $k \in \mathbb{N}$, we define

$$V_k = w^{-k-1}\mathcal{B}_c = \{\psi = w^{-k-1}\varphi : \varphi \in \mathcal{B}_c\} \tag{5}$$

with the following topology: $\psi_j \to 0$ in V_k if $w^{k+1}\psi_j \to 0$ in $\mathcal{B}_c$. The dual of V_k is

$$V'_k = w^{k+1}\mathcal{D}'_{L^1} = \{f \in S' : w^{-k-1}f \in \mathcal{D}'_{L^1}\}. \tag{6}$$

A sequence $f_j \in V'_k$ strongly converges to zero in V'_k if $w^{-k-1}f_j$ converges to zero in the strong dual topology of $\mathcal{D}'_{L^1}$. In the sequel, $\mathcal{D}'_{L^1}$ and V'_k are supposed to carry the strong dual topologies $\beta(\mathcal{D}'_{L^1}, \mathcal{B}_c)$, $\beta(V'_k, V_k)$ unless otherwise stated. For the definitions of these topologies and weaker ones, we refer to [Tre] for instance.

(c) The indexed Hilbert transforms $S_{\eta.m}$, $m \geq 1$, defined by (2) and (3) depend on the way we distribute 1 into $\xi + \eta$. To abbreviate, we write $\xi \in U(0)$ if $\xi(x) \in \mathcal{D}$, $0 \leq \xi(x) \leq 1$, $\xi(x) = 1$ in a neighborhood of the origin. We write $\eta \in U(\infty)$ if $\eta = 1 - \xi$ for some $\xi \in U(0)$. It is useful to remark that for every given $\eta \in U(\infty)$, there exists $\eta_1 \in U(\infty)$ such that $\eta\eta_1 = \eta$ on $\mathbb{R}$. When no confusion may arise, we write $S_m f$ instead of $S_{\eta.m} f$. In this case, $S_m f$ must be understood as an abbreviation for $S_{\eta.m} f$ where $\xi \in U(0)$, $\eta = 1 - \xi$ are given and fixed. As a remark, we shall see that $S_{\eta_1.m} - S_{\eta_2.m}$ is a specific polynomial of degree $\leq m - 1$ (Corollary 1).

We now mention several preliminary lemmas. The first two are proven in [CL2]; see [CL2, page 244] for Lemma 2.

LEMMA 1

For every $g \in \mathcal{E}', s \in \mathbb{N}, s \neq 0$,

$$x^s Hg = H(t^s g) + \sum_{q=0}^{s-1} x^q M(t^{s-1-q}g).$$

LEMMA 2

If $f \in V_0', m \geq 1, \eta \in U(\infty)$, *then*

$$x^m H\left(\frac{\eta f}{t^m}\right) = H(\eta f) + \sum_{k=0}^{m-1} x^k M\left(\frac{\eta f}{t^{k+1}}\right).$$

COROLLARY 1

Suppose $m \geq 1$ *and* $f \in V_m'$. *Let* $\xi_i \in U(0), \eta_i = 1 - \xi_i, i = 1, 2$. *Then,*

$$S_{\eta_1.m}f - S_{\eta_2.m}f = P_{m-1}(x; \eta_1 - \eta_2) \tag{7}$$

where $P_{m-1}(x; \eta_1 - \eta_2) = \sum_{k=0}^{m-1} x^k M\left(\frac{(\eta_1-\eta_2)f}{t^{k+1}}\right)$.

This follows from the fact that the left side of (7) is equal to

$$H((\xi_1 - \xi_2)f) + x^m H\left(\frac{(\eta_1 - \eta_2)f}{t^m}\right)$$

$$= H((\xi_1 - \xi_2)f) + H((\eta_1 - \eta_2)f) + P_{m-1}(x; \eta_1 - \eta_2)$$

$$= P_{m-1}(x; \eta_1 - \eta_2).$$

Lemma 2 is indeed applicable since $(\eta_1 - \eta_2)f \in \mathcal{E}'(\mathbb{R} \setminus \{0\})$.

LEMMA 3

Let $f \in V_s', 0 \leq s < k$. *Then,*

$$S_k f - S_s f = x^s \sum_{j=0}^{k-s-1} x^j M\left(\frac{\eta f}{t^{s+j+1}}\right).$$

PROOF Let $g = \left(\frac{\eta_1 f}{t^s}\right)$ where $\eta_1 \in U(\infty)$ is chosen, so that $\eta_1 \eta = \eta$ on $\mathbb{R}$. Using Lemma 2, we obtain

$$S_k f - S_s f = x^s \left\{ x^{k-s} H\left(\frac{\eta g}{t^{k-s}}\right) - H(\eta g) \right\}$$

$$= x^s \sum_{j=0}^{k-s-1} x^j M\left(\frac{\eta g}{t^{j+1}}\right). \qquad \blacksquare$$

The assertion follows.

9.3 Continuity

THEOREM 1

Let $k \in \mathbb{N}, \xi \in U(0), \eta = 1 - \xi$. Define

$$T_k \psi = \xi H \psi + \left(\frac{\eta}{x^k}\right) H(t^k \psi) \tag{8}$$

for $\psi \in V_{k+1}$; $T_0 \psi = H \psi$ for $\psi \in V_1$. Then,

(i) *for every $f \in V_k'$,*

$$_{V_{k+1}'} \langle S_k f, \psi \rangle_{V_{k+1}} = -_{V_k'} \langle f, T_k \psi \rangle_{V_k}, \qquad \forall \psi \in V_{k+1}.$$

(ii) *S_k is continuous from V_k' into V_{k+1}'.*

PROOF (a) Suppose $k = 0$. First, we prove that H is a continuous map from V_1 into V_0.

Assume $\xi \in U(0)$ is fixed, $\eta = 1 - \xi$ and ψ_j is a converging sequence to zero in V_1. Writing $\psi_j = \xi \psi_j + \left(\frac{\eta}{x^2}\right) b_j$ with $b_j = x^2 \psi_j$, we have

$$x H \psi_j = x H(\xi \psi_j) + x H \left(\frac{\eta b_j}{t^2}\right)$$

where by Lemma 1, $x H(\xi \psi_j) = H(t \xi \psi_j) + M(\xi \psi_j)$.

Clearly, $\xi \psi_j$ and $t \xi \psi_j$ converge to zero in $\mathcal{D}(K)$, $K = \sup \xi$. Therefore, $\lim M(\xi \psi_j) = 0$ as $j \to \infty$, where the limit can be interpreted as a limit of constant functions in $\mathcal{B}_c$. Furthermore, $H(t \xi \psi_j)$ converges to zero in $\mathcal{D}_{L^2}$ and, in particular, in the topology of $\mathcal{B}_c$.

From these facts, it results that $x H(\xi \psi_j)$ converges to zero in $\mathcal{B}_c$. On the other hand, by Lemma 2,

$$x H \left(\frac{\eta b_j}{t^2}\right) = H \left(\frac{\eta b_j}{t}\right) + M \left(\frac{\eta b_j}{t^2}\right).$$

Since $\left(\frac{\eta b_j}{t}\right) = \eta t \ \psi_j$ converges to zero in $w^{-1} \mathcal{B}_c = V_0$, it also converges to zero in the topology of $\mathcal{D}_{L^2}$. This yields $\lim_{j \to \infty} H \left(\frac{\eta b_j}{t}\right) = 0$ in $\mathcal{D}_{L^2}$ and, in particular, in $\mathcal{B}_c$. Noting now that $M \left(\frac{\eta b_j}{t^2}\right)$ converges to zero in $\mathcal{B}_c$, we obtain $\lim_{j \to \infty} \left\{x H \left(\frac{\eta b_j}{t^2}\right)\right\} = 0$ in $\mathcal{B}_c$ and from what precedes, $\lim_{j \to \infty} (x H \psi_j) = 0$ in $\mathcal{B}_c$. As a consequence, $H \psi_j$ converges to zero in V_0 because $w H \psi_j = \xi w H \psi_j + \left(\frac{\eta w}{x}\right) (x H \psi_j)$, where $\left(\frac{\eta w}{x}\right) \in \mathcal{B}$ and $\lim(\xi w H \psi_j) = 0$ in $\mathcal{D}(K)$ as $j \to \infty$. The continuity of $H : V_1 \to V_0$ follows.

Consider now the transpose H' of H. For each $f \in V_0'$,

$$_{V_1'} \langle H' f, \psi \rangle_{V_1} = -_{V_0'} \langle f, H \psi \rangle_{V_0}, \qquad \psi \in V_1,$$

and H' is continuous from V_0' into V_1' (see [H] and [Tre] for instance). In view of the definition of Hf for $f \in V_0'$, we have

$$_{S'}\langle Hf, \varphi \rangle_S = -_{V_0'}\langle f, H\varphi \rangle_{V_0} \qquad \text{for all } \varphi \in S.$$

Since S is dense in V_1, this implies $H'f = -Hf$ for all $f \in V_0'$. The assertion of continuity of the $H : V_0' \to V_1'$ follows.

(b) Suppose $k \geq 1$. If ψ_j converges in V_{k+1}, then $t^k \psi_j$ converges in V_1. From the first part of the proof, it follows that $H(t^k \psi_j)$ converges in V_0 and, as a consequence, $\left(\frac{\eta}{x^k}\right) H(t^k \psi_j)$ converges in V_k. Since $\xi H \psi_j$ converges in $\mathcal{D}$, this proves that T_k is a continuous map from V_{k+1} into V_k.

Notice now that for each $f \in V_k'$, the following equalities hold for all $\varphi \in S$:

$$_{V_k'}\langle f, T_k \varphi \rangle_{V_k} = \langle f, \xi H\varphi \rangle + _{V_0'}\left\langle \frac{\eta f}{x^k}, H(t^k \varphi) \right\rangle_{V_0}$$

$$= -\langle H(\xi f), \varphi \rangle - \left\langle H\left(\frac{\eta f}{x^k}\right), t^k \varphi \right\rangle$$

$$= -\langle S_k f, \varphi \rangle.$$

Since S is dense in V_{k+1}, we conclude that $S_k f = -T_k' f$ for every $f \in V_k'$. This and what precedes, yield the continuity assertion for $S_k, k \geq 1$. ∎

REMARK 1   (a) For each $k \in \mathbb{N}$, the above continuity property of $S_k : V_k' \to V_{k+1}'$ holds for the strong dual topologies $\beta(V_k', V_k)$ on V_k', $\beta(V_{k+1}', V_{k+1})$ on V_{k+1}', as well as for the weak dual topologies $\sigma(V_k', V_k)$ on V_k', $\sigma(V_{k+1}', V_{k+1})$ on V_{k+1}'. Other topologies can be considered too (see [H] and [Tre], for instance).

(b) For each $k \geq 1$ and $0 \leq s \leq k - 1$, the restriction $S_k \mid V_s'$ of S_k on the subspace V_s' of V_k' is a continuous mapping from V_s' into V_k'. To show this, suppose f_m tends to zero in V_s'. It results from Lemma 3 that $S_k f_m$ differs from $S_s f_m$ by a polynomial $P_{k-1}(\cdot; f_m)$ of degree inferior or equal to $k - 1$, each coefficient of which tends to zero. This implies that $P_{k-1}(\cdot; f_m)$ converges to zero in V_k'. On the other hand, Theorem 1 implies that $S_s f_m$ converges to zero in V_{s+1}' and thus in V_k'. The announced continuity property of $S_k \mid V_s'$ follows. As a particular case, note, for instance, that for each $s \geq 0$, both S_s and $S_{s+1} \mid V_s'$ are continuous mappings from V_s' into V_{s+1}'. Of course, a remark similar to (a) above, can be made for the topologies on both spaces. ∎

COROLLARY 2

If f_j converges to f in S', then there exists $N \in \mathbb{N}$ such that $S_N f_j$ converges to $S_N f$ in V_{N+1}'. As a consequence, for each $m \geq N$, $S_m f_j$ converges to $S_m f$ in S'.

PROOF   Since $f \in S'$, there exists $p \in \mathbb{N}$ such that $f \in V_p'$. Let $g_j = f_j - f$. By [S2, page 240], there exists $r \in \mathbb{R}$ such that $\frac{g_j}{w^r}$ strongly converges to zero in $\mathcal{D}_{L^\infty}'$.

Since $w^{-\sigma} \in \mathcal{D}_{L^1}$ for every $\sigma > 1$, it results from [S2, page 203] that $\frac{g_j}{w^{\sigma+r}}$ strongly converges to zero in $\mathcal{D}'_{L^1}$ whenever $\sigma > 1$. Choose now $\sigma = 1 + \epsilon, 0 < \epsilon < 1$, so that $r + \epsilon \in \mathbb{N}$ and set $N = \max\{p, r+\epsilon\}$. Then, f_j strongly converges to f in V'_N as $j \to \infty$. By Theorem 1, this implies that $S_N f_j$ converges to $S_N f$ in V'_{N+1}. From Remark 1(b), it then follows that for each $m \geq N + 1$, $S_m f_j$ converges to $S_m f$ in V'_m, and a fortiori in S', as $j \to \infty$. ∎

COROLLARY 3

Let $k \in \mathbb{N}$. If $f \in V'_k$ and $\xi_j \in AU$, then $\lim S_k(\xi_j f) = S_k f$ in the strong topology of V'_{k+1} as $\xi_j \to 1$ in $\mathcal{B}_c$.

PROOF The assertion will follow from Theorem 1 if we show that $\lim(\xi_j f) = f$ in the strong topology of V'_k, as $\xi_j \to 1$ in $\mathcal{B}_c$. To that purpose, we note that $h = w^{-k-1} f$ can be written under the form $h = \sum_{q=0}^M D^{s(q)} F_q$ for some $M \in \mathbb{N}$, $F_q \in L^1$, $s(q) \in \mathbb{N}$. [S2, Theorem XXV, page 201]. Hence, for any bounded set B of $\mathcal{B}_c$,

$$
\sup_{b \in B} |\langle \eta_j h, b \rangle| = \sup_{b \in B} \left| \left\langle \sum_{q=0}^M D^{s(q)} F_q, \eta_j b \right\rangle \right|
$$

$$
\leq \sum_{q=0}^M \sup_{b \in B} \left| \int_{\mathbb{R}} F_q \cdot D^{s(q)}(\eta_j b) dx \right|
$$

where $\eta_j = 1 - \xi_j$. For every $\epsilon > 0$, there exists a compact interval $K = K(\epsilon)$ such that $\int_{\mathbb{R}/K} |F_k| dx < \epsilon$ for all $0 \leq q \leq M$. Next, there exists $J = J(K) \in \mathbb{N}$ such that $[D^{s(q)}(\eta_j b)](x) = 0$ for $x \in K, 0 \leq q \leq M$ and all $j \geq J$. Therefore, the sum in the right side of the above estimate is dominated by $\epsilon \sum_{q=0}^M \sup_{b \in B} \|D^{s(q)}(\eta_j b)\|_\infty \leq \epsilon C$ for all $j > J$, where C depends only on the bounded set B under consideration. This shows $\eta_j f$ converges to zero in the strong topology of V'_k. The result then follows from Theorem 1. ∎

9.4 Inversion

The main result of this section is shown in Theorem 2.

THEOREM 2

Suppose $\xi \in U(0), \eta = 1 - \xi$.

(i) If $f \in V_0'$, then

$$S_{\eta.1} H f = -f - \left(\frac{1}{\pi}\right)_{V_0'} \left\langle f, H\left(\frac{\eta}{t}\right)\right\rangle_{V_0} \quad \text{in } V_1'. \tag{9}$$

(ii) If $k \geq 1$ and $f \in V_k'$, then

$$S_{\eta.k+1} S_{\eta.k} f = -f - \left(\frac{1}{\pi}\right) \sum_{r=0}^{k-1} c_r(\eta; k; f) x^r$$

$$- \left(\frac{1}{\pi}\right)_{V_k'} \left\langle f, T_{\eta.k}\left(\frac{\eta}{t^{k+1}}\right)\right\rangle_{V_k} x^k \tag{10}$$

in V_{k+1}', where

$$T_{\eta.k} \psi = \xi H \psi + \left(\frac{\eta}{x^k}\right) H(t^k \psi), \qquad T_{\eta.0} = T_0 = H \tag{11}$$

and

$$c_r(\eta; k; f) =_{V_k'} \left\langle f, \xi H\left(\frac{\eta}{t^{r+1}}\right) - \left(\frac{\eta}{x^k}\right) H(t^{k-r-1}\xi)\right\rangle_{V_k}. \tag{12}$$

PROOF   As before, we write S_m instead of $S_{\eta.m}$ and T_m instead of $T_{\eta.m}$.
(i) Suppose $f \in V_0'$, $\xi_j \in AU$. Then, $H(H(\xi_j f)) = -\xi_j f$ and, by Lemma 3,

$$S_1 H(\xi_j f) = -\xi_j f + \left(\frac{1}{\pi}\right) \left\langle H(\xi_j f), \frac{\eta}{x}\right\rangle$$

$$= -\xi_j f - \left(\frac{1}{\pi}\right) \left\langle \xi_j f, H\left(\frac{\eta}{t}\right)\right\rangle.$$

From Corollary 3 and Theorem 1, it follows that the left side of the above equality converges to $S_1 H f$ in V_2'. Since $H\left(\frac{\eta}{t}\right) \in V_0$ and $\lim(\xi_j f) = f$ in V_0', the second term in the right side converges to $-\left(\frac{1}{\pi}\right)\langle f, H\left(\frac{\eta}{t}\right)\rangle$ in V_1'. Assertion (i) follows.

Note: The above argument shows that $\lim S_1 H(\xi_j f) = S_1 H f$ holds, in fact, in the topology of V_1', which is stronger than that of V_2'.

(ii) We first prove the assertion for $k = 1$. Suppose $f \in V_1'$. Then, $\xi_j f \in V_0'$ and Lemma 3 implies $S_1(\xi_j f) = H(\xi_j f) + M\left(\frac{\eta \xi_j f}{x}\right)$. Since the right side of this equality belongs to V_1', we can apply S_1 to both sides. In view of the proof of 1 above, this yields

$$S_1 S_1(\xi_j f) = S_1 H(\xi_j f) + M\left(\frac{\eta \xi_j f}{x}\right) S_{\eta}, (1)$$

$$= -\xi_j f - \left(\frac{1}{\pi}\right)\langle \xi_j f, H\left(\frac{\eta}{t}\right) - \left(\frac{\eta}{x}\right) S_{\eta}, (1)\rangle,$$

where the test function in the last term is equal to $\xi H\left(\frac{\eta}{t}\right) - \left(\frac{\eta}{x}\right) H\xi$ and belongs to V_1. Taking the limit as $\xi_j \to 1$ in $\mathcal{B}_c$, we obtain

$$\lim(S_1 S_1)(\xi_j f) = -f - \left(\frac{1}{\pi}\right), \qquad c_0(\eta; 1; f) \text{ in } V_1'.$$

Now, by Lemma 3,

$$S_2 S_1(\xi_j f) = S_1 S_1(\xi_j f) + \left(\frac{1}{\pi}\right) x \left\langle S_1(\xi_j f), \left(\frac{\eta}{t^2}\right)\right\rangle.$$

On one hand, Theorem 1 implies $\lim S_2 S_1(\xi_j f) = S_2 S_1 f$ in V_3', as $\xi_j \to 1$ in $\mathcal{B}_c$. On the other hand,

$$\left\langle S_1(\xi_j f), \left(\frac{\eta}{t^2}\right)\right\rangle = -\left\langle \xi_j f, T_1\left(\frac{\eta}{t^2}\right)\right\rangle,$$

as this can be seen by using the definitions of S_1, T_1 and the fact that $\xi_j f \in \mathcal{E}'$, $T_1\left(\frac{\eta}{t^2}\right) \in V_1$. In addition, the right side converges to $-\left\langle f, T_1\left(\frac{\eta}{t^2}\right)\right\rangle$. From this and what precedes, the result is

$$S_2 S_1 f = -f - \left(\frac{1}{\pi}\right), \qquad c_0(\eta; 1; f) - \left(\frac{1}{\pi}\right), \qquad \left\langle f, T_1\left(\frac{\eta}{t^2}\right)\right\rangle x,$$

which completes the proof of (ii), for $k = 1$.

We now use an induction argument. Suppose Assertion (ii) is true for some $k \geq 1$. Let $f \in V_{k+1}'$. Then, $\xi_j f \in V_0' \subset V_k'$ and $S_{k+1}(\xi_j f) \in V_{k+1}'$. By applying Lemma 3 to $S_{k+1}(\xi_j f)$, we thus have

$$S_{k+1} S_{k+1}(\xi_j f) = S_{k+1} S_k(\xi_j f) + M\left(\frac{\eta \xi_j f}{t^{k+1}}\right) S_{\eta,k+1}(t^k).$$

In view of the hypothesis we have just made,

$$S_{k+1} S_k(\xi_j f) = -(\xi_j f) - \left(\frac{1}{\pi}\right) \sum_{r=0}^{k-1} c_r(\eta; k; \xi_j f)x^r - \left(\frac{1}{\pi}\right)\left\langle \xi_j f, T_k\left(\frac{\eta}{t^{k+1}}\right)\right\rangle x^k.$$

Simple calculations show that, for $0 \leq r \leq k - 1$,

$$c_r(\eta; k; \xi_j f) = c_r(\eta; k+1; \xi_j f) + \left\langle \xi_j f, \frac{\eta}{x^{k+1}}\right\rangle M(t^{k-r-1}\xi),$$

which yields

$$\sum_{r=0}^{k-1} [c_r(\eta; k; \xi_j f) - c_r(\eta; k+1; \xi_j f)]x^r = \left\langle \xi_j f, \frac{\eta}{x^{k+1}}\right\rangle \{x^k H\xi - H(t^k \xi)\}.$$

On the other hand,

$$\left\langle \xi_j f, T_k\left(\frac{\eta}{t^{k+1}}\right)\right\rangle = c_r(\eta; k+1; \xi_j f) + S_{\eta,1}(1)\left\langle \xi_j f, \frac{\eta}{x^{k+1}}\right\rangle$$

and

$$S_{\eta.k+1}(t^k) = H(\xi t^k) - x^k H\xi + x^k S_{\eta.1}(1).$$

From these observations, we deduce

$$S_{k+1}S_{k+1}(\xi_j f) = -\xi_j f - \left(\frac{1}{\pi}\right)\sum_{r=0}^{k} c_r(\eta; k+1; \xi_j f)x^r.$$

Now,

$$S_{k+2}S_{k+1}(\xi_j f) = S_{k+1}S_{k+1}(\xi_j f) + \left(\frac{1}{\pi}\right)\left\langle S_{k+1}(\xi_j f), \frac{\eta}{t^{k+2}}\right\rangle x^{k+1},$$

where

$$\left\langle S_{k+1}(\xi_j f), \frac{\eta}{t^{k+2}}\right\rangle = -\left\langle \xi_j f, T_{k+2}\left(\frac{\eta}{t^{k+2}}\right)\right\rangle.$$

The last equality was obtained by using the definitions of S_{k+1} and T_{k+1} and the fact that $\xi_j f$ has a compact support. Since $T_{k+1}\left(\frac{\eta}{t^{k+2}}\right) \in V_{k+1}$, we may write

$$\lim_{j\to\infty}\left\langle \xi_j f, T_{k+1}\left(\frac{\eta}{t^{k+2}}\right)\right\rangle = \left\langle f, T_{k+1}\left(\frac{\eta}{t^{k+2}}\right)\right\rangle.$$

For similar reasons,

$$\lim_{j\to\infty} c_r(\eta; k+1; \xi_j f) = c_r(\eta; k+1; f)$$

for $0 \le r \le k$. Assertion (ii) thus follows for the index $k+1$, instead of k, which ends the proof of the theorem. ∎

COROLLARY 4

Let $\xi \in U(0)$, $\eta = 1 - \xi$, $k \ge 1$. Then, for all $\psi \in S$,

$$T_{\eta.k}T_{\eta.k+1}\psi = -\psi - \sum_{r=0}^{k-1}M(t^r\psi)\left[\xi H\left(\frac{\eta}{t^{r+1}}\right) - \left(\frac{\eta}{x^k}\right)H(t^{k-r-1}\xi)\right]$$

$$- M(t^k\psi)\cdot T_{\eta.k}\left(\frac{\eta}{t^{k+1}}\right).$$

If $k = 0$, $HT_{\eta.1}\psi = -\psi - M(\psi)\cdot H\left(\frac{\eta}{t}\right)$.

This is a consequence of Theorems 1 and 2.

9.5   Inversion in subspaces

The main result in this section is Theorem 3. We first prove a preliminary lemma.

LEMMA 4

Suppose $f \in V_0', u \in V_0$. Then, $H(uHf + fHu) = Hf \cdot Hu - fu$ in V_1.

PROOF Note first that $fHu \in V_0'$ and $uHf \in V_0'$. As a consequence, $uHf + fHu \in V_0'$. Hence, for each $\psi \in S$,

$$_{S'}\langle H(uHf + fHu), \psi \rangle_S$$
$$= -_{V_0'}\langle (uHf + fHu), H\psi \rangle_{V_0}$$
$$= -_{V_1'}\langle Hf, uH\psi \rangle_{V_1} -_{V_0'} \langle f, Hu \cdot H\psi \rangle_{V_0}$$
$$= _{V_0'}\langle f, H(uH\psi) - Hu \cdot H\psi \rangle_{V_0},$$

where the last equality follows from Theorem 1 since $uH\psi \in V_1$.

On the other hand, for every $\psi \in S$,

$$_{S'}\langle Hf \cdot Hu - fu, \psi \rangle_S = _{S'}\langle Hf, \psi Hu \rangle_S - _{S'}\langle f, u\psi \rangle_S$$
$$= -_{V_0'}\langle f, H(\psi Hu) + u\psi \rangle_{V_0}.$$

The assertion then follows from the known identity

$$HuH\psi - u\psi = H(\psi Hu + uH\psi), \qquad \psi \in S, \qquad u \in V_0$$

(see [Tri] and [CL1], for instance). ∎

THEOREM 3

(i) Suppose $f \in V_0'$ and $Hf \in V_0'$. Then

$$H(Hf) = -f \tag{13}$$

and, for every $u \in \{u \in V_0 : Hu \in V_0\}$,

$$\langle Hf, u \rangle = -\langle f, Hu \rangle;$$
$$\langle Hf, Hu \rangle = \langle f, u \rangle. \tag{14}$$

(ii) Let $\xi \in U(0), \eta = 1 - \xi, k \geq 1$. If $f \in V_k'$ and $S_{\eta.k}f \in V_k'$, then

$$S_{\eta.k}S_{\eta.k}f = -f - \left(\frac{1}{\pi}\right) \sum_{r=0}^{k-1} c_r(\eta; k; f)x^r \tag{15}$$

and, for every $\psi \in \{\psi \in V_k : T_{\eta.k}\psi \in V_k\}$,

$$_{V_k'}\langle S_{\eta.k}f, \psi \rangle_{V_k} = -_{V_k'}\langle f, T_{\eta.k}\psi \rangle_{V_k}. \tag{16}$$

In what precedes, $c_r(\eta; k; f)$ and $T_{\eta.k}$ are defined by (11) and (12), respectively.

PROOF (i) In view of the hypotheses, the distributions $F = fHu + uHf$ and $G = Hf \cdot Hu - fu$ belong to $\mathcal{D}'_{L^1}$. Therefore, their Fourier transforms ΦF, ΦG are continuous functions on $\mathbb{R}$ [S2, page 256]. Furthermore, since $F \in \mathcal{D}'_{L^2}$, we have $\Phi H F = \sigma \Phi F$ where $\sigma(x) = -i \operatorname{sign} x$ [S2, pages 259, 270]. By Lemma 4, $G = HF$. Hence $\Phi G = \sigma \Phi F$ and by what precedes, $\Phi G(0) = \Phi F(0) = 0$. This means that $\langle fHu + uHf, 1 \rangle = 0$ and $\langle Hf \cdot Hu - fu, 1 \rangle = 0$, which yields (14).

To prove (13), recall that $S_1 Hf = H(Hf) + M\left(\frac{\eta Hf}{t}\right)$ in view of Lemma 3 applied with $Hf \in V'_0$ instead of f.

On the other hand, Theorem 2 implies $S_1 Hf = -f - M\left(fH\left(\frac{\eta}{t}\right)\right)$. This yields

$$H(Hf) + f = \left(\frac{1}{\pi}\right)\left\{\left\langle Hf, \frac{\eta}{x}\right\rangle + \left\langle f, H\left(\frac{\eta}{t}\right)\right\rangle\right\}.$$

Since $\left(\frac{\eta}{x}\right) \in V_0$ and $H\left(\frac{\eta}{t}\right) \in V_0$, (14) shows that the right side of the above equality is equal to zero. Assertion (4.1) follows.

(ii) Since $S_k f \in V'_k$, Lemma 3 implies

$$S_{k+1}(S_k f) = S_k(S_k f) + \left(\frac{1}{\pi}\right)\left\langle S_k f, \frac{\eta}{x^{k+1}}\right\rangle x^k.$$

From this and Theorem 2(ii), we deduce

$$S_k(S_k f) = -f - \left(\frac{1}{\pi}\right)\sum_{r=0}^{k-1} c_r(\eta; k; f)x^r - \left(\frac{x^k}{\pi}\right)R(\eta; k; f),$$

where

$$R(\eta; k; f) = {}_{V'_k}\left\langle f, T_k\left(\frac{\eta}{t^{k+1}}\right)\right\rangle_{V_k} + {}_{V'_k}\left\langle S_k f, \frac{\eta}{x^{k+1}}\right\rangle_{V_k}.$$

Since $\frac{\eta}{x^{k+1}}$ and $T_k\left(\frac{\eta}{t^{k+1}}\right)$ both belong to V_k, we shall have $R(\eta; k; f) = 0$ if (16) is true, and in this case, (15) will follow. To prove (16), note that $f \in V'_1$ implies $\left(\frac{\eta f}{x}\right) \in V'_0$. Moreover, $S_k f \in V'_k$ implies $x^k H(\eta/t^k) \in V'_k$ and, as a consequence, $H\left(\frac{\eta}{t^k}\right) \in V'_0$.

On the other hand, if $\psi \in V_k$, then $x^k \psi \in V_0$. If $T_1 \psi \in V_k$, then $H(x^k \psi) \in V_0$. Therefore,

$$_{V'_k}\langle S_k f, \psi \rangle_{V_k} = {}_{\mathcal{D}'_{L^2}}\langle H(\xi f), \psi \rangle_{\mathcal{D}_{L^2}} + {}_{V'_0}\left\langle H, \left(\frac{\eta f}{t^k}\right), x^k \psi \right\rangle_{V_0}$$

$$= -\langle \xi f, H\psi \rangle - {}_{V'_0}\left\langle \frac{\eta f}{t^k}, H(x^k \psi) \right\rangle_{V_0},$$

where the last term was obtained by using the first part of the theorem. The result then follows from the definition of T_k. ∎

The following corollaries of Theorems 2 and 3 deal with the convergence of some products of operators.

COROLLARY 5

(i) Let $k \geq 0$. Suppose $f \in V_k'$ converges to f in V_k', then

$$\lim(S_{k+1}S_k f_j) = S_{k+1}S_k f \text{ in } V_{k+1}', \text{ as } j \to \infty.$$

This holds when V_k' and V_{k+1}' both carry their strong dual topology (respectively, their weak dual topology).

(ii) Let $k \geq 0$. Suppose f_j and f belong to $\{g \in V_k' : S_k g \in V_k'\}$. Assume f_j converges to f in the strong topology (respectively, in the weak topology) of V_k'. Then, $\lim(S_k S_k f_j) = S_k S_k f$ strongly (respectively, weakly) in V_k', as $j \to \infty$.

PROOF (i) In case $k \geq 1$, the assertion results from Theorem 2(ii) and the fact that $\lim_{j \to \infty} c_r(\eta; k; f_j) = c_r(\eta; k; f)$ for $0 \leq r \leq k - 1$ and $\lim_{j \to \infty} \langle f_j, T_k \left(\frac{\eta}{j^{k+1}}\right)\rangle = \langle f, T_k \left(\frac{\eta}{j^{k+1}}\right)\rangle$, since the involved test functions belong to V_k. In case $k = 0$, the proof is similar.

(ii) In case $k = 0$, the assertion is a direct consequence of Theorem 3(i) and the hypothesis of convergence of f_j. In case $k \geq 1$, we use (15) and remark that $\lim c_r(\eta; k; f_j) = c_r(\eta; k; f)$ as $j \to \infty$, for each $0 \leq r \leq k - 1$. ∎

As a consequence of Corollary 5, we obtain the following associativity property.

COROLLARY 6

Let $k \geq 0$. If $f \in V_k'$ then for any $\xi \in U(0)$, $\eta = 1 - \xi$,

$$S_{\eta,k+1}(S_{\eta,k+1}S_{\eta,k}f) = (S_{\eta,k+1}S_{\eta,k+1})(S_{\eta,k}f). \tag{17}$$

PROOF Let $f_j = \xi_j f$ where $\xi_j \in AU$. Then, with the usual abbreviations, $S_{k+1}(S_{k+1}S_k f_j) = (S_{k+1}S_{k+1})(S_k f_j)$ for each j. By Corollary 5(i), $(S_{k+1}S_k f_j)$ converges to $S_{k+1}S_k f$ in V_{k+1}' as $\xi_j \to 1$ in $\mathcal{B}_c$. Theorem 1 then implies that $\lim_{j \to \infty} S_{k+1}(S_{k+1}S_k f_j) = S_{k+1}(S_{k+1}S_k f)$ in V_{k+2}'. On the other hand, since $F_j = S_k(\xi_j f)$ converges to $F = S_k f$ in V_{k+1}', and $S_{k+1}F_j$ and $S_{k+1}F$ both belong to V_{k+1}', Corollary 5(ii) implies $\lim_{j \to \infty}(S_{k+1}S_{k+1}F_j) = S_{k+1}S_{k+1}F$. The assertion follows. ∎

Note: One can prove (17) only by doing algebraic calculations. Since $S_{k+1}S_k f \in V_{k+1}'$, one can apply the operator S_{k+1} to both members of (10). On the other hand, since $F = S_k f \in V_{k+1}'$ is such that $S_{k+1}F \in V_{k+1}'$, one can evaluate $S_{k+1}S_{k+1}F$ by using Theorem 3. Explicit calculations of $c_r(\eta; k+1; S_k f)$ for $0 \leq r \leq k$ then lead to the announced identity.

References

[Cl1] C. Carton-Lebrun, *An extension to BMO functions of some product properties of Hilbert transforms,* J. Approx. Theory 49 (1987), 75–78.

[Cl2] C. Carton-Lebrun, *A real variable definition of the Hilbert transform on D′ and S′,* Applicable Anal. 29 (1988) 235–251.

[Cl3] C. Carton-Lebrun, *Continuity and inversion of the Hilbert transform of distributions,* J. Math. Anal. Applic. 161 (1991), 274–283.

[F] C. Fefferman, *Characterization of bounded mean oscillation,* Bull. Amer. Math. Soc. 77 (1971), 587–588.

[H] J. Horvath, *Topological vector spaces and distributions,* Vol. I, Addison-Wesley, Reading, MA, 1966.

[S1] L. Schwartz, *Causalité et analyticité,* An. Acad. Brasil. Cien. 34 (1962), 13–21.

[S2] L. Schwartz, *Théorie des distributions,* Hermans, Paris, 1978.

[Tre] F. Treves, *Topological vector spaces, distributions and kernels,* Academic Press, New York, 1967.

[Tri] F. G. Tricomi, *Integral equations,* Interscience, New York, 1965.

10

Failure of an Endpoint Estimate for Integrals Along Curves

Michael Christ[1]

10.1 Introduction

Let $m \geq 2$ be an integer and consider the integral operator

$$Af(x) = \int_{\mathbb{R}} f(x - \gamma(t))\sigma(t)\,dt, \tag{1}$$

acting on functions f defined in $\mathbb{R}^2$, where $\gamma(t) = (t, t^m)$ and $\sigma \in C_0^\infty(\mathbb{R})$ with $\sigma(0) \neq 0$. It is well known that such operators are smoothing to a certain degree. Denote by L_s^p the space of L^p functions having s derivatives in L^p, for $1 < p < \infty$ and $\mathbb{R} \ni s \geq 0$. The fundamental result concerning the smoothing properties of A, in the scale of spaces L_s^p, is that when γ has nonvanishing curvature, that is, when $m = 2$, A maps L^p to L_s^p, where $s = p^{-1}$ for $2 \leq p \leq \infty$, and this value of $s = s(p)$ is best possible for all p in that range. For $1 \leq p < 2$, the degree of smoothing on L^p is, of course, identical to the degree of smoothing on $L^{p'}$, where p' always denotes the exponent conjugate to p.

More generally, one can ask when operators such as A map L^p to L_s^q; the most delicate case is often $q = p$. Some recent results may be found in [PS].

The purpose of this note is to determine the optimal result for $m > 2$. That case is distinguished by the vanishing curvature of γ, at an isolated point, $t = 0$.

For $2 \leq p < \infty$, define

$$s(p, m) = \min(p^{-1}, m^{-1}).$$

THEOREM 1
For $2 \leq p < \infty$ and $m > 2$, A maps L^p to L_s^p, if and only if either

$$\begin{cases} p \neq m \text{ and } s \leq s(p, m), \text{ or} \\ p = m \text{ and } s < s(p, m). \end{cases}$$

[1]Research supported by the National Science Foundation.

As will be seen, the three cases $s < s(p, m)$, $p > m$, and $p = 2$ are rather easy. The case $2 < p < m$, $s = s(p, m)$ is subtler, but follows from a result of Grafakos [G] (and its proof). What is new here is simply the observation that the endpoint result with $s = s(p, m)$ fails for $p = m$.

These endpoint problems are closely related to the work of Grafakos [G], who built on [C] to analyze operators

$$\mathcal{A}_\alpha f(x) = \int_\mathbb{R} f(x - \gamma(t)) |t|^\alpha \, dt$$

for $m = 2$, and showed that for $2 \le p < \infty$, $|\partial_2|^s \circ \mathcal{A}_\alpha$ maps the Lorentz space $L^{p \cdot p'}$ to L^p, where $s = p^{-1}$, $\alpha = 2s - 1 = 2p^{-1} - 1$. This value of α is forced by homogeneity considerations, once the optimal $s = p^{-1}$ is chosen; it causes $\partial_2^s \circ \mathcal{A}_\alpha$ to be formally of order zero in the sense of scaling. $|\partial_2|^s$ is the Fourier multiplier operator defined by

$$(|\partial_2|^s f)^\wedge(\xi) = |\xi_2|^s \widehat{f}(\xi);$$

we will also make use of $|\partial_1|^s$, defined in the same way except that ξ_2 is replaced by ξ_1. Since $L^{p \cdot p'}$ is contained properly in L^p for $p > 2$, Grafakos' work left open the question of L^p boundedness.

[G] is concerned with certain convolution kernels that are homogeneous of the critical degree -3, with respect to the natural dilation group preserving the curve (t, t^2); the behavior of the kernel in a compact neighborhood of $\mathbb{R}^2 \backslash \{0\}$ (essentially) determines the kernel everywhere, and it is irrelevant whether the curve has nonvanishing curvature at the origin. Defining more generally

$$\mathcal{A}_{m,\alpha} f(x) = \int_\mathbb{R} f(x - \gamma(t)) |t|^\alpha \, dt$$

with $\gamma(t) = (t, t^m)$, the same method proves that $|\partial_2|^s \circ \mathcal{A}_{m,\alpha}$ maps $L^{p \cdot p'}$ to L^p for $2 \le p < \infty$, where $s = p^{-1}$ and $\alpha = \alpha(s, m) = ms - 1$. In particular, taking $p = m$, one finds that

$$f \mapsto |\partial_2|^{m^{-1}} \int f(x - \gamma(t)) \, dt$$

maps $L^{m,m'}$ to L^m. It follows quickly that the operator A in (1) maps $L^{m,m'}$ to L^m_s with $s = m^{-1}$, a result only slightly weaker than the desired smoothing estimate $L^m \mapsto L^m_s$.

Our only contribution here is the

Observation: $|\partial_2|^{1/p} \circ \mathcal{A}_{m,mp^{-1}-1}$ *does not map* L^p *to* L^p, *for* $2 < p < \infty$ *and* $m \ge 2$.

It is not relevant that m be an integer; the same applies to all real $m > 1$, taking the curve to be either of $(t, \pm|t|^m)$ for $t < 0$.

I am indebted to C. D. Sogge for pointing out the question answered by the Theorem.

10.2 The positive results

We take for granted the result proven in detail in [G], only for $m = 2$, that $|\partial_2|^{1/p} \circ \mathcal{A}_{m,mp^{-1}-1}$ maps $L^{p \cdot p'}$ to L^p for $2 \leq p < \infty$. An almost immediate corollary is that $|\partial_2|^{1/m} \circ A$ maps $L^{m,m'}$ to L^m. On the other hand, van der Corput's lemma implies that $|\partial_2|^{1/m} \circ A$ is bounded on L^2. Therefore interpolation by the real method yields L^p boundedness of $|\partial_2|^{1/m} \circ A$, for $2 < p < m$. But A is given by convolution with a distribution whose Fourier transform decays faster than any power of $|\xi|$ in a conic neighborhood of the axis $\xi_2 = 0$, which implies that A maps L^p to L^p_{m-1} for $2 \leq p < m$.

Next consider $m < p < \infty$ and $s = m^{-1}$. Fix $\varphi \in C_0^\infty(\mathbb{R}^+)$ such that $\sum_{-\infty}^{\infty} \varphi(2^j t) \equiv 1$ on $\mathbb{R}^+$. The operator

$$f \mapsto |\partial_2|^{1/p} \circ \int f(x - \gamma(t)) \, \sigma(t) \varphi(2^j t) \, dt$$

has L^p operator norm equal to exactly $2^{j((m/p)-1)}$ times the operator norm of

$$f \mapsto |\partial_2|^{1/p} \circ \int f(x - \gamma(t)) \sigma(2^{-j} t) \varphi(t) \, dt,$$

and these are bounded on L^p uniformly for $j \geq 0$, since the functions $\sigma(2^{-j} t)$ are uniformly C^∞, and since the curve (t, t^m) has nonvanishing curvature on the support of φ. Therefore setting $\tilde{\varphi}(t) = \sum_0^\infty \varphi(2^j t)$ and summing a convergent geometric series, we find that the operator

$$f \mapsto |\partial_2|^{1/p} \circ \int f(x - \gamma(t)) \, \tilde{\varphi}(t) \sigma(t) \, dt$$

is bounded on L^p for $m < p < \infty$. Since $(1 - \tilde{\varphi}) \cdot \sigma$ is supported in a compact region in which the curve γ has nonvanishing curvature, the associated operator maps L^p to L^p_{p-1} for all $2 \leq p < \infty$, and hence is harmless, and the result for $m < p < \infty$ is proven.

Boundedness from L^p to L^p_s for $s < s(p, m)$, now follows by interpolation, although it can also be proven more simply.

10.3 Proof of the negative result

The lack of L^p boundedness for $p = m$ in the Theorem follows directly from the failure of $|\partial_2|^{1/m} \circ \mathcal{A}_{m,0}$ to be bounded on L^p. Indeed, by summing a convergent series, as was done two paragraphs above, we may replace A by an operator of the same form, in which σ is replaced by a C_0^∞ function, which is identically equal to 1 in some neighborhood of the origin. Then L^m boundedness of $|\partial_2|^{1/m} \circ A$ would imply L^m boundedness of $|\partial_2|^{1/m} \circ \mathcal{A}_{m,0}$ by a routine homogeneity argument. Therefore it suffices to establish the Observation.

Fix m and set

$$B_0 f(x) = \int f(x - \gamma(t))\phi(t)\,dt$$

where $\phi \in C_0^\infty(\mathbb{R}^+)$ is supported in $\left(\frac{1}{2}, \frac{3}{2}\right)$, $\phi \geq 0$, and $\phi(1) \neq 0$. Fix $\eta \in C_0^\infty(\mathbb{R}^2)$, nonnegative and not vanishing identically, and supported, where $|x| < 1$. Define $\eta_r(x) = \eta(\delta_r x)$ where $\delta_r x = (rx_1, r^m x_2)$. Write

$$g(x) = e^{zx_1} \eta_r(x)$$

where the parameters $z, r \in \mathbb{R}^+$ are understood. We assume always that r is large, and that $|z| \geq 1$. Fix a positive, but sufficiently small, constant c_0.

LEMMA 1
Assume that $1 < p < \infty$.

(i)   *For all $N \geq 0$,*

$$\|B_0 g\|_p \leq C_N (1 + (r^{-m}|z|))^{-N} r^{-m} r^{-1/p}. \tag{2}$$

(ii) *For all $\beta \geq 0$, for $i = 1, 2$,*

$$\| |\partial_i|^\beta B_0 g \|_p \leq C_N (r^m + |z|)^\beta (1 + (r^{-m}|z|))^{-N} r^{-m} r^{-1/p}. \tag{3}$$

(iii) *There exists $\varepsilon_0 > 0$ such that for all $\beta \geq 0$, when $z = c_0 r^m$ and r is sufficiently large,*

$$\| |\partial_1|^\beta B_0 g \|_p \geq \varepsilon_0 r^{m\beta} r^{-m} r^{-1/p}. \tag{4}$$

(iv) *There exists $\varepsilon_0 > 0$ such that when $z = c_0 r^m$ and r is sufficiently large,*

$$\frac{\| |\partial_2|^{1-p^{-1}} B_0 g \|_p}{\|g\|_p} \geq \varepsilon_0. \tag{5}$$

PROOF   We have

$$B_0 g(x) = e^{izx_1} \int e^{-izt} \eta(r(x_1 - t), r^m(x_2 - t^m))\,\phi(t)\,dt. \tag{6}$$

This is supported in the set of all points within distance Cr^{-1} of a compact portion of the curve γ, and is everywhere $O(r^{-m})$, so (2) holds for small $|z|$. When $|z| \geq r^m$, the extra factor of $r^{Nm}|z|^{-N}$ in (2) is obtained via repeated integration by parts. From Leibniz's rule, (3) follows for integral-order derivatives, and the general case then follows by complex interpolation. If $z = c_0 r^m$, then there is no significant cancellation in the integral (6), and $B_0 g$ is bounded below by $\varepsilon_0 r^{-m}$

on a tubular neighborhood T of a fixed compact subset of γ, of width cr^{-1}, so (4) holds when $\beta = 0$. When k is a positive integer, we have for large r

$$\left| \frac{\partial^k}{\partial x_1^k} B_0 g(x) \right| \geq \varepsilon_0 r^{mk} r^{-m} \text{ for } x \in T$$

by Leibniz's rule, whence (4) holds for all integers β. Moreover, there is an upper bound of the same form for all x, with ε_0 replaced by another finite constant. Fixing $\beta \in \mathbb{R}^+$ and two integers such that $\beta < k < n$, interpolating by the complex method between L_β^p, L_k^p and L_n^p, and invoking the lower bound for the L_k^p norm and the upper bound for the L_n^p norm, we obtain the desired lower bound for the L_β^p norm. (5) follows at once, with $|\partial_2|^{1/m}$ replaced by $|\partial_1|^{1/m}$. But B_0 is defined by convolution with a measure whose Fourier transform decays faster than any power of ξ in a conic neighborhood of the union of the two axes $\xi_1 = 0$, $\xi_2 = 0$, and outside such a conic neighborhood $|\partial_1|^s$ is equivalent to $|\partial_2|^s$, therefore (5) holds. ∎

To establish the Observation, fix $m > 1$ and $p \in (2, \infty)$. Set $q = p' \in (1, 2)$; we will demonstrate that $|\partial_2|^{p^{-1}} \circ \mathcal{A}_{m,mp^{-1}-1}$ does not map L^q to L^q. Let $N \in \mathbb{N}$ be a large parameter, then choose a very sparse sequence of positive integers k_n and set

$$f_n(x) = \exp(i\zeta_n x_1) \eta(x)$$

where

$$\zeta_n = c_0 2^{(m-1)k_n},$$

and

$$F = \sum_{n=1}^N f_n.$$

By the Littlewood–Paley theory we obtain

$$\|F\|_q \sim N^{1/2}.$$

Write $\mathcal{A}_{m,mp^{-1}-1} = \sum_{-\infty}^\infty T_j$ where

$$T_j f(x) = \int_{\mathbb{R}} f(x - \gamma(t)) |t|^{mp^{-1}-1} \varphi(2^j |t|) \, dt$$

and where $\varphi \in C_0^\infty(\mathbb{R}^+)$ is nonnegative and $\sum_{-\infty}^\infty \varphi(2^j t) \equiv 1$ on $\mathbb{R}^+$. It will suffice to show that $\| |\partial_2|^{p^{-1}} \mathcal{A}_{m,mp^{-1}-1} F \|_q \geq \varepsilon N^{1/q}$ for some fixed $\varepsilon > 0$.

Now by rescaling,

$$\frac{\| |\partial_2|^{p^{-1}} T_{-k_n} f_n \|_q}{\| f_n \|_q} = \frac{\| |\partial_2|^{p^{-1}} B_0 g \|_q}{\| g \|_q}$$

where $g(x) = \exp(i c_0 r^m x_1) \eta_r(x)$ for $r = 2^{k_n}$. Thus by (5),

$$\| |\partial_2|^{p^{-1}} T_{-k_n} f_n \|_q \geq \varepsilon_0 > 0.$$

Moreover, these functions have pairwise disjoint supports, if the sequence k_n is reasonably sparse. Therefore

$$\left\| \sum_n |\partial_2|^{p^{-1}} T_{-k_n} f_n \right\|_q \geq \varepsilon N^{1/q}$$

as desired, and the Observation results, provided that there is no significant interference from terms $T_j f_n$ with $j \neq -k_n$. But that follows immediately, provided the sequence $\{k_n\}$ is chosen to be sufficiently sparse, from parts (2) and (3) of the lemma, noting that if η has sufficiently small support and if φ is chosen to be supported in $\left(\frac{1}{2}, \frac{3}{2}\right)$ and to satisfy $\varphi(1) \neq 0$, then for $j \neq -k_n$, $|\partial_2|^{p^{-1}} T_j F$ vanishes identically in the tubular region where $|\partial_2|^{p^{-1}} T_{-k_n} f_n$ is large.

Question: For which Lorentz classes does $|D|^{1/m} \circ A$ map $L^{m'}$ to $L^{m',r}$? It does so for all $r \geq m$, as a consequence of [G], as remarked above. Our analysis actually shows that it fails to do so for any $r < 2$; indeed it fails to map the smaller space $L^{m',1}$ to $L^{m',r}$. A. Seeger has outlined an unpublished argument showing that $L^{m'}$ is mapped to $L^{m',2}$.

References

[C] M. Christ, *Weak type* (1, 1) *bounds for rough operators,* Annals of Math. 128 (1988), 19–42.

[G] L. Grafakos, *Endpoint bounds for an analytic family of Hilbert transforms,* Duke Math. J. 62 (1991), 23–59.

[PS] D. H. Phong and E. M. Stein, *Radon transforms and torsion,* International Mathematics Research Notices 4 (1991), 49–60.

11

Spline Wavelet Bases of Weighted Spaces

J. García-Cuerva[1]
K. S. Kazarian

11.1 Introduction

The purpose of this paper is to give conditions on the weight w for the systems of m-spline wavelets to be bases or unconditional bases of some weighted spaces. We want to present these results here just with the essential details, so that the general philosophy can be easily transmitted. Complete details will be given in the forthcoming papers [G–K1] and [G–K2]. In the general framework of wavelets in $\mathbb{R}^n$, it is quite clear that sufficient conditions follow from variations of the theory of regular singular integrals. We do this in Section 11.3, and show that for any $m \geq 1$, $w \in A_p$ is enough to guarantee that the m-spline wavelets form an unconditional basis of $L^p(w)$. However, for $H^p(w)$, there is an interplay between the regularity (i.e., order) of the splines and the critical index of the weight (in terms of its membership in the A_p classes), so that when p decreases to 0 and the critical index of w increases, we are forced to take more and more regular splines. This agrees with what happens in the unweighted situation, see [St]. The whole theory in Section 11.3 depends on a simple atomic estimate plus the extrapolation theorem of [G1]. In Section 11.4 we take up the problem of necessity, which we study only in dimension one for the spaces $L^p(w)$, $1 \leq p < \infty$. Again the A_p condition plays a role, but a new phenomenon arises, which suggests the existence of other types of weights, without singularities at infinity, that we plan to investigate elsewhere. For the Haar system (i.e., $m = 0$), the characterization of these weights is entirely similar to the one obtained in [K2] for the interval [0, 1].

[1] The first author was supported by DGICYT Spain, under Grant PB90-187. The second author was supported by Ministerio de Educación, Spain, under Sabattical Grant SB90-82.

1980 *Mathematics Subject Classification* (1985) *Revision* 42B20, 42B25, 42C10, 46B15.

The history of the discovery by Maurey ([Mau]) of the fact that H^1 has an unconditional basis, and the subsequent construction of concrete bases by Carleson ([Ca]) and Wojtaszczyk ([W]), can be seen in [Me]. For the spline bases in the unweighted case, after the research of Ciesielski (see [C1] and [C2]) and Bochkariev (see [Bo]), we have to cite [St] and further work by Chang and Ciesielski ([C–C]) and also by Sjölin and Strömberg ([S–S], [S–S1]).

We devote Section 11.5 to clarify some point connected to the so called "remarkable identity" of Meyer ([Me], Chapter 2, Section 6). The result we obtain may have an interest in itself, since it identifies precisely those functions of slow growth that are orthogonal to all the m-spline wavelets.

11.2   Basic definitions

Let E be a separable F-space (i.e., a complete invariant metric topologic vector space). The sequence of vectors $\{e_k\}_{k=1}^\infty$ is called a (Schauder) basis of the space E if for every $e \in E$, there exists a unique sequence of numbers $\{\xi_k\}_{k=1}^\infty$ such that the sequence of partial sums

$$S_N(e) = \sum_{k=1}^N \xi_k e_k$$

converges to e in the metric of the space E. A basis $\{e_k\}_{k=1}^\infty$ of E is called *unconditional* if it remains a basis after every rearrangement of its elements. It is a well known fact, essentially due to Banach (see [B]), that in order for $\{e_k\}_{k=1}^\infty$ to be a basis of E, the fundamental property one must check (apart from the completeness of the system and the existence of a biorthogonal system in the dual) is the uniform boundedness in E of the partial sum operators. Likewise, for $\{e_k\}_{k=1}^\infty$ to be an unconditional basis of E, one must establish the uniform boundedness of the operators $S_{N,\epsilon}$ given by

$$S_{N,\epsilon}(e) = \sum_{k=1}^N \epsilon_k \xi_k e_k$$

where $\epsilon = \{\epsilon_k\}_{k=1}^\infty$ is any sequence of numbers ± 1.

We shall always work on $\mathbb{R}^n$, and we shall be concerned mainly with the spaces $L^p(w)$, $1 \le p < \infty$ and $H^p(w)$, $0 < p \le 1$. A basic role will be played by the classes A_p of weights. These classes were introduced by Muckenhoupt in [Mu], and their theory was further developed in [C–F]. See also [G–R].

DEFINITION 1   *Given a weight $w \ge 0$ on $\mathbb{R}^n$, we shall denote by $H^p(w)$ the space consisting of those $f \in S'(\mathbb{R}^n)$, the space of tempered distributions, for* which the maximal function

$$\phi^*(f)(x) = \sup_{t>0} |\phi_t \star f(x)| \in L^p(w)$$

where $\phi \in S(\mathbb{R}^n)$, is a fixed Schwartz function with $\int_{\mathbb{R}^n} \phi \neq 0$.

Under minimal conditions on w (in particular for $w \in A_\infty$), the definition does not depend on the choice of ϕ (see [S–T]), and we can write

$$\|f\|_{H^p(w)} = \|\phi^*(f)\|_{L^p(w)},$$

obtaining a norm if $p \geq 1$ and a p-norm otherwise.

For $w \in A_\infty$ we define

$$q_w = \inf\{q > 1 : w \in A_q\},$$

the critical index of w.

Also for $0 < p \leq 1$ we write

$$N_p(w) = \left[n \left(\frac{q_w}{p} - 1 \right) \right],$$

i.e., the largest integer $\leq n \left(\frac{q_w}{p} - 1 \right)$.

DEFINITION 2   *Given a weight $w \geq 0$ on $\mathbb{R}^n$, and a number p with $0 < p \leq 1$, a p-atom with respect to w will be a function a supported in a cube Q, and such that*

$$\|a\|_\infty \leq w(Q)^{-1/p} \tag{1}$$

and

$$\int_{\mathbb{R}^n} x^\alpha a(x)\, dx = 0 \text{ for every multiindex } \alpha \text{ such that } |\alpha| \leq N_p(w). \tag{2}$$

These p-atoms with respect to w are the basic building blocks of $H^p(w)$, as stated in the next proposition, whose proof can be seen in [G] and [S–T]. In those references our p-atoms are called (p, ∞)-atoms, since other (p, q)-atoms are considered in them, which we shall not need to consider in this chapter.

PROPOSITION 1

Let $w \in A_\infty$ be a weight in $\mathbb{R}^n$, and let $0 < p \leq 1$. A tempered distribution f on $\mathbb{R}^n$ belongs to the space $H^p(w)$, if and only if, f can be written as a series

$$f = \sum_j l_j a_j \tag{3}$$

convergent in the sense of distributions, where each a_j is a p-atom with respect to w, and the coefficients l_j satisfy

$$\sum_j |l_j|^p < \infty. \tag{4}$$

Besides, the infimum of the sums like (4), taken over all possible decompositions like (3), is equivalent to the p-norm $\| f \|^p_{H^p(w)}$.

Next, we shall define the m-splines on $\mathbb{R}$. Let m be an integer ≥ 0. Let

$$V_0 = \{ f \in L^2(\mathbb{R}) \cap C^{m-1}(\mathbb{R}) \text{ such that the restriction of } f \text{ to each}$$
$$\text{interval }]n, n+1[\text{ is a polynomial of degree } \leq m \},$$

where by $C^r(\mathbb{R})$ we denote the class of functions on $\mathbb{R}$ whose derivatives of order r are continuous and by $C^{-1}(\mathbb{R})$ we denote the class of piecewise continuous functions on $\mathbb{R}$.

Then we obtain a multiscale analysis $\{V_j\}_{j \in \mathbb{Z}}$ of $L^2(\mathbb{R})$ in the sense of Mallat [Ma] and Meyer [Me], simply by defining $V_j \subset L^2(\mathbb{R})$ in this way:

$$f(2x) \in V_{j+1} \iff f(x) \in V_j. \tag{5}$$

This multiscale analysis is m-*regular*, and therefore we can find an m-regular analyzing wavelet, that is, a function $\psi \in V_1$, $\psi \perp V_0$, such that

$$|D^\alpha \psi(x)| \leq C_{N,\alpha} (1 + |x|)^{-N} \quad \forall \alpha \tag{6}$$

such that $0 \leq \alpha \leq m$ and $\forall N \in \mathbb{N}$, and $\{\psi(x-k)\}_{k \in \mathbb{Z}}$ is an orthonormal basis of W_0, the orthogonal complement of V_0 in V_1. See [Me] and [D]. Since

$$L^2(\mathbb{R}) = \bigoplus_{j \in \mathbb{Z}} W_j,$$

it turns out that the system

$$\psi_{j,k}(x) = 2^{j/2} \psi(2^j x - k), \qquad j, k \in \mathbb{Z}$$

is an orthonormal basis of $L^2(\mathbb{R})$. In $\mathbb{R}^n$ one can define m-splines by starting with the space

$$V_0(\mathbb{R}^n) = V_0 \otimes V_0 \otimes \ldots \otimes V_0 \qquad (n \text{ times}).$$

The only difference is that when $n > 1$, one analyzing wavelet is not enough; but we can always find $2^n - 1$ functions ψ_η, $\eta \in E = \{0, 1\}^n \setminus \{0\}$ such that each ψ_η satisfies

$$|\partial^\alpha \psi_\eta(x)| \leq C_{N,\alpha} (1 + |x|)^{-N} \quad \forall \alpha \tag{7}$$

with $0 \leq |\alpha| \leq m$ and $\forall N \in \mathbb{N}$, and the system

$$\psi_{\eta,j,k}(x) = 2^{nj/2} \psi_\eta(2^j x - k), \qquad \eta \in E, j \in \mathbb{Z}, k \in \mathbb{Z}^n$$

is an orthonormal basis of $L^2(\mathbb{R}^n)$. The details are given in [Me] and [D].

Given a function f, its wavelet expansion will be

$$\sum_{\eta \in E; j \in \mathbb{Z}, k \in \mathbb{Z}^n} \langle f, \psi_{\eta,j,k} \rangle \psi_{\eta,j,k}.$$

We are actually interested in the partial sum operators

$$T_\Omega f(x) = \sum_{(\eta,j,k) \in \Omega} \langle f, \psi_{\eta,j,k} \rangle \psi_{\eta,j,k}(x)$$

for $\Omega \subset E \times \mathbb{Z} \times \mathbb{Z}^n$ finite sets; or even in the operators

$$T_{\Omega,\epsilon} f(x) = \sum_{(\eta,j,k) \in \Omega} \epsilon_{\eta,j,k} \langle f, \psi_{\eta,j,k} \rangle \psi_{\eta,j,k}(x),$$

where $\epsilon_{\eta,j,k} = \pm 1$. Since

$$T_{\Omega,\epsilon} f(x) = \int_{\mathbb{R}^n} K_{\Omega,\epsilon}(x,y) f(y) \, dy,$$

where

$$K_{\Omega,\epsilon}(x,y) = \sum_{(\eta,j,k) \in \Omega} \epsilon_{\eta,j,k} \psi_{\eta,j,k}(x) \psi_{\eta,j,k}(y)$$

then by (7), we obtain Proposition 2.

PROPOSITION 2
The kernels $K_{\Omega,\epsilon}(x,y)$ associated to the multiscale analysis by m-splines (or to any m-regular multiscale analysis) in $\mathbb{R}^n$ satisfy the following estimates:

$$\left| \partial_y^\alpha K_{\Omega,\epsilon}(x,y) \right| \le C |x-y|^{-n-|\alpha|}, \qquad 0 \le |\alpha| \le m.$$

11.3 Sufficient conditions for unconditional basisness in $H^p(w)$

We start by considering the $H^p(w) \to L^p(w)$ boundedness of the modified partial sum operators. This is obtained very easily from the Calderón–Zygmund theory, which we shall sketch below. Then the $H^p(w) \to H^p(w)$ boundedness will be obtained by applying the maximal function ϕ^* and realizing that the operators obtained satisfy the same estimates.

We define a class of *regular singular integrals* to which the modified partial sum operators belong.

DEFINITION 3 *Given a kernel $K(x,y)$, $x \in \mathbb{R}^n$, $y \in \mathbb{R}^n$, $x \ne y$ and given $\gamma \in \mathbb{R}$, $\gamma > 0$, we say that K is γ-regular with respect to y if:*

(i)   *K has continuous derivatives $\partial_y^\alpha K(x, y)$ for every multiindex α with order $|\alpha| < \gamma$ and they satisfy*

$$\left|\partial_y^\alpha K(x, y)\right| \le C|x - y|^{-n-|\alpha|}.$$

(ii)   *For those derivatives of the highest order, that is, for those corresponding to $\gamma - 1 \le |\alpha| < \gamma$, we have*

$$\left|\partial_y^\alpha K(x, y) - \partial_y^\alpha K(x, y')\right| \le C\frac{|y - y'|^{\gamma-|\alpha|}}{|x - y|^{n+\gamma}},$$

provided $2|y - y'| < |x - y|$.

Of course, there is a parallel notion of γ-regularity with respect to x, which amounts to saying that the dual kernel $K^*(x, y) = K(y, x)$ satisfies the conditions above.

Observation: *If $\gamma < \gamma'$ and $K(x, y)$ is γ'-regular with respect to y, then $K(x, y)$ is also γ-regular with respect to y.*

DEFINITION 4   *We shall say that T is a singular integral operator with kernel $K(x, y)$ if for every L^∞ function f with bounded support and for almost every x in the complement of that support*

$$Tf(x) = \int_{\mathbb{R}^n} K(x, y) f(y) \, dy.$$

For example, the operators $T_{\Omega, \epsilon}$ associated to the wavelet expansion by m-splines, or in general, to the expansion in wavelets corresponding to an m-regular multiscale analysis, are singular integral operators with m-regular symmetric kernels.

The starting point for the Calderón–Zygmund theory is the following simple estimate.

THEOREM 1
Let T be a singular integral operator with kernel $K(x, y)$ γ-regular with respect to y. Let f be an L^∞ function supported in a cube Q with center y_0, such that

$$\int_{\mathbb{R}^n} f(x)x^\alpha \, dx = 0 \text{ for every multiindex } \alpha \text{ with order } |\alpha| < \gamma.$$

Then for almost every $x \notin \widetilde{Q}$, the $2\sqrt{n}$-dilated of Q, we have

$$|T(f)(x)| \le C\left(\frac{|Q|}{|x - y_0|^n}\right)^{1+(\gamma/n)} \|f\|_\infty.$$

PROOF   We just need to subtract from $K(x, y)$ its Taylor polynomial of degree $[\gamma] - 1$ in y around y_0.   ∎

Next, by the observation made after Definition 3, we obtain the following.

COROLLARY 1

Let T be a singular integral operator with kernel $K(x, y)$ γ-regular with respect to y. Let $f \in L_N^\infty(Q)$ (that is, f is an essentially bounded function supported in the cube Q, and with moments vanishing up to order N). Then for a.e. $x \notin \widetilde{Q}$

$$|Tf(x)| \leq C \left(\frac{|Q|}{|x - y_0|^n} \right)^{1 + \frac{\min(N+1,\gamma)}{n}} \|f\|_\infty,$$

and, integrating against $w \in A_\infty$:

COROLLARY 2

Let T be a singular integral operator with kernel $K(x, y)$ γ-regular with respect to y. Let $w \in A_\infty$ and let f be a p-atom with respect to w, supported in a cube Q. Suppose $0 < p \leq 1$ is such that $\frac{q_w}{p} < 1 + \frac{\gamma}{n}$. Then

$$\int_{\mathbb{R}^n \setminus \widetilde{Q}} |Tf(x)|^p w(x) \, dx \leq C \text{ independent of } f.$$

DEFINITION 5   *A singular integral operator that is bounded in $L^{p_0}(\mathbb{R}^n)$ for some p_0 such that $1 < p_0 < \infty$ will be called a Calderón–Zygmund operator.*

PROPOSITION 3

Let T be a Calderón–Zygmund operator bounded in $L^{p_0}(\mathbb{R}^n)$ for a given p_0 such that $1 < p_0 < \infty$. Suppose that T has a kernel that is γ-regular in y. Then for every $w \in A_1 \cap A_\infty^{1/p_0'}$, T maps $H^1(w)$ boundedly into $L^1(w)$.

PROOF   Let f be a 1-atom with respect to w. By applying Corollary 2 with $q_w = 1$ and $p = 1$, we have

$$\int_{\mathbb{R}^n \setminus \widetilde{Q}} |Tf(x)| w(x) \, dx \leq C.$$

We need a similar estimate over $\widetilde{Q}$. We shall obtain it by using the boundedness of T on $L^{p_0}(\mathbb{R}^n)$ plus the fact that $w^{p_0'} \in A_\infty$. Note that this is equivalent to saying that w satisfies a reverse Hölder's inequality with exponent p_0' (see [S–W] or [J–N]). Then

$$\int_{\widetilde{Q}} |Tf(x)| w(x) \, dx \leq \left(\int_{\widetilde{Q}} |Tf(x)|^{p_0} \, dx \right)^{1/p_0} \left(\int_{\widetilde{Q}} w(x)^{p_0'} \, dx \right)^{1/p_0'}$$

$$\leq Cw(Q)^{-1} |Q| \left(\frac{1}{|\widetilde{Q}|} \int_{\widetilde{Q}} w(x)^{p_0'} \, dx \right)^{1/p_0'} \leq C. \quad \blacksquare$$

Now we can combine in a single statement all the boundedness properties of the regular Calderón–Zygmund operators. We shall use the following extrapolation theorem of [G1].

THEOREM 2   (Extrapolation)

Let T be an operator, either linear or sublinear, requiring also in this second case that $Tf \geq 0$ for every f. Then

(i) *If T is bounded from wL^∞ to $BMO(w)$ for every $w \in A_1^\alpha$ and some α such that $0 < \alpha \leq 1$, with norm depending only on the A_1 constant of $w^{1/\alpha}$; it follows that T is bounded in $L^q(w)$ for every $q > \frac{1}{\alpha}$ and every $w \in A_{\alpha q}$.*

(ii) *If T is bounded from $H^1(w)$ to $L^1(w)$ for every $w \in A_1$, with norm depending only on the A_1 constant of w, it follows that T is bounded in $L^p(w)$ for every p such that $1 < p < \infty$ and every $w \in A_p$.*

Part (i) is Corollary 1.15(b) of [G1] and part (ii) is Corollary 2.7 of [G1].

THEOREM 3

Let T be a Calderón–Zygmund operator with a kernel γ-regular in y and ϵ-regular in x. Then

(i) *If $1 < p < \infty$ and $w \in A_p$, T is bounded in $L^p(w)$.*

(ii) *If $0 < p \leq 1$ and $w \in A_\infty$ are such that $\frac{q_w}{p} < 1 + \frac{\gamma}{n}$, T is bounded from $H^p(w)$ to $L^p(w)$.*

In all cases the norm of T depends only on the constants for the kernel and the weight.

PROOF   By applying Proposition 3 to the dual operator T^*, we see that T^* is bounded from $H^1(w)$ to $L^1(w)$ for every $w \in A_1 \cap A_\infty^\alpha$ and some $\alpha, 0 < \alpha < 1$. In particular T^* maps $H^1(w)$ boundedly into $L^1(w)$ for every $w \in A_1^\alpha$. This implies that T is bounded from wL^∞ to $BMO(w)$ for every $w \in A_1^\alpha$. Then part (i) of Theorem 2 above can be used to prove that T is bounded in $L^q(w)$ for every $q > \frac{1}{\alpha}$ and every $w \in A_{\alpha q}$.

Let us prove (ii) first. We assume $0 < p \leq 1$ and $w \in A_\infty$ such that $\frac{q_w}{p} < 1 + \frac{\gamma}{n}$. If f is a p-atom with respect to w having support in a cube Q, we know that

$$\int_{\mathbb{R}^n \setminus \widetilde{Q}} |Tf(x)|^p w(x)\, dx \leq C.$$

We need a similar estimate over $\widetilde{Q}$. We shall get it by taking q so large that $\alpha q > q_w$. Then $w \in A_{\alpha q}$ and, consequently, T is bounded in $L^q(w)$. Thus

$$\int_{\widetilde{Q}} |Tf(x)|^p w(x)\, dx = \int_{\widetilde{Q}} |Tf(x)|^p w(x)^{p/q} w(x)^{1-(p/q)}\, dx$$

$$\leq \left(\int_{\widetilde{Q}} |Tf(x)|^q w(x)\, dx \right)^{p/q}$$

$$\cdot \left(\int_{\widetilde{Q}} w(x)^{(1-(p/q))(q/p)'}\, dx \right)^{1/(q/p')}$$

$$\leq C w(Q)^{-1} w(\widetilde{Q})^{p/q} w(\widetilde{Q})^{1-(p/q)} = C.$$

This completes the proof of (ii).

Note that this implies, in particular, that T is bounded from $H^1(w)$ to $L^1(w)$ for every $w \in A_1$. By part (ii) of Theorem 2, this implies (i). ∎

According to Proposition 2, the modified partial sum operators $T_{\Omega,\epsilon}$ associated to the expansion by m-spline wavelets are Calderón–Zygmund operators with m-regular kernels. Consequently, they are uniformly bounded from $H^p(w)$ to $L^p(w)$ provided $\frac{q_w}{p} < 1 + \frac{m}{n}$. We are going to see that they are actually bounded in $H^p(w)$ with the same restriction on the indices. To do that, we must look at the operators

$$\phi^*(T_{\Omega,\epsilon}f)(x) \leq \sum_{(\eta,j,k)\in\Omega} |\langle f, \psi_{\eta,j,k}\rangle| \phi^*(\psi_{\eta,j,k})(x).$$

Now it turns out that these operators satisfy precisely the same basic estimate as the corresponding Calderón–Zygmund operators; namely, Proposition 4.

PROPOSITION 4

If f is an L^∞ function supported in a cube Q, centered at x_0 and such that

$$\int_{\mathbb{R}^n} f(x) x^\alpha\, dx = 0 \quad \forall \alpha \text{ satisfying } 0 \leq |\alpha| \leq m-1,$$

then for every $x \notin \widetilde{Q}$, we have

$$\phi^*(T_{\Omega,\epsilon}f)(x) \leq C \left(\frac{|Q|}{|x-x_0|^n} \right)^{1+(m/n)} \|f\|_\infty.$$

We omit the proof, which is based upon estimates for the maximal function of the wavelets and of the wavelet coefficients of f. Estimates of this type were obtained for dimension 1 in [St] and will be fully carried out in [G-K2]. Note that this is the same estimate as the one in Theorem 1. After it, everything works in the same way as before and we arrive at the main result.

THEOREM 4

For $m \geq 1$, the m-spline wavelets in $\mathbb{R}^n$ are an unconditional basis for

(i) $L^p(w)$, $1 < p < \infty$ with $w \in A_p$.
(ii) $H^p(w)$, $0 < p \leq 1$ with $w \in A_\infty$ such that $\frac{q_w}{p} < 1 + \frac{m}{n}$.

For example, Strömberg's modified Franklin system ($m = 1$, $n = 1$) is a unconditional basis of $H^1(w)$ provided $q_w < 2$, that is, provided $w \in A_2$.

11.4   Necessary conditions for basisness in $L^p(w)$

Here the setting will be $\mathbb{R}$. We want to study necessary conditions on the weight w so that the system $\{\psi_{j,k}\}_{j,k\in\mathbb{Z}}$, for some enumeration of the indices, is a basis in the weighted space $L^p(w)$, $1 \leq p < \infty$, where ψ is an m-regular analyzing wavelet associated to the splines of order m. The basic result needed to analyze the situation is the following.

THEOREM 5

Let Φ be a locally integrable function on $\mathbb{R}$ such that

$$\int_{\mathbb{R}} \frac{|\Phi(x)|}{(1 + |x|)^N}\, dx < \infty \quad \text{for some } N > 0 \tag{7}$$

and

$$\int_{\mathbb{R}} \Phi(t)\psi_{j,k}(t)\, dt = 0 \quad \text{for every } j, k \in \mathbb{Z}. \tag{8}$$

Then on each half-line $\mathbb{R}_- =\,] - \infty, 0]$ and $\mathbb{R}_+ = [0, +\infty[$, the function Φ is almost everywhere equal to some polynomial of degree $\leq m$.

The detailed proof will appear in [G-K1].

Denote by U_m the set of those locally integrable functions which satisfy (7) and (8) for a given m. By Theorem 5 one can easily conclude that U_m is a linear space of dimension, at most $2m + 2$. Let

$$(x)_+ = \begin{cases} x & \text{if } x \geq 0 \\ 0 & \text{if } x < 0. \end{cases}$$

We can easily see that the functions $1, x, \ldots, x^{m-1}, (x)_+^m, (-x)_+^m$ are orthogonal to the functions $\psi_{j,k}$ ($j, k \in \mathbb{Z}$); so that the dimension of U_m is at least $m + 2$. The dimension is actually $m + 2$, but this point is not important here and we postpone its discussion until Section 11.5.

We omit the proof of the following lemma because it is entirely similar to that of the corresponding result in [K] pages 38–40.

LEMMA 1

Let $w \geq 0$ be a locally integrable function satisfying as Φ condition (7) in Theorem 5 and let $1 \leq p < \infty$.

For the system $\{\psi_{j,k}\}_{j,k \in \mathbb{Z}}$ of m-spline wavelets to be complete and/or minimal in the space $L^p(w)$, it is necessary and sufficient that the following conditions (i) and/or (ii) respectively are fulfilled:

(i)   A function of the form $w^{-1}u$, where $u \in U_m$, belongs to $L^{p'}(w)$ ($p^{-1} + p'^{-1} = 1$ and $p' = \infty$ for $p = 1$), if and only if, $u = 0$ a.e..

(ii) For every $j, k \in \mathbb{Z}$ there exists a uniquely determined function $u_{j,k} \in U_m$, such that $\xi_{j,k} = (\psi_{j,k} + u_{j,k})w^{-1} \in L^{p'}(w)$.

DEFINITION 6 We will say that the weight function $w \geq 0$ has a singularity of order p at $+\infty$(resp. $-\infty$) if for every $a > 0$ (resp. $b < 0$)

$$\int_a^{+\infty} w^{-1/(p-1)} dt = +\infty \qquad \left(resp. \int_{-\infty}^b w^{-1/(p-1)} dt = +\infty \right).$$

THEOREM 6

Let $1 \leq p < \infty$ and let $w \geq 0$ be a nonnegative function that satisfies the same condition (7) as Φ in Theorem 5 and also has singularities of order p at both $+\infty$ and $-\infty$. Assume also that ψ is the m-spline wavelet considered above with $m \geq 1$, so that $\psi \in C(\mathbb{R})$. Then if the system $\{\psi_{j,k}\}_{j,k \in \mathbb{Z}}$ for some enumeration of the indices constitutes a Schauder basis in the space $L^p(w)$, we necessarily have $w \in A_p$.

PROOF By Theorem 5 and our assumptions on the function w, it is obvious that

$$fw^{-1} \in L^{p'}(w) \text{ for } f \in U_m \text{ iff } f = 0 \quad a.e.$$

Hence by Lemma 1 we determine that the conjugate system of the system $\{\psi_{j,k}\}_{j,k \in \mathbb{Z}}$ is $\{\psi_{j,k}w^{-1}\}_{j,k \in \mathbb{Z}}$. The coefficients of the expansion of the function $f \in L^p(w)$ with respect to the system $\{\psi_{j,k}\}_{j,k \in \mathbb{Z}}$, are defined by the following equations:

$$a_{j,k}(f) = \int_{\mathbb{R}} f(t)\psi_{j,k}(t)\,dt, \qquad j, k \in \mathbb{Z}. \tag{9}$$

By Banach theorem we conclude that there is a number $C > 0$ independent of f such that:

$$|a_{j,k}(f)|\,\|\psi_{j,k}\|_{L^p(w)} \leq C\|f\|_{L^p(w)} \text{ for every } j, k \in \mathbb{Z}. \tag{10}$$

From (9) and (10) we easily obtain

$$\left(\int_{\mathbb{R}} |\psi_{j,k}|^{p'} w^{1-p'} dt \right)^{1/p'} \|\psi_{j,k}\|_{L^p(w)} \le C \text{ if } p > 1 \qquad (11)$$

and

$$\|\psi_{j,k} w^{-1}\|_{L^\infty} \|\psi_{j,k}\|_{L^1(w)} \le C \text{ if } p = 1 \qquad (12)$$

and this gives almost immediately the condition A_p.   ∎

11.5   On Meyer's remarkable identity for splines

We are going to prove that the dimension of the space U_m of Section 11.4 is actually $m + 2$. This is related to what Meyer calls remarkable identity ([Me], Chapter 2, Section 6). We need to recall some definitions. Every time that we have an m-regular multiscale analysis, say in $\mathbb{R}$, we have a scaling function ϕ, satisfying the same conditions (6) as ψ, and such that the translates $\phi(x - k)$, $k \in \mathbb{Z}$ form an orthonormal basis of V_0. Consequently, the projection on V_0 is given by the kernel

$$E(x, y) = \sum_{k \in \mathbb{Z}} \phi(x - k)\overline{\phi(y - k)}$$

and the projection on V_j by the kernel

$$E_j(x, y) = 2^j E(2^j x, 2^j y).$$

Then Meyer's remarkable identity is the following theorem.

THEOREM 7

$$\int_{\mathbb{R}} E(x, y)y^\alpha \, dy = x^\alpha, \qquad \alpha = 0, 1, \ldots m.$$

It follows from the discussion of Section 11.4 that for m-spline wavelets, this theorem is also true for $(\pm x)_+^m$. We shall see next that this is optimal for this case, and it will follow that the dimension of U_m is $m + 2$. More precisely, we have the following proposition.

PROPOSITION 5
Let $m \in \mathbb{N}$ and let ψ be an m-regular analyzing wavelet in $\mathbb{R}$; then none of the functions $(x)_+^l, (-x)_-^l, l = 0, 1, \ldots m - 1$ is orthogonal to all $\{\psi_{j,k}\}_{j,k \in \mathbb{Z}}$.

PROOF   In our proof we use essentially the following.

Claim:
None of these identities hold

$$\int_{\mathbb{R}} (\pm y)_+^l E(x, y) \, dy \equiv \text{constant}, \qquad l = 0, 1, 2, \ldots$$

Proof of Claim: We first consider the case $l = 0$. Suppose $\int_0^\infty E(x, y) dy$ is a constant. If the constant is 0, using the identity

$$E(x - k, y - k) = E(x, y), \qquad k \in \mathbb{Z} \tag{13}$$

we immediately see a contradiction with

$$\int_{\mathbb{R}} E(x, y) \, dy = 1.$$

Now suppose that the constant is different from 0; then using (13), we can prove that

$$\text{constant} \equiv \int_{-k}^\infty E(x, y) \, dy \longrightarrow 0, \qquad k \to -\infty,$$

a contradiction.

Let us proceed to the case $l \in \mathbb{N}$. If we have, for instance,

$$\text{constant} = \int_0^\infty y^l E(x, y) \, dy$$

$$= \int_0^\infty y^l E(x - k, y - k) \, dy = \int_{-k}^\infty (z + k)^l \, E(\xi, z) \, dz$$

using (6) for ϕ, one can easily deduce that the last integral, at least on some points x_k tends to ∞ as x_k^l, which is a contradiction. This finishes the proof of the claim.

Now suppose that, say, $(x)_+^l$ is orthogonal to all the $\{\psi_{j,k}\}_{j,k \in \mathbb{Z}}$. Then, since

$$\sum_{k \in \mathbb{Z}} \psi(x - k)\overline{\psi(y - k)} = E_1(x, y) - E(x, y),$$

we easily find for every $j \in \mathbb{Z}$

$$\int_0^\infty y^l E(x, y) \, dy \equiv \int_0^\infty y^l E_j(x, y) \, dy = 2^{-lj} \int_0^\infty y^l E(2^j x, y) \, dy.$$

Then we differentiate l times with respect to x, obtaining

$$\int_0^\infty y^l \partial_x^l E(x, y) \, dy \equiv \int_0^\infty y^l \partial_x^l E(2^j x, y) \, dy. \tag{14}$$

The right side depends on j, and from the continuity of $\partial_x^l E(2^j x, y)$, we see that the left side is identically constant. If that constant is 0, then integrating the right side in x, we obtain

$$2^{-j} \int_0^\infty y^l \partial_x^{l-1} E(2^j x, y)\, dy \equiv \text{constant}.$$

Letting $j \to \infty$, we obtain

$$\int_0^\infty y^l \partial_x^{l-1} E(2^j x, y)\, dy = 0.$$

By iteration we arrive at a contradiction with our claim.

Finally, if the constant in (14) is not 0, then we integrate l times in x and arrive at

$$\int_0^\infty y^l E(x, y)\, dy \equiv a_0 x^l + a_1 x^{l-1} + \cdots + a_l.$$

As in the last part of the proof of the claim, we conclude that $a_0 = 1$; hence, using Meyer's identity we obtain

$$-\int_{-\infty}^0 y^l E(x, y)\, dy = a_1 x^{l-1} + \cdots + a_l.$$

The same strategy used in the proof of the claim leads us to a contradiction.   ∎

REMARK   It is clear that an analogue of Proposition 5 holds in higher dimensions.   ∎

References

[B]   S. Banach, *Théorie des opérations linéaires,* 1932, English translation: Elsevier 1987.

[Bo]   S. V. Bochkariev, *Existence of bases in the space of analytic functions and some properties of the Franklin system,* Mat. Sbornik 98 (1974), 3–18.

[Ca]   L. Carleson, *An explicit unconditional basis in H^1,* Bull. des Sciences Math. 104 (1980), 405–416.

[C-C]   A. Chang and Z. Ciesielski, *Spline characterizations of H^1,* Studia Math 75 (1983), 183–192.

[C1]   Z. Ciesielski, *Properties of the orthonormal Franklin system,* Studia Math. 23 (1963), 141–157.

[C2]   _____, *Properties of the orthonormal Franklin system II,* Studia Math. 27 (1966), 289–323.

[C-F]   R. Coifman and C. Fefferman, *Weighted norm inequalities for maximal functions and singular integrals,* Studia Math. 51 (1974), 241–250.

[D]   G. David, *Wavelets and singular integrals on curves and surfaces,* Lecture Notes in Mathematics 1465, Springer–Verlag, 1991.

[G]   J. García-Cuerva, *Weighted H^p spaces,* Dissertationes Mathematicae 162 (1979), 1–63.

[G1]   _____, *Extrapolation of weighted norm inequalities from endpoint spaces to Banach lattices,* Journal of the London Mathematical Society (2) 46 (1992), 280–294.

[G-K1]   J. García-Cuerva and K. S. Kazarian, *Spline wavelet bases of weighted L^p spaces,* $1 \le p < \infty$, Proc. Amer. Math. Soc., to appear,

[G-K2]   _____, *Calderón-Zygmund operators and unconditional bases of weighted Hardy spaces,* to appear.

[G-R]   J. García-Cuerva and José-Luis Rubio de Francia, *Weighted norm inequalities and related topics,* North Holland Math Studies 114 1985.

[J-N]   R. Johnson and C. J. Neugebauer, *Homeomorphisms preserving A_p,* Revista Matemática Iberoamericana 3 (1987), 249–273.

[K]   K. S. Kazarian, *On the multiplicative completion of some orthonormal systems to bases in L^p,* $1 \le p < \infty$, Analysis Math. 4 (1978), 37–52, (Russian).

[K2]   _____ *On bases and unconditional bases in the spaces $L^p(d\mu)$,* $1 \le p < \infty$, Studia Math. 71 (1982), 227–249.

[Ma]   S. G. Mallat, *Multiresolution approximation and wavelet orthonormal bases of $L^2(\mathbb{R})$,* Trans. Amer. Math. Soc. 315 (1989), 69–87.

[Mau]   B. Maurey, *Isomorphismes entre espaces H^1,* Acta Math. 145 (1980), 79–120.

[Me]   Y. Meyer, *Ondelettes et Opérateurs,* vol. I and II, Hermann, Paris, 1990.

[Mu]   B. Muckenhoupt, *Weighted norm inequalities for the Hardy maximal function,* Trans. Amer. Math. Soc. 165 (1972), 207–226.

[S-S]   P. Sjölin and J. O. Strömberg, *Spline systems as bases in Hardy spaces,* Israel Jour. of Math. 45 (1983), 147–153.

[S-S1]   _____, *Basis properties of Hardy spaces,* Ark. Mat. 21 (1983), 111–125.

[St]   J. O. Strömberg, *A modified Franklin system and higher order spline systems on $\mathbb{R}^n$ as unconditional bases for Hardy spaces,* Proc. Conf. in Honor of Antoni Zygmund, W. Beckner, A. P. Calderón, R. Fefferman and P. W. Jones (eds), Wadsworth, NY, 1981, 475–493, vol 2.

[S-T]   J. O. Strömberg and A. Torchinsky, *Weighted Hardy spaces,* Lecture Notes in Mathematics 1381, Springer–Verlag, 1989.

[S-W]   J. O. Strömberg and R. Wheeden, *Fractional integrals on weighted H^p and L^p spaces,* Trans. Amer. Math. Soc. 287 (1985), 293–321.

[W]   P. Wojtaszczyk, *The Franklin system is an unconditional basis in H^1,* Ark. Mat. 20 (1982), 293–300.

12

A Note on Hardy's Inequality in Orlicz Spaces

Hans P. Heinig[1]

12.1 Introduction

Let T be the integral operator defined by

$$(Tf)(x) = \int_0^x f(t)\, dt, \qquad x > 0.$$

If u and v are nonnegative locally integrable (weight) functions on $(0, \infty)$, then the inequality

$$\left\{ \int_0^\infty u(x) |(Tf)(x)|^q\, dx \right\}^{1/q} \le C \left\{ \int_0^\infty v(x) |f(x)|^p\, dx \right\}^{1/p}, \qquad (1)$$

$0 < p, q < \infty$, $1 < p < \infty$, $C > 0$, a constant, together with its discrete analogue is Hardy's inequality in weighted spaces. Conditions on the weight functions u, v can be given that are equivalent to the validity of (1) (cf., [O–K], [S] and the literature cited there), whereas the weight characterization for the discrete analogue may be found in [B] and [S]. In the case $1 < p \le q < \infty$, the inequality (1) was generalized (in [H–M]) in the sense that conditions on weight functions u_0, u_1, v_0, and v_1 were given, which are equivalent to the validity of the modular inequality

$$Q^{-1} \left\{ \int_0^\infty Q \left[u_1(x) \int_0^x f(t)\, dt \right] u_0(x)\, dx \right\}$$

$$\le C P^{-1} \left\{ \int_0^\infty P[v_1(x) f(x)] v_0(x)\, dx \right\} \qquad (2)$$

[1]This Research was supported by the Natural Sciences and Engineering Research Council of Canada under grant A-4837.

and its dual. Here P and Q are certain Young functions. Of course if $P(x) = |x|^p$, $Q(x) = |x|^q$, $1 < p \leq q < \infty$, the result reduces to the weighted $L^p - L^q$ case, and for $u_1 = v_1 = 1$ the result goes back to Lai [L].

In this chapter we shall prove the discrete analogue of (2) and provide conditions on the weight sequences that are equivalent to the corresponding modular inequality. Although such a result is not unexpected, its proof requires an extension of (3), (Theorem 1) where the term $u_0(x)\,dx$ is replaced by $d\mu(x)$, with μ a positive regular Borel measure, which is of independent interest.

12.2   Hardy inequalities

We first require some notation and preliminaries.

A function P is called a Young function[2] if it has the form $P(x) = \int_0^{|x|} \phi(s)\,ds$, where ϕ is nondecreasing, right continuous on $(0, \infty)$, $\phi(0) = 0$, and $\phi(s) > 0$ if $s > 0$. If ϕ^{-1} denotes the right continuous inverse of ϕ, then the complementary Young function of P is $\widetilde{P}(x) = \int_0^{|x|} \phi^{-1}(s)\,ds$. The inequality

$$t \leq P^{-1}(t)\widetilde{P}^{-1}(t) \leq 2t \tag{3}$$

always holds ([K–R, 2.10]), and from it one easily obtains

$$P\left[\frac{\widetilde{P}(t)}{t}\right] \leq \widetilde{P}(t), \qquad t > 0. \tag{4}$$

Now if P satisfies the Δ_2-condition ($P \in \Delta_2$), that is, there is a constant D dependent on P, such that $P(2t) \leq DP(t)$ holds for all $t > 0$, then it follows from (3) that

$$P\left[\frac{t}{\widetilde{P}^{-1}(t)}\right] \leq t \leq DP\left[\frac{t}{\widetilde{P}^{-1}(t)}\right], \qquad t > 0. \tag{5}$$

If P and Q are Young functions, then we write $P \prec Q$, if there is a constant $C > 0$, such that

$$\sum_i Q \circ P^{-1}(a_i) \leq CQ \circ P^{-1}\left(\sum_i a_i\right) \tag{6}$$

holds for all nonnegative sequences $\{a_i\}$. If $Q \circ P^{-1}$ is convex, then $P \prec Q$. For other conditions that imply $P \prec Q$ see [H–M, Lemma 1.1].

---

[2] Strictly speaking this is an N-function [K–R] but we use the term Young function to simplify the statements of our theorems.

Recall that a μ-measurable function f defined on X belongs to the Orlicz space $L_P(\mu)$, where P is a Young function, if its (Luxemburg) norm

$$\|f\|_{P(\mu)} = \inf\left\{\lambda > 0 : \int_X P\left[\frac{|f(x)|}{\lambda}\right] d\mu(x) \le 1\right\}$$

is finite. For properties of these spaces and Young functions we refer to [K–R].

In the sequel A, B, C, and D denote positive constants, which may be different at different places with D always being the constant arising from the Δ_2-condition. χ_E is the characteristic function of a set E, and we write $\chi_r = \chi_{(0,r)}$ and $\chi^r = \chi_{(r,\infty)}$. Throughout, $f \ge 0$ is assumed and inequalities (such as (2)) are interpreted to mean that if the right side is finite, the left side is also, and the inequality holds.

After these preliminaries, the first result is the following.

THEOREM 1
Suppose P, Q are Young functions such that $P \prec Q$ and $Q \in \Delta_2$. Let u, v_0, and v_1 be nonnegative locally integrable weight functions and μ a positive regular Borel measure. Then there is a constant $A > 0$, such that

$$Q^{-1}\left\{\int_0^\infty Q\left[u(x)\int_0^x f(t)\,dt\right]d\mu(x)\right\}$$
$$\le AP^{-1}\left\{\int_0^\infty P[v_1(x)f(x)]v_0(x)\,dx\right\} \tag{7}$$

holds, if and only if, there is a constant $B > 0$ such that for every $\varepsilon > 0$ and $r > 0$

$$Q^{-1}\left\{\int_r^\infty Q\left[u(x)\left\|\frac{\chi_r}{v_0 v_1 \varepsilon}\right\|_{\widetilde{P}(\varepsilon v_0)}\right]d\mu(x)\right\} \le BP^{-1}\left\{\frac{1}{\varepsilon}\right\} \tag{8}$$

is satisfied.

For the dual operator the inequality

$$Q^{-1}\left\{\int_0^\infty Q\left[u(x)\int_x^\infty f(t)\,dt\right]d\mu\right\}$$
$$\le AP^{-1}\left\{\int_0^\infty P[v_1(x)f(x)]v_0(x)\,dx\right\} \tag{9}$$

holds if and only if

$$Q^{-1}\left\{\int_0^r Q\left[u(x)\left\|\frac{\chi^r}{v_0 v_1 \varepsilon}\right\|_{\widetilde{P}(\varepsilon v_0)}\right]d\mu(x)\right\} \le BP^{-1}\left\{\frac{1}{\varepsilon}\right\}. \tag{10}$$

The proof follows along the lines of corresponding results of [H–M] and [L]. The sufficiency part is given here for completeness, and the necessity is essentially that given in the proof of Theorem 3.

PROOF   Let $\{x_k\}_{k \in \mathbb{Z}}$ be a positive increasing sequence such that

$$\int_0^{x_k} f(t)\,dt = 2^k.$$

Then, since $Q \in \Delta_2$

$$Q^{-1}\left\{\int_0^\infty Q\left[u(x)\int_0^x f(t)\,dt\right]d\mu(x)\right\}$$

$$\leq Q^{-1}\left\{\sum_{k \in \mathbb{Z}}\int_{x_k}^{x_{k+1}} Q\left[u(x)\int_0^{x_{k+1}} f(t)\,dt\right]d\mu(x)\right\}$$

$$= Q^{-1}\left\{\sum_{k \in \mathbb{Z}}\int_{x_k}^{x_{k+1}} Q\left[u(x)2^2 \cdot 2^{k-1}\right]d\mu(x)\right\}$$

$$\leq Q^{-1}\left\{\sum_{k \in \mathbb{Z}} D\int_{x_k}^{x_{k+1}} Q\left[u(x)\int_{x_{k-1}}^{x_k} f(t)\,dt\right]d\mu(x)\right\}. \qquad (11)$$

Let $f_k(x) = \chi_{(x_{k-1},x_k)}(x)f(x)$ and choose $\varepsilon_k > 0$, so that

$$\int_0^\infty P[v_1(x)f_k(x)]\varepsilon_k v_0(x)\,dx = 1.$$

Then $\|f_k v_1\|_{P(\varepsilon v_0)} \leq 1$, and applying Hölder's inequality, it follows that

$$\int_{x_{k-1}}^{x_k} f \leq 2\|f_k v_1\|_{P(\varepsilon_k v_0)}\left\|\frac{\chi_{x_r}}{v_0 v_1 \varepsilon_k}\right\|_{\widetilde{P}(\varepsilon_k v_0)}$$

$$\leq 2\left\|\frac{\chi_{x_r}}{v_0 v_1 \varepsilon_k}\right\|_{\widetilde{P}(\varepsilon_k v_0)}.$$

Since $Q \in \Delta_2$, (8) and the fact that $P \prec Q$, the right side of (11) is not larger than

$$Q^{-1}\left\{D\sum_{k \in \mathbb{Z}}\int_{x_k}^\infty Q\left[u(x)\left\|\frac{\chi_{x_r}}{v_0 v_1 \varepsilon_k}\right\|_{\widetilde{P}(\varepsilon_k v_0)}\right]d\mu(x)\right\}$$

$$\leq Q^{-1}\left\{D\sum_{k \in \mathbb{Z}}Q\left[BP^{-1}\left(\frac{1}{\varepsilon_k}\right)\right]\right\}$$

$$\leq Q^{-1}\left\{D\sum_{k \in \mathbb{Z}}Q \circ P^{-1}\left(\frac{1}{\varepsilon_k}\right)\right\}$$

$$\leq Q^{-1}\left\{DQ \circ P^{-1}\left(\sum_{k \in \mathbb{Z}}\frac{1}{\varepsilon_k}\right)\right\}$$

$$\leq AP^{-1}\left\{\sum_{k \in \mathbb{Z}}\int_0^\infty P[v_1(x)f_k(x)]v_0(x)\,dx\right\}$$

$$= A P^{-1} \left\{ \sum_{k \in \mathbb{Z}} \int_{x_{k-1}}^{x_k} P[v_1(x) f(x)] v_0(x) \, dx \right\}$$

$$= A P^{-1} \left\{ \int_0^\infty P[v_1(x) f(x)] v_0(x) \, dx \right\}.$$

Here concavity of Q^{-1} was used to obtain the estimate.

Note that if $u(x) = u_1(x)$ and $d\mu(x) = u_0(x) \, dx$, u_0 a weight function, one obtains [H–M, Theorem 2.1, and Corollary 2.2]. ∎

We now apply this result to obtain the discrete Hardy inequality.

THEOREM 2

Let P and Q be Young functions, such that $P \prec Q$ and $Q \in \Delta_2$. If $\{u_n^0\}_{n \in \mathbb{N}}$, $\{u_n^1\}_{n \in \mathbb{N}}$, $\{v_n^0\}_{n \in \mathbb{N}}$ and $\{v_n^1\}_{n \in \mathbb{N}}$ are sequences of positive numbers, then there is a constant $A > 0$, such that

$$Q^{-1} \left\{ \sum_{n=1}^\infty Q \left[u_n^1 \sum_{k=1}^n a_k \right] u_n^0 \right\} \le A P^{-1} \left\{ \sum_{n=1}^\infty P \left[a_n v_n^1 \right] v_n^0 \right\} \tag{12}$$

holds for all nonnegative sequences $\{a_n\}$, if and only if there is a constant $B > 0$, such that

$$Q^{-1} \left\{ \sum_{n=m}^\infty Q \left[u_n^1 \left\| \frac{\chi_m}{v^0 v^1 \varepsilon} \right\|_{\widetilde{P}(\varepsilon v^0)} \right] u_n^0 \right\} \le B P^{-1} \left\{ \frac{1}{\varepsilon} \right\} \tag{13}$$

is satisfied for every $m \in \mathbb{N}$ and $\varepsilon > 0$.

For the dual operator, the inequality

$$Q^{-1} \left\{ \sum_{n=1}^\infty Q \left[u_n^1 \sum_{k=n}^\infty a_k \right] u_n^0 \right\} \le A P^{-1} \left\{ \sum_{n=1}^\infty P \left[a_n v_n^1 \right] v_n^0 \right\} \tag{14}$$

holds, if and only if,

$$Q^{-1} \left\{ \sum_{n=1}^m Q \left[u_n^1 \left\| \frac{\chi^m}{v^0 v^1 \varepsilon} \right\|_{\widetilde{P}(\varepsilon v^0)} \right] u_n^0 \right\} \le B P^{-1} \left\{ \frac{1}{\varepsilon} \right\}. \tag{15}$$

Note that in this discrete case the Luxemburg norm takes the form

$$\left\| \frac{\chi_m}{v^0 v^1 \varepsilon} \right\|_{\widetilde{P}(\varepsilon v_0)} = \inf \left\{ \lambda > 0 : \sum_{n=1}^m \widetilde{P} \left(\frac{1}{\lambda v_n^0 v_n^1 \varepsilon} \right) \varepsilon v_0 \le 1 \right\},$$

and similarly for $\left\| \frac{\chi^m}{v^0 v^1 \varepsilon} \right\|_{\widetilde{P}(\varepsilon v_0)}$.

PROOF Let $\mu(x) = \sum_{n=1}^\infty u^0(x) \delta_n(x)$, where δ_n is the Dirac measure concentrated at $n \in \mathbb{N}$, and $u^0(x)$, a weight function such that $u^0(n) = u_n^0$. Now define

f, v_0, v_1 by $f(x) = a_n$, $v_1(x) = v_n^1$ and $v_0(x) = v_n^0$ for $x \in (n-1, n]$, $n = 1, 2, \ldots$. Let $r > 0$. Then there is an $m \in \mathbb{N}$, such that $m - 1 < r \leq m$, and hence condition (8) is implied by

$$Q^{-1} \left\{ \sum_{n=m}^{\infty} Q \left[u(n) \left\| \frac{\chi_r}{v_0 v_1 \varepsilon} \right\|_{\widetilde{P}(\varepsilon v_0)} \right] u_n^0 \right\} \leq B P^{-1} \left\{ \frac{1}{\varepsilon} \right\}.$$

But since

$$\left\| \frac{\chi_r}{v_0 v_1 \varepsilon} \right\|_{\widetilde{P}(\varepsilon v_0)} \leq \left\| \frac{\chi_m}{v_0 v_1 \varepsilon} \right\|_{\widetilde{P}(\varepsilon v_0)}$$

$$= \inf \left\{ \lambda > 0 : \sum_{k=1}^{m} \widetilde{P} \left(\frac{1}{\lambda v_k^0 v_k^1 \varepsilon} \right) \varepsilon v_k^0 \leq 1 \right\},$$

condition (13) (with $u_n^1 = u(n)$) implies (8) and therefore (12) follows from Theorem 1. The proof that (15) implies (14) is similar and is hence omitted.

Out of necessity we proceed as in [H–M, Theorem 2.1].

Fix $\varepsilon > 0$ and $m \in \mathbb{N}$ and define $a_n = \widetilde{P} \left[\frac{1}{\varepsilon v_n^1 v_n^0} \right] \varepsilon v_n^0$, $n = 1, 2, \ldots, m$, and zero if $n > m$. Then by (12) and (4)

$$Q^{-1} \left\{ \sum_{n=m}^{\infty} Q \left[u_n^1 \sum_{k=1}^{m} \widetilde{P} \left(\frac{1}{\varepsilon v_k^1 v_k^0} \right) \varepsilon v_k^0 \right] u_n^0 \right\}$$

$$\leq A P^{-1} \left\{ \sum_{n=1}^{m} P \left[v_n^1 v_n^0 \varepsilon \widetilde{P} \left(\frac{1}{\varepsilon v_n^1 v_n^0} \right) \right] v_n^0 \right\}$$

$$\leq A P^{-1} \left\{ \sum_{n=1}^{m} \widetilde{P} \left(\frac{1}{\varepsilon v_n^1 v_n^0} \right) v_n^0 \right\}.$$

Now choose $\eta > 0$, so that

$$\sum_{n=1}^{m} \widetilde{P} \left(\frac{1}{\varepsilon v_n^1 v_n^0} \right) v_n^0 \varepsilon \eta = 1,$$

then

$$\left\| \frac{\chi_m}{\varepsilon \eta v^0 v^1} \right\|_{\widetilde{P}(\varepsilon v^0)} \leq \frac{1}{\eta},$$

and hence

$$\sum_{n=m}^{\infty} Q \left[u_n^1 \sum_{k=1}^{m} \widetilde{P} \left(\frac{1}{\varepsilon v_k^1 v_k^0} \right) \varepsilon v_k^0 \right] u_n^0 \leq Q \left[A P^{-1} \left(\frac{1}{\varepsilon \eta} \right) \right].$$

But this implies

$$Q^{-1} \left\{ \sum_{n=m}^{\infty} Q \left[u_n^1 \left\| \frac{\chi_m}{\varepsilon \eta v^0 v^1} \right\|_{\widetilde{P}(\varepsilon v^0)} \right] u_n^0 \right\} \leq A P^{-1} \left\{ \frac{1}{\varepsilon \eta} \right\}.$$

Let $\frac{1}{\eta} = g(\varepsilon)$, then g is continuous, taking on values from 0 to ∞; therefore $\eta = \frac{1}{g(\varepsilon)}$. Hence given $\delta > 0$, there is an $\varepsilon > 0$ such that $\eta\varepsilon = \frac{\varepsilon}{g(\varepsilon)} = \delta$. Substituting $\eta\varepsilon = \delta$ into the last inequality one obtains (13) with $\varepsilon = \delta$.

The necessity for the dual is proven in the same way. ∎

COROLLARY 1

Let $P \in \Delta_2$ be a Young function, then

$$P^{-1}\left\{\sum_{n=1}^{\infty} P\left[\frac{1}{n}\sum_{k=1}^{n} a_k\right]\right\} \leq A P^{-1}\left\{\sum_{n=1}^{\infty} P[a_n]\right\} \tag{16}$$

holds for all nonnegative sequences $\{a_n\}$, if and only if, for every $\varepsilon > 0$ and $m \in \mathbb{N}$, there is a $B > 0$, such that

$$P^{-1}\left\{\sum_{n=m}^{\infty} P\left[\frac{1}{n\varepsilon \widetilde{P}^{-1}(\frac{1}{m\varepsilon})}\right]\right\} \leq B P^{-1}\left(\frac{1}{\varepsilon}\right). \tag{17}$$

PROOF Let $u_n^1 = \frac{1}{n}$, $u_n^0 = v_n^0 = v_n^1 = 1$ for $n = 1, 2, \ldots$, and $Q = P$ in Theorem 2, then (16) is satisfied, if and only if,

$$P^{-1}\left\{\sum_{n=m}^{\infty} P\left[\frac{1}{n}\left\|\frac{X_m}{\varepsilon}\right\|_{\widetilde{P}(\varepsilon)}\right]\right\} \leq B P^{-1}\left(\frac{1}{\varepsilon}\right)$$

holds for all $m \in \mathbb{N}$ and $\varepsilon > 0$. But since

$$\left\|\frac{X_m}{\varepsilon}\right\|_{\widetilde{P}(\varepsilon)} = \inf\left\{\lambda > 0 : \sum_{k=1}^{m} \widetilde{P}\left(\frac{1}{\lambda\varepsilon}\right)\varepsilon \leq 1\right\}$$

$$= \inf\left\{\lambda > 0 : m\varepsilon\widetilde{P}\left(\frac{1}{\lambda\varepsilon}\right) \leq 1\right\}$$

$$= \inf\left\{\lambda > 0 : \frac{1}{\lambda\varepsilon} \leq \widetilde{P}^{-1}\left(\frac{1}{m\varepsilon}\right)\right\} = \frac{1}{\varepsilon\widetilde{P}^{-1}\left(\frac{1}{m\varepsilon}\right)},$$

this is precisely (17). ∎

REMARK 1 (i) If $P(x) = \frac{|x|^p}{p}$, then $\widetilde{P}(x) = \frac{|x|^{p'}}{p'}$, $p' = \frac{p}{(p-1)}$. An easy calculation shows that (17) holds in this case, and one obtains the classical Hardy inequality.

(ii) If $P(x) = x^p \ln^{\gamma}(x+1)$, $p > 1$, $\gamma > 0$, then $P \in \Delta_2$ and P^{-1} exists. Let $a = \widetilde{P}^{-1}\left(\frac{1}{m\varepsilon}\right)$, then (17) is of the form

$$\sum_{n=m}^{\infty} P\left[\frac{1}{n\varepsilon a}\right] \leq \frac{C}{\varepsilon}.$$

But

$$\sum_{n=m}^{\infty} \left(\frac{1}{n\varepsilon a}\right)^P \ln^\gamma \left(\frac{1}{n\varepsilon a}+1\right) \leq \left(\frac{1}{\varepsilon a}\right)^P \ln^\gamma \left(\frac{1}{m\varepsilon a}+1\right) \sum_{n=m}^{\infty} n^{-P}$$

$$\leq C m \left(\frac{1}{m\varepsilon a}\right)^P \ln^\gamma \left(\frac{1}{m\varepsilon a}+1\right)$$

$$= C m P \left(\frac{1}{m\varepsilon \widetilde{P}^{-1}\left(\frac{1}{m\varepsilon}\right)}\right) \leq \frac{C}{\varepsilon},$$

where the last inequality follows from (5). Hence (6) holds also for this Young function P.

(iii) If $P(x) = x \ln(x+1)$, $x > 0$, then the inequality (16) fails to be satisfied. For otherwise (17) would hold for all $m \in \mathbb{N}$ and $\varepsilon > 0$. But (17) can be written in this case as

$$\sum_{n=m}^{\infty} \frac{1}{na} \ln\left(\frac{1}{na}+1\right) \leq \frac{B}{\varepsilon} \tag{18}$$

where $a = \varepsilon \widetilde{P}^{-1}\left(\frac{1}{m\varepsilon}\right)$. Now the left side is larger than

$$\int_m^\infty \frac{1}{xa} \ln\left(\frac{1}{xa}+1\right) dx = \frac{1}{a} \int_0^{1/ma} \frac{\ln(t+1)}{t} dt$$

$$\geq \frac{1}{a} \left[\ln\left(\frac{1}{ma}+1\right)\right]^2$$

$$= m^2 a \left[\frac{1}{ma} \ln\left(\frac{1}{ma}+1\right)\right]^2$$

$$= m^2 \varepsilon \widetilde{P}^{-1}\left(\frac{1}{m\varepsilon}\right) \left[P\left(\frac{1}{m\varepsilon \widetilde{P}\left(\frac{1}{m\varepsilon}\right)}\right)\right]^2$$

$$\geq m^2 \varepsilon \widetilde{P}^{-1}\left(\frac{1}{m\varepsilon}\right) \left[\frac{1}{D} \cdot \frac{1}{m\varepsilon}\right]^2$$

$$= \left(\frac{1}{D}\right) \frac{1}{\varepsilon} \widetilde{P}^{-1}\left(\frac{1}{m\varepsilon}\right),$$

where the last inequality follows from (5). But then (18) implies $\widetilde{P}^{-1}\left(\frac{1}{m\varepsilon}\right) \leq C$, which is impossible. ∎

Since modular inequalities imply Orlicz norm inequalities, one obtains from Theorem 2 the following.

COROLLARY 2

Suppose P, Q, $\{u_n^0\}$, $\{u_n^1\}$, $\{v_n^0\}$ and $\{v_n^1\}$ are as in Theorem 2. If T is defined by

$$(Ta)(n) = \sum_{k=1}^{n} a_k,$$

then (13) implies $\|u^1 Ta\|_{Q(u^0)} \le C\|v^1 a\|_{P(v^0)}$.

Similarly if $(T^*a)(n) = \sum_{k=n}^{\infty} a_k$, then (15) implies that $\|u^1 T^*a\|_{Q(u^0)} \le C\|v^1 a\|_{P(v^0)}$.

PROOF Since (13) implies (12), then (12) shows that

$$\|u^1 Ta\|_{Q(u^0)} = \inf \left\{ \lambda > 0 : \sum_{n=1}^{\infty} Q\left[\frac{u_n^1 (Ta)(n)}{\lambda} \right] u_n^0 \le 1 \right\}$$

$$\le \inf \left\{ \lambda > 0 : Q\left[AP^{-1} \left(\sum_{n=1}^{\infty} P\left[\frac{a_n v_n^1}{\lambda} \right] v_n^0 \right) \right] \le 1 \right\}$$

$$\le \inf \left\{ \lambda > 0 : \sum_{n=1}^{\infty} P\left[\frac{a_n v_n^1}{\lambda_n} \right] v_n^0 \le P\left[\frac{Q^{-1}(1)}{A} \right] \right\}.$$

Let $\eta = P\left[\frac{Q^{-1}(1)}{A} \right]$. If $\eta \ge 1$, the last infimum is not larger than $\|v^1 a\|_{P(v^0)}$, and the result follows in this case. If $\eta < 1$, then the convexity of P shows that $P\left[\frac{a_n v_n^1}{\lambda} \right] \le \eta P\left[\frac{a_n v_n^1}{\lambda \eta} \right]$ so that in this case the above infimum is not larger than

$$\inf \left\{ \frac{\eta \lambda}{\eta} > 0 : \sum_{n=1}^{\infty} P\left[\frac{v_n^1 a_n}{\eta \lambda} \right] v_n^0 \le 1 \right\} \le \frac{1}{\eta} \|v^1 a\|_{P(v^0)},$$

and the result follows with $C \le \max \left(1, \frac{1}{P\left(\frac{Q^{-1}(1)}{A} \right)} \right)$. ∎

References

[B]   G. Bennett, *Some elementary inequalities III,* Quart. J. Math. Oxford, Ser. (2) 42 (1991), 149–174.

[H–M] H. P. Heinig and L. Maligranda, *Interpolation with weights in Orlicz spaces,* Boll. della Unione Math. Ital. (7) 8-B (1994), 37–55.

[K–R] M. A. Krasnosel'skii and Ya. B. Rutickii, *Convex functions and Orlicz spaces,* Noordhoff, Groningen, 1961.

[L]   Q. Lai, *Two weight mixed Φ-inequalities for the Hardy and Hardy-Littlewood operator,* J. Lond. Math. Soc. (2) 42 (1992), 301–318.

[O–K] B. Opic and A. Kufner, *Hardy-type inequalities,* Pitman Research Notes in Math. Series, vol. 219, Longman Scientific and Technical, 1990.

[S]   G. Sinnamon, *Spaces defined by their level function and their dual.* Studia Math. (to appear).

13

A Characterization of Commutators of Parabolic Singular Integrals

Steve Hofmann[1]

ABSTRACT   *We characterize those functions $A(x, t)$ for which one has L^2 boundedness of the "parabolic" Calderón Commutator*

$$T[A]f(x, t) \equiv \text{P.V.} \int_{\mathbb{R}} \int_{\mathbb{R}^{n-1}} K(x - y, t - s)[A(x, t) - A(y, s)] f(y, s) \, dy \, ds,$$

where $K(\lambda x, \lambda^2 t) \equiv \lambda^{-n-2} K(x, t)$, and where K is smooth away from the origin, and satisfies an appropriate cancellation condition on the sphere.

13.1   Introduction

In this chapter we characterize those bilinear singular integrals that are the parabolic analogues of the first "Calderón commutator," and are bounded on $L^2(\mathbb{R}^n)$. In contrast to the "elliptic" case (see Calderón [Ca], and also, for example, David and Journé [DJ] and the survey article of Coifman and Meyer [CM]), the theory of multilinear singular integrals having "parabolic" homogeneity is not nearly so well understood. It is not the mixed homogeneity, per se, that causes problems, as this phenomenon has been well understood for some time (see e.g., Jones [J], Fabes and Riviere [FR1], [FR2], Riviere [R], and Coifman and Weiss [CW]). Rather, the fundamental difficulty lies in the fact that, unlike their elliptic counterparts, the multilinear operators that arise in this setting are rough in a sense which will be explained below. The first Calderón commutator is the bilinear operator defined by

$$T[A]f(x) \equiv \text{P.V.} \int_{\mathbb{R}} \frac{A(x) - A(y)}{(x - y)^2} f(y) \, dy. \tag{1}$$

The famous and elegant result of Calderon [Ca] is that $T[A]$ defines a bounded

---

[1]The author was partially supported by the NSF, and by the Wright State University Research Council.

This paper is dedicated to S. Pichorides and A. Zygmund.

operator on $L^2(R)$, if and only if, A satisfies a Lipschitz condition, and furthermore

$$\|T[A]\|_{op} \approx \|A'\|_\infty.$$

The well known n-dimensional version of (1) is also described in [Ca]. The operators that we shall consider in this chapter are those of the form

$$T[A]f(z) \equiv \text{P.V.} \int_{\mathbb{R}^n} K(z-w)[A(z) - A(w)]f(w)\,dw, \qquad (2)$$

where $z \equiv (x,t)$, $w \equiv (y,s) \in \mathbb{R}^{n-1} \times \mathbb{R}$, and where K satisfies the homogeneity condition

$$K(\lambda x, \lambda^2 t) \equiv \lambda^{-n-2} K(x,t), \qquad (3)$$

as well as certain smoothness and cancellation conditions to be described below. For the sake of simplicity, we shall restrict our attention to operators invariant with respect to the parabolic dilations

$$z \to \delta_\lambda z \equiv \lambda^\alpha z \equiv (\lambda x, \lambda^2 t) \qquad (4)$$

(we use the multiindex notation $\alpha \equiv (1, 1, \ldots, 1, 2)$), which case is already typical, but the same methods also apply to the dilations $z \to (\lambda^{\alpha_1} z_1, \lambda^{\alpha_2} z_2, \ldots, \lambda^{\alpha_n} z_n)$, $1 \le \alpha_1 \le \alpha_2 \le \cdots \le \alpha_n$. We leave the details of this more general case to the interested reader.

Fabes and Riviere [FR1] obtained the first L^p boundedness result for operators of the type (2), but only by imposing more smoothness on A than one would like, especially in contrast to [Ca].

In the special case that $K(x,t)$ in (2) equals $\frac{1}{t}$ times the Gaussian, Lewis and Murray ([LM1] and [LM2]) have recently obtained results under conditions on A that are much closer to being optimal. They consider functions $A(x,t)$ having (as we shall see) the "right"amount of smoothness in x (namely Lipschitz), and with the following nearly sharp condition, motivated by the one-dimensional results of Murray ([Mu1] and [Mu2]), in the t variable: namely that the one-dimensional half-order differentiation operator, defined by $(|D|^{1/2} f)^\wedge(\tau) \equiv |\tau|^{1/2} \hat{f}(\tau)$, acting on $A(x,t)$ in the time variable, yields a function belonging to one-dimensional BMO, uniformly in x. Lewis and Murray have extended this result to the higher order commutators as well, permitting them to obtain norm estimates for the caloric double layer potential on domains with a time dependent boundary, given by the graph of such a function $A(x,t)$.

Furthermore, it has been realized, in retrospect, by Lewis, Murray, and myself, that the estimates for the commutators, obtained via perturbation in [LM2], were obtained for A, satisfying a condition equivalent to that of the present chapter, in slightly disguised form (our condition is described in Theorems 1 and 2 below, and in the adjacent remarks). Our purpose here is to establish the necessity of our condition, and also to provide a direct proof (i.e., without perturbation) of the sufficiency.

The fundamental difficulty presented by the parabolic commutators is this: viewed as multilinear singular integrals, they are rough operators. It is true that, as an operator acting on f, the kernel in (2) is a "standard" Calderón–Zygmund kernel, but for fixed $f \in L^\infty$, viewed as acting on $a \equiv |D_p|A$, (here $|D_p|$ is an appropriate "parabolic" derivative, see (9) below), the operator in (2) has a "rough" kernel. The Calderon commutator (1) and its higher order analogues present no such problem. It turns out that for the first-order commutators discussed here, in the special case $f \equiv 1$ we obtain a "nice" kernel acting on a, and after [DJ], the case $f \equiv 1$ is all that matters. The second-order commutator, while a bit worse, can still be handled by the arguments given below, with only a bit more technical difficulty, although we shall not bother to do this here. The full magnitude of the "roughness" problem presented by these parabolic operators does not appear until one considers the third-order commutator, and in that case, some further ideas are required even when $f \equiv 1$. We plan to discuss the higher order commutators in a future paper. For now, we content ourselves with the simpler first-order case, our main purpose here being to obtain the sharp condition on A.

Acknowledgements I am grateful to J. Lewis and M. Murray for introducing me to the problem, for describing to me their joint work, and for several helpful and encouraging conversations. I also thank R. Craighead for some useful comments.

13.2 Preliminaries and statement of results

Our results will be obtained in $\mathbb{R}^n \equiv \mathbb{R}^{n-1} \times \mathbb{R}$, endowed with the group of parabolic dilations (4). Associated to this dilation group is a nonisotropic "norm," which we denote $\|z\|$, defined as the unique positive solution of the identity

$$1 = \sum_{j=1}^{n-1} \frac{x_j^2}{\|(x,t)\|^2} + \frac{t^2}{\|(x,t)\|^4}, \tag{5}$$

and having the dilation invariance property

$$\|\delta_\lambda z\| \equiv \lambda \|z\|.$$

With this "norm", $\mathbb{R}^n$ is a space of homogeneous type, in the sense of Coifman and Weiss [CW], with homogeneous dimension

$$d \equiv |\alpha| \equiv n + 1.$$

One has the polar decomposition

$$z \equiv \delta_\rho \sigma,$$

with $\sigma \in S^{n-1}$, $\rho \equiv \|z\|$, and $dz \equiv \rho^{d-1} d\rho\, J(\sigma)\, d\sigma$, where $J(\sigma)$ is a smooth, nonnegative function of $\sigma \in S^{n-1}$, which is even in each of $\sigma_1, \sigma_2, \ldots, \sigma_n$

separately (see, e.g., [FR1] or [R]). Suppose that

$$K(\lambda^\alpha z) \equiv \lambda^{-d-1} K(z)$$

(i.e., as in (3)). We will assume that $K \in C^\infty(\mathbb{R}^n \setminus \{0\})$ (this assumption could almost certainly be relaxed considerably; we impose it here to simplify certain arguments by exploiting symbolic calculus results of [FR1]; in any case, this assumption is valid for most applications). With $\rho \equiv \|z\|$, set

$$K(z) \equiv \rho^{-d-1} K(\rho^{-\alpha} z) \equiv \rho^{-d-1} \Omega(z) \tag{6}$$

(so that, of course, Ω is parabolically homogeneous of degree zero). We shall require the following cancellation condition

$$\int_{S^{n-1}} \Omega(\sigma) \sigma_j J(\sigma) \, d\sigma \equiv 0, \qquad 1 \le j \le n-1. \tag{7}$$

Note that (7) cannot, in general, be relaxed, as may be seen by considering the convolution operators that arise by taking $A(x, t) \equiv x_j$ (see [FR2] for the theory of convolution operators in this setting). Of particular interest is the special case

$$K(x, t) \equiv t^{-1-(n/2)} \exp\left(-\frac{|x|^2}{4t}\right) \chi\{t > 0\}.$$

We define (as usual) the parabolic fractional integral of order 1, which we denote I_p, by

$$(I_p f)^\wedge(\zeta) \equiv \|\zeta\|^{-1} \widehat{f}(\zeta), \tag{8}$$

and its inverse by

$$(|D_p| f)^\wedge(\zeta) \equiv \|\zeta\| \widehat{f}(\zeta) \tag{9}$$

Here $\zeta \equiv (\xi, \tau) \in \mathbb{R}^{n-1} \times \mathbb{R}$.

By a (parabolic) cube $Q \subseteq R^n$, we mean an interval of the form

$$Q \equiv Q_r \equiv I \times \widetilde{I}, \tag{10}$$

where $I \subseteq \mathbb{R}^{n-1}$ is a cube with side length r, and $\widetilde{I}$ is a one-dimensional interval of length r^2. The (parabolic) "ball" of radius r and center z is defined by

$$B_r(z) \equiv \{w \in \mathbb{R}^n : \|z - w\| < r\}. \tag{11}$$

As usual, the parabolic BMO space is defined as the collection of all locally L^1 functions modulo constants with norm

$$\|b\|_* \equiv \sup_{\text{parabolic } Q} \frac{1}{|Q|} \int_Q |b - m_Q(b)|,$$

where $m_Q(b) \equiv \frac{1}{|Q|} \int_Q b$.

Following Strichartz [Stz], we define the parabolic BMO Sobolev space

$$I_p(BMO) \equiv \{A(x, t) : A = I_p a, \qquad a \in BMO\}.$$

(Elements in this space are well-defined modulo 1st degree polynomials in x.)
We shall characterize the functions A, for which the operators in (2) are bounded
on L^2, as the subset of I_p (BMO) consisting of those A that are Lipschitz in x.
Specifically, we have the following

THEOREM 1
Suppose that $A = I_p a$, where $a \in BMO$, and that is A Lipschitz in x (uniformly
in t) with

$$\|\nabla_x A(x, t)\|_{L^\infty(\mathbb{R}^n)} \le B.$$

We assume that K satisfies (3) and (7) and that $K \in C^\infty(\mathbb{R}^n \backslash \{0\}$. Then the operator
$T[A]$ defined in (2) is bounded on $L^2(\mathbb{R}^n)$ with

$$\|T[A]\|_{op} \le C(n, K)(B + \|a\|_*).$$

We remark that weak $(1, 1)$ and L^p bounds for $T[A]$ then adhere to the
usual Calderón–Zygmund program (see, e.g., [FR2], [R], or [CW]), since we will
observe below that the hypotheses of Theorem 1 imply, in particular,
that $|A(z) - A(w)| \le \|z - w\|$; hence the kernel $K(z - w)[A(z) - A(w)]$
is "standard."
   We also remark that by (5), we may write

$$|D_p| \equiv \sum_{j=1}^{n-1} \frac{\partial}{\partial x_j} R_j + D_n R_n,$$

where

$$\widehat{R_j}(\xi, \tau) \equiv \frac{\xi_j}{\|(\xi, \tau)\|}, \qquad 1 \le j \le n - 1$$

$$\widehat{R_n}(\xi, \tau) \equiv \frac{\tau}{\|(\xi, \tau)\|^2}$$

define L^p and BMO bounded singular integrals, and where the half-order differentiation operator D_n is defined by

$$\widehat{D_n}(\xi, \tau) \equiv \frac{\tau}{\|(\xi, \tau)\|} \equiv (R_n|D_p|)^\wedge(\xi, \tau).$$

Thus in the bound for $\|T[A]\|_{op}$ we have the equivalence

$$\|\nabla_x A\|_\infty + \|a\|_* \approx \|\nabla_x A\|_\infty + \|D_n A\|_*.$$

Conversely, the sufficient conditions of Theorem 1 are also necessary in the
following sense.

THEOREM 2

There exists a kernel $K \in C^\infty(\mathbb{R}^n \setminus \{0\})$, satisfying (3) and (7), such that if the corresponding operator $T[A]$, defined as in (2), is bounded on $L^2(\mathbb{R}^n)$, then

(i) $|A(x,t) - A(y,s)| \leq C\|T[A]\|_{op}\|(x,t) - (y,s)\|$

  (so in particular $\|\nabla_x A\|_{L^\infty(\mathbb{R}^n)} \leq C\|T[A]\|_{op}$), and

(ii) $\||D_p|A\|_* \leq C\|T[A]\|_{op}$ *(so that $A \in I_p(BMO)$).*

In fact, we shall see that we may take $T[A] = [H^{1/2}, A]$, where $H \equiv \Delta - \frac{\partial}{\partial t}$. This particular commutator arises in the series expansion of certain caloric layer potentials (see Section 13.4).

13.3   Proof of Theorem 1

The proof will be a direct application of the T1 Theorem of David and Journé [DJ], or rather the nonisotropic version of their theorem. Such a generalization of the T1 Theorem, in the even more general homogeneous space setting, was proven by Coifman (unpublished), but the reader could easily provide his or her own proof in the special case of parabolic homogeneity in $\mathbb{R}^n$ by adapting the argument in [CM, pages 13–16] (see also [LM1, pages 12–19]. The T1 Theorem will be applied in the same way that the original T1 Theorem can be applied to treat the "elliptic" Calderón commutators (see e.g., [CM, page 13]), but with just a few additional (and relatively minor) technical difficulties.

We first prove a lemma that is a transparent extension to the parabolic setting of [Stz, Theorem 3.4, pages 553–554].

LEMMA 1

Suppose $A \in I_p$ (BMO). Then

$$|A(x,t) - A(x,s)| \leq C\|\,|D_p|A\|_*|s-t|^{1/2}.$$

PROOF   We follow [Stz]. Let $A = I_p a$, $a \in BMO$. Then up to a 1st degree polynomial in x,

$$A(x,t) \equiv \int\!\!\int a(y,s)\big[k(x-y,t-s) - k(y,s)$$

$$+ x \cdot \nabla_y k(y,s)\chi_{\{\|(y,s)\|>1\}}\big]\,dy\,ds,$$

for an appropriate even $k \in C^\infty(\mathbb{R}^n \setminus \{0\})$, where

$$k(\lambda^\alpha z) \equiv \lambda^{-d+1}k(z). \tag{12}$$

By the (parabolic) $H^1 - BMO$ duality (see, e.g., Calderón and Torchinsky [CT1], [CT2]), and translation invariance, it is sufficient to prove that, as a function of (y, s),

$$\|k(y, s - h) - k(y, s)\|_{H^1} \leq C|h|^{1/2}. \tag{13}$$

By an easy computation using the smoothness and homogeneity of k, we have

$$\int\int |k(y, s - h) - k(y, s)| \, dy \, ds \leq C|h|^{1/2}. \tag{14}$$

Furthermore, Coifman and Dahlberg [CD] have shown that there exists a finite collection of parabolic singular integrals $T_1, \ldots, T_N$ (we recall that the identity may also be viewed as a (parabolically) homogeneous singular integral with C^∞ symbol $m(\zeta) \equiv 1$), such that

$$\|f\|_{H^1} \approx \sum_{j=1}^{N} \|T_i f\|_{L^1}.$$

But by the symbolic calculus of parabolic kernels (see, e.g., [FR1], the composition $I_p T_i$ also satisfies (12) and (14). This proves the lemma. ∎

Thus, by assuming in addition that A is Lipschitz in the space variable, we find in particular that

$$|A(z) - A(w)| \leq C(\| \, |D_p|A\|_* + \|\nabla_x A\|_{L^\infty(\mathbb{R}^n)})\|z - w\|, \tag{15}$$

and therefore the kernel $K(z-w)[A(z)-A(w)]$ is a standard (parabolic) Calderón–Zygmund kernel. Furthermore, this kernel and its transpose, are of essentially the same form, so by using the $T1$ Theorem it is enough to verify that $T1 \in BMO$, and that T satisfies the weak boundedness property (WBP) (see (17) and (18) below). Only the verification of WBP presents any difficulty, so we shall dispose of $T1$ first. In the sequel, we will, as is common, make the *a priori* assumption that $A \in C^\infty$, but of course we obtain bounds depending only on $\|\nabla_x A\|_\infty$ and $\| \, |D_p|A\|_*$. The smoothness assumption is justified by taking a regularization of the identity, which may be removed by standard limiting arguments. We give a formal proof of the fact that $T1 \in BMO$. A rigorous approach is discussed at the end of the argument.

We write $T[A]1$ in parabolic polar coordinates, and then integrate by parts to obtain

$$T[A]1(x, t) \equiv \text{P.V.} \int_{S^{n-1}} \Omega(\sigma) \int_0^\infty [A(x, t) - A(x - \rho\sigma', t - \rho^2\sigma_n)]\frac{d\rho}{\rho^2} J(\sigma) \, d\sigma$$

$$\equiv \text{P.V.} \int_{S^{n-1}} \Omega(\sigma)\sigma' \cdot \int_0^\infty \nabla_x A(x - \rho\sigma', t - \rho^2\sigma_n)\frac{d\rho}{\rho} J(\sigma) \, d\sigma$$

$$+ \text{P.V.} \, 2 \int_{S^{n-1}} \Omega(\sigma) \sigma_n \int_0^\infty \frac{\partial}{\partial t} A(x - \rho \sigma', t - \rho^2 \sigma_n) \, d\rho \, J(\sigma) \, d\sigma$$

$$\equiv I + II,$$

where $\sigma \equiv (\sigma', \sigma_n) \in S^{n-1}$ and $\sigma' \equiv (\sigma_1, \sigma_2, \ldots, \sigma_{n-1})$. Now, $\nabla_x A \in L^\infty(\mathbb{R}^n)$, so by (7), I is just a "nice" parabolic singular integral operator acting on a bounded function. Thus, $\|I\|_* \le C \|\nabla_x A\|_\infty$. To handle II, we recall that $|D_p| A \in BMO$; i.e., $A = I_p a, a \in BMO$, so that

$$\frac{\partial}{\partial t} A = \frac{\partial}{\partial t} I_p a \equiv D_n a. \tag{16}$$

Thus $(D_n a)^\wedge(\xi, \tau) \equiv \frac{\tau}{\|(\xi, \tau)\|} \widehat{a}(\xi, \tau)$.

Also, by following the symbolic calculus argument of {FR1, pages 109–110], the reader may readily verify that the operator defined by

$$\widetilde{I}_p f(z) \equiv \int \frac{\widetilde{\Omega}(z - w)}{\|z - w\|^{d-1}} f(w) \, dw,$$

where $\widetilde{\Omega}(\sigma) \equiv 2 \Omega(\sigma) \sigma_n$ is parabolically homogeneous of degree zero, and is "parabolically smoothing," i.e.,

$$(\widetilde{I}_p f)^\wedge(\zeta) \equiv \frac{m(\zeta)}{\|\zeta\|} \widehat{f}(\zeta),$$

where $m \in C^\infty(\mathbb{R}^n \setminus \{0\})$ and $m(\lambda^\alpha \zeta) \equiv m(\zeta)$. Thus, viewing Π as an operator on a, we have

$$\widehat{\Pi}(\xi, \tau) \equiv m(\xi, \tau) \frac{\tau}{\|(\xi, \tau)\|^2} \widehat{a}(\xi, \tau).$$

Again, by symbolic calculus [FR1], the latter expression defines a "nice" parabolic singular integral that is bounded on BMO, so that $\|\Pi\|_* \le C \|a\|_*$. This concludes the formal treatment of T1. One can make all of this rigorous as follows. On each ball, one writes $1 = \chi_{10B} + 1 - \chi_{10B}$, where χ is a "smoothed out" characteristic function. The part corresponding to $1 - \chi$ can be treated by the "standard" smoothness estimates for the kernel, and $T[A] \chi_{10B}$ can be treated by integration by parts, as in the proof of WBP below. The boundary terms that arise in the integration by parts are zero.

We now turn to a discussion of WBP. Let $A(z, r)$ denote the space of all $\varphi \in C_0^\infty$, supported in $B_r(z) \equiv \{w : \|z - w\| < r\}$, and satisfying

(i) $\|\varphi\|_\infty \le 1$

(ii) $|\varphi(u) - \varphi(w)| \le \dfrac{\|u - w\|}{r} \left(\text{in particular, } \|\nabla_x \varphi\|_\infty \le \dfrac{1}{r} \right)$ $\qquad$ (17)

(iii) $\displaystyle \sup_{|\gamma| + \beta \le 2} \left\{ r^{2\beta + |\gamma|} \left\| \left(\frac{\partial}{\partial x} \right)^\gamma \left(\frac{\partial}{\partial t} \right)^\beta \varphi \right\|_\infty \right\} \le 1.$

Then we state that T satisfies WBP, if for all $z \in \mathbb{R}^n$, $r > 0$, and for all $\psi, \varphi \in A(z, r)$, we have

$$|\langle \psi, T\varphi \rangle| \leq Cr^d. \tag{18}$$

REMARK 1   Usually WBP is stated only for functions satisfying (17), (i) and (ii). An inspection of the proof of the T1 Theorem shows that the higher order smoothness condition, (17) (iii), is no loss of generality.   ∎

PROOF   By dilation invariance, we may take $r = 1$. Then φ, ψ are supported in the parabolic ball $B \equiv B_1(z_0)$.

We write $T[A]\varphi$ in polar coordinates and integrate by parts so that

$$T[A]\varphi(x, t)$$

$$= \text{P.V.} \int_{S^{n-1}} \Omega(\sigma) \int_0^\infty [A(x, t) - A(x - \rho\sigma', t - \rho^2\sigma_n)]$$

$$\cdot \varphi(x - \rho\sigma', t - \rho^2\sigma_n) \frac{d\rho}{\rho^2} J(\sigma) \, d\sigma$$

$$\equiv \text{P.V.} \int_{S^{n-1}} \Omega(\sigma)\sigma' \cdot \int_0^\infty \nabla_x A(x - \rho\sigma', t - \rho^2\sigma_n)$$

$$\cdot \varphi(x - \rho\sigma', t - \rho^2\sigma_n) \frac{d\rho}{\rho} J(\sigma) \, d\sigma$$

$$+ 2 \, \text{P.V.} \int_{S^{n-1}} \Omega(\sigma)\sigma_n \int_0^\infty \frac{\partial}{\partial t} A(x - \rho\sigma', t - \rho^2\sigma_n)$$

$$\cdot \varphi(x - \rho\sigma', t - \rho^2\sigma_n) \, d\rho \, J(\sigma) \, d\sigma$$

$$+ \int_{S^{n-1}} \Omega(\sigma)\sigma' \cdot \int_0^\infty [A(x, t) - A(x - \rho\sigma', t - \rho^2\sigma_n)]$$

$$\cdot \nabla_x \varphi(x - \rho\sigma', t - \rho^2\sigma_n) \frac{d\rho}{\rho} J(\sigma) \, d\sigma$$

$$+ 2 \int_{S^{n-1}} \Omega(\sigma)\sigma_n \int_0^\infty [A(x, t) - A(x - \rho\sigma', t - \rho^2\sigma_n)]$$

$$\cdot \frac{\partial}{\partial t} \varphi(x - \rho\sigma', t - \rho^2\sigma_n) \, d\rho \, J(\sigma) \, d\sigma$$

$$\equiv I + II + III + IV.$$

By (3) the term I is a "nice" L^2 bounded parabolic singular integral operator, acting on $(\nabla_x A)\varphi$. Thus by Schwarz

$$|\langle \psi, I \rangle| \leq C\|(\nabla_x A)\varphi\|_2 \leq C\|\nabla_x A\|_\infty.$$

To handle III and IV, we note that since $(x, t) \in B$, and also $(x - \rho\sigma', t - \rho^2\sigma_n) \in B$, we have $\rho \leq C$. Thus by (15)

$$|III| \leq C(\| \, |D_\rho|A\|_* + \|\nabla_x A\|_\infty) \int_0^C \|\nabla_x \varphi\|_\infty \, d\rho,$$

and the desired bound for $|\langle \psi, III \rangle|$ follows naturally, by (17) (ii). A similar computation disposes of IV, by (17) (iii).

Finally, we consider II, which in rectangular coordinates equals $\int k(z - w) \frac{\partial}{\partial t} A \cdot (w)\varphi(w) \, dw$, where

$$k(\lambda^\alpha z) \equiv \lambda^{-d+1} k(z), \qquad \text{and } k \in C_0^\infty(\mathbb{R}^n \backslash \{0\}). \tag{19}$$

Then, as before, the operator defined by $\tilde{I}_\rho f \equiv k * f$ is "parabolically smoothing," i.e.,

$$k(\zeta) \equiv \frac{m(\zeta)}{\|\zeta\|}, \quad \text{where } m(\lambda^\alpha \zeta) \equiv m(\zeta), \qquad m \in C^\infty(\mathbb{R}^n \backslash \{0\}).$$

Also (see (16)) $\frac{\partial}{\partial t} A \equiv D_n a$, where $D_n \equiv \frac{\partial}{\partial t} I_\rho$, $a \in BMO$, and

$$D_n a(w) \equiv \text{P.V.} \int D_n(w - v)a(v) \, dv. \tag{20}$$

The kernel $D_n(w)$ is odd, has vanishing first moments in the space variable on the sphere, belongs to $C^\infty(\mathbb{R}^n \backslash \{0\})$, and satisfies the homogeneity property $D_n(\lambda^\alpha w) \equiv \lambda^{-d-1} D_n(w)$ (see [FR1, Lemmas 1 and 2, pages 111–113]). ∎

We note that given our *a priori* regularization of A, and hence of a, the principal value in (20) is well defined. In fact, since

$$D_n 1(w) \equiv \lim_{\epsilon \to 0, r \to \infty} \int_{\epsilon < \|w - v\| < r} D_n(w - v)1 \, dv = 0,$$

we have

$$D_n a(w) \equiv \text{P.V.} \int D_n(w - v)[a(v) - m_B a]\eta(v) \, dv$$

$$+ \int D_n(w - v)[a(v) - m_B a][1 - \eta(v)] \, dv$$

$$\equiv D_n a_1(w) + D_n a_2(w).$$

where $\eta \in C_0^\infty(B_{20}(z_0))$, $\eta \equiv 1$ on $B_{10}(z_0)$. Then $D_n a_1$ is defined by [FR1], and for $w \in B$, we have

$$|D_n a_2(w)| \leq C \int \frac{1}{1 + \|z_0 - v\|^{d+1}} |a(v) - m_B a| \, dv \leq C\|a\|_*. \tag{21}$$

The last inequality follows by the parabolic version of a well known result of Fefferman and Stein [FS] (the proof is the same as in the elliptic case). We then write

$$II(z) = \int k(z - w) \text{ (P.V.)} \int D_n(w - v)a_1(v)[\varphi(w) - \varphi(v)]\, dv\, dw$$

$$+ \int k(z - w) \text{ (P.V.)} \int D_n(w - v)a_1(v)\varphi(v)\, dv\, dw$$

$$+ \int k(z - w)\varphi(w)D_na_2(w)\, dw$$

$$\equiv II_1 + II_2 + II_3.$$

Now $II_2 \equiv k * D_n(a_1\varphi)$ defines, by symbolic calculus [FR1], a "nice" parabolic singular integral acting on $a_1\varphi$, and furthermore $\|a_1\varphi\|_2 \leq C\|a\|_*$. The desired estimate for $|\langle \psi, II_2 \rangle|$ follows easily.

Also since $z \in \text{sup } \varphi \leq B$, by (19) and (21) we have

$$|II_3| \leq C\|a\|_* \int_{\|z - w\| \leq C} \|z - w\|^{-d+1}\, dw = C\|a\|_*.$$

Again $\langle \psi, II_3 \rangle$ is easily handled. Finally, the inner integral in II_1 is the commutator

$$[D_n, \varphi]a_1(w).$$

By the higher order smoothness of φ, (17) (iii), $[D_n, \varphi]$ is bounded on L^2 [FR1, pages 116–117]. Since $k(z - w) \in L^1 + L^2$, we find that $k * [D_n, \varphi]a$ belongs to $L^2 + L^\infty$, with bound $C\|a\|_*$, and WBP follows.

13.4 Proof of Theorem 2

The proof is based on some ideas of Murray [Mu1, pages 206–207]. In our case the proof depends on the nonisotropic analogue of a well known result of Meyers [Ms].

PROPOSITION 1
Suppose that for all $r > 0$, and for every parabolic cube J with $|J| = r^d$, there is a constant a_J such that

$$\int_J |A(z) - a_J|\, dz \leq Br^{d+1},$$

for some fixed constant B. Then

$$|A(z) - A(w)| \leq CB\|z - w\|.$$

The proof of the proposition is a verbatim repetition of Meyers' argument. One need only change the usual dilations to parabolic ones, and use the parabolic version of the Calderón–Zygmund decomposition lemma (see, e.g., [R]).

We prove Theorem 2 (i) first. Fix

$$K(x, t) \equiv t^{-1-(n/2)} \exp\left(\frac{-|x|^2}{4t}\right) \chi \ \{t > 0\}, \tag{22}$$

which is easily seen to verify the hypotheses of the theorem. Then the corresponding operator $T[A]$, defined for this K as in (2), is the first-order commutator that arises in the series expansion of the caloric double layer potential on a domain with a time-dependent boundary $(x, t, A(x, t) \in \mathbb{R}^{n+1}$. We shall also see that $T[A] \equiv [\sqrt{H}, A]$, where $H \equiv \Delta - \frac{\partial}{\partial t}$. We are assuming that $T[A]$ is bounded on L^2, so by duality, so is its transpose $(T[A])^*$, defined by

$$(T[A])^* f(z) \equiv \text{P.V.} \int K(w - z)[A(w) - A(z)] f(w) \, dw.$$

We also have, therefore, the L^2 boundedness of $T[A] - (T[A])^*$. The kernel of the latter is

$$h(x, t, y, s) \equiv |t - s|^{-1-(n/2)} \exp\left(\frac{-|x - y|^2}{4|t - s|}\right)[A(x, t) - A(y, s)]$$
$$\equiv k(x - y, t - s)[A(x, t) - A(y, s)]. \tag{23}$$

We base the proof of Theorem 2 (i) on the idea of the heuristic argument in [Mu, Lemma 2.2], although we shall actually give the details here. It is enough to verify the hypothesis of Proposition 1 with a constant

$$B = C\|T[A] - (T[A])^*\|_{op} \leq C\|T[A]\|_{op}.$$

Let J be an arbitrary parabolic cube of the form $Q \times [a, a + r^2]$, where Q is an $n - 1$ dimensional cube with side length r. Let I be a parabolic cube of the same size, but translated "upward" by a distance of $(Nr)^2$, N a large number to be determined; i.e., $I \equiv Q \times [a + (Nr)^2, a + (Nr)^2 + r^2]$, with the same $Q \subseteq \mathbb{R}^{n-1}$. By Schwarz's inequality,

$$\frac{1}{|J|} \int_J \left| \int_I h(z, w) \, dw \right| \, dz \leq C\|T[A]\|_{op}. \tag{24}$$

On the other hand, the inner integral on the left side of (24) equals

$$[A(z) - m_I(A)] \int_I k(z - w) \, dw + \int_I [m_I(A) - A(w)]k(z - w) \, dw$$

$$\equiv F(z) + G(z),$$

where k is defined in (23).

Since

$$\frac{1}{2} \leq \exp\left(\frac{-|x - y|^2}{4|t - s|}\right) \leq 1 \text{ for } (x, t) \in J, (y, s) \in I$$

and N sufficiently large, we have

$$|F(z)| \geq C_0 \frac{|A(z) - m_I(A)|}{N^{d+1}r}.$$

Furthermore, since $A - m_I(A)$ has mean value zero on I,

$$G(z) = \int_I [m_I(A) - A(w)]\{k(z - w) - k(z - w_I)\} \, dw,$$

where w_I is the center of I. Thus

$$|G(z)| \leq \frac{C_1}{N^{d+2}r} \frac{1}{|I|} \int_I |m_I(A) - A(w)| \, dw.$$

Now suppose for the moment that

Case 1

$$\frac{1}{|J|} \int_J |A(z) - m_I(A)| \, dz \geq \frac{1}{2} \frac{1}{|I|} \int_I |A(w) - m_I(A)| \, dw.$$

If we choose N large enough, depending only on C_0, then the left side of (24) is greater than or equal to

$$\frac{1}{|J|} \int_J (|F| - |G|) \geq \frac{C}{|J|r} \int_J |A - m_I(A)|,$$

and Theorem 2 (i) follows in this case, with $a_J \equiv m_I(A)$, since $|J| = r^d$.

On the other hand, suppose that we have

Case 2

$$\frac{1}{|J|} \int_J |A - m_I(A)| \leq \frac{1}{2} \frac{1}{|I|} \int_I |A - m_I(A)|.$$

Then

$$\frac{1}{|J|} \int_J |A - m_J(A)| \leq \frac{2}{|J|} \int_{|J|} |A - m_I(A)|$$

$$\leq \frac{1}{|I|} \int_I |A - m_I(A)|$$

$$\leq 2\frac{1}{|I|} \int_I |A - m_J(A)|.$$

But now we just repeat the previous argument, with the roles of I and J reversed, so that

$$\int_J |A - m_J(A)| \leq 2 \int_I |A - m_J(A)| \leq Cr^{d+1} \|T[A]\|_{op},$$

and Theorem 2 (i) again follows, this time with $a_J \equiv m_J(A)$.

Next, we verify Theorem 2 (ii). We continue to assume that $T[A]$ is bounded on L^2, where as above $T[A] = [K, A]$, and K denotes convolution with the kernel $K(x, t)$ in (22). By Theorem 2 (i), $K(z - w)[A(z) - A(w)]$ satisfies the standard size and smoothness conditions of parabolic Calderón–Zygmund theory, with bound $C \|T[A]\|_{op}$. Hence, T[A] maps L^∞ boundedly into BMO, and in particular $\|T[A]1\|_* \leq C \|T[A]\|_{op}$. We now claim that $T[A] \equiv C[H^{1/2}, A]$, where $H = \Delta - \frac{\partial}{\partial t}$. Here, $(H^{1/2}f)^\wedge(\xi, \tau) \equiv (|\xi|^2 - i\tau)^{1/2} \widehat{f}(\xi, \tau)$, for an appropriately chosen branch of the square root. Let us assume the claim for the moment, and deduce the theorem. Since $H^{1/2}1 = 0$, we have $[H^{1/2}, A]1 = H^{1/2}A$, so that

$$\|H^{1/2}A\|_* \leq C \|T[A]\|_{op}. \tag{25}$$

But $|D_p| = SH^{1/2}$, where S is the BMO bounded singular integral operator with symbol

$$\widehat{S}(\xi, \tau) \equiv \frac{\|(\xi, \tau)\|}{(|\xi|^2 - i\tau)^{1/2}}.$$

Thus Theorem 2 (ii) follows from (25).

It remains only to establish the claim. Let $W(x, t)$ denote the Gaussian

$$W(x, t) \equiv t^{-n/2} \exp\left(\frac{-|x|^2}{4t}\right) \chi\{t > 0\}. \tag{26}$$

(We have ignored a multiplicative constant). Then

$$\widehat{W}(\xi, \tau) \equiv \int_0^\infty e^{it\tau} t^{-1/2} t^{-(n-1)/2} \int_{\mathbb{R}^{n-1}} e^{-\frac{|x|^2}{4t}} e^{i\xi \cdot x} \, dx \, dt$$

$$= C \int_0^\infty e^{-t(|\xi|^2 - i\tau)} t^{-1/2} \, dt$$

$$= C(|\xi|^2 - i\tau)^{-1/2},$$

where the second equality holds by a well known identity, and the third equality is obtained by choosing an appropriate branch of the square root and shifting the contour of integration. A routine computation shows that

$$\left(\Delta - \frac{\partial}{\partial t}\right) W(x, t) \equiv CK(x, t), \qquad (x, t) \neq (0, 0)$$

where K is the kernel of (22). (To avoid possible confusion, we note that W is the projection onto $\mathbb{R}^n$ of the fundamental solution for the heat equation in $\mathbb{R}^{n+1}$, so we do not obtain zero away from the origin by applying the n-dimensional heat operator to W). But

$$(HWf)^\wedge(\xi, \tau) = C(|\xi|^2 - i\tau)^{1/2}\widehat{f}(\xi, \tau),$$

and the claim follows.

References

[Ca]   A. P. Calderón, *Commutators of singular integral operators,* Proc. Nat. Acad. Sci., USA 74 (1977), 1324–1327.

[CT1]   ———, and A. Torchinsky, *Parabolic maximal functions associated with a distribution,* Adv. Math. 16 (1975), 1–64.

[CT2]   ———, *Parabolic maximal functions associated with a distribution II,* Adv. Math 24 (1977), 101–171.

[CD]   R. Coifman and B. Dahlberg, *Singular integral characterizations of non-isotropic H^p spaces and the F. and M. Riesz theorem,* Harmonic Analysis in Euclidean Spaces, G. Weiss and S. Wainger, Eds., Proc. Symp. Pure Math., Vol. 35, Amer. Math. Soc., Providence, RI, 1979, pp. 231–234.

[CM]   R. Coifman and Y. Meyer, *Non-linear harmonic analysis, operator theory, and P.D.E.,* Beijing Lectures in Harmonic Analysis, E.M. Stein, Ed., Princeton Univ. Press, Princeton, N.J. 1986, pp. 3–45.

[CW]   R. Coifman and G. Weiss, *Analyse harmonique non-commutative sur certains espaces homogènes,* Lecture Notes in Math., Vol. 242, Springer-Verlag, Berlin, 1971.

[DJ]   G. David and J. L. Journé, *A boundedness criterion for generalized Calderón-Zygmund operators,* Ann. of Math. 120 (1984), 371–397.

[FR1]   E. B. Fabes and N. M. Riviere, *Symbolic calculus of kernels with mixed homogeneity,* Singular Integrals, A. P. Calderón, Ed., Proc. Symp. Pure Math., Vol. 10, Amer. Math. Soc., Providence, 1967, pp. 106–127.

[FR2]   ———, *Singular Integrals with mixed homogeneity,* Studia Math. 27 (1966), 19–38.

[FS]   C. Fefferman and E. M. Stein, *H^p spaces of several variables,* Acta. Math. 129 (1972), 137–193.

[J]   B. F. Jones, *A class of singular integrals,* Amer. J. Math. 86 (1964), 441–462.

[LM1]   J. L. Lewis and M. A. M. Murray, *The method of layer potentials for the heat equation in time-varying domains I: singular integrals,* preprint.

[LM2] _____, *The method of layer potentials for the heat equation in time-varying domains II: the David buildup scheme,* preprint.

[Ms] N. Meyers, *Mean oscillation over cubes and Hölder continuity,* Proc. Amer. Math. Soc. 15 (1964), 717–721.

[Mu1] M. A. M. Murray, *Commutators with fractional differentiation and BMO Sobolev spaces,* Indiana Univ. Math. J. 34 (1985), 205–215.

[Mu2] _____, *Multilinear singular integrals involving a derivative of fractional order,* Studia Math. 87 (1987), 139–165.

[R] N. M. Riviere, *Singular integrals and multiplier operators,* Ark. Math. 9 (1971), 243–278.

[Stz] R. Strichartz, *Bounded mean oscillation and Sobolev spaces,* Indiana Univ. Math. J. 29 (1980), 539–558.

14

Inequalities for Classical Operators in Orlicz Spaces

M. Krbec

14.1 Introduction

This chapter deals with several aspects of the Orlicz spaces theory connected with boundedness of classical operators. Aside from what follows more or less directly by interpolation of L^p spaces, there are a lot of interesting results and unsolved problems in this area in both reflexive and nonreflexive spaces.

A classical theory of Orlicz spaces is surveyed in the Krasnoselskii and Rutitskii monograph [KR], interpolation properties are investigated in the Krein, Petunin, and Semenov book [Kre], and further in a series of papers; one can consult the bibliography on interpolation spaces collected by Maligranda and Persson [MP]. Generalized modular spaces (Musielak–Orlicz spaces) are the topic of Musielak [Mu]. Among recent surveys the monographs by Bennet and Sharpley [BS], and Ren and Rao [RR] should be named. The study of classical operators begins with the Torchinsky paper [T] on Riesz potentials in nonweighted spaces; further on, there is a deep result on A_p weights and maximal functions in reflexive Orlicz spaces, described by Kerman and Torchinsky [KT]. Since then, many papers have appeared, devoted to various topics in this area. The recent monograph by Kokilashvili and the author [KK], and its bibliography can be used as a reference.

During the 1960's when the theory of monotone operators was flourishing, the Orlicz spaces were used by various authors to solve boundary and initial value problems. Whereas the modification to reflexive spaces was more of a technical problem (sometimes very complicated), the challenge was the existence theory for equations with slowly or rapidly increasing coefficients, requiring use of spaces where bounded sets are not relatively weakly compact. For the existence theorems see Gossez [Gos], who developed the concept of quadruples of Orlicz spaces.

A careful study of the boundedness of classical operators in Orlicz spaces soon reveals one substantial problem, namely, a deterioration in proximity of L^1. To get a flavor for what is going on near L^1, let us consider the Orlicz–Sobolev space

$$W^1 L \log^+ L(\Omega) = \{ f \in L \log^+ L(\Omega); \ \nabla f \in L \log^+ L(\Omega) \}$$

where Ω is a bounded domain in $\mathbb{R}^n$, for simplicity with a Lipschitzian boundary $\partial\Omega$ (i.e., Ω can be reduced to a unit square $(0, 1)^n$) and $L \log^+ L(\Omega)$ is the Zygmund space, consisting of all f with

$$\int\limits_\Omega |u(x)| \log^+ |u(x)| \, dx < \infty;$$

$W^1 L \log^+ L(\Omega)$ is then equipped with an appropriate Orlicz or Luxemburg norm $u \mapsto \|u\| + \|\nabla u\|$. The question then will be about the traces of functions from $W^1 L \log^+ L(\Omega)$ on $\partial\Omega$. Of course the traces exist, and it is easy to show that $\text{Tr} : W^1 L \log^+ L(\Omega) \to L \log^+ L(\partial\Omega)$; also one can consider the norm

$$\|u\|_{\text{Tr}\,(W^1 L \log^+ L(\partial\Omega))} = \inf\{\|f\|_{W^1 L \log^+ L(\Omega)}; \ \text{Tr} \, f = u \text{ a.e. on } \partial\Omega\}.$$

Nevertheless, no intrinsic norm seems to be known, e.g., analogous to that in Sobolev spaces with fractional derivatives. This problem was only solved for Orlicz–Sobolev spaces with lower indices greater than 1; i.e., we have

$$\text{Tr} \, W^1 L^\Phi(\Omega) = \left\{ f \in L^\Phi(\partial\Omega); \ \text{there is } \varepsilon > 0 \text{ such that} \right.$$

$$\left. \int\limits_{\partial\Omega} \int\limits_{\partial\Omega} \Phi \left(\frac{\varepsilon |f(x) - f(y)|}{|x - y|} \right) \frac{dx \, dy}{|x - y|^{n-2}} < \infty \right\}$$

for classical Young functions Φ growing faster at ∞ than $t \mapsto t^{1+\varepsilon}$ for some $\varepsilon > 0$. Different proofs were given by Lacroix [L] and Palmieri [P]. This is a natural analogue of $(1 < p < \infty)$

$$\text{Tr} \, W^{1,p}(\Omega)$$
$$= W^{1,1-1/p,(\partial\Omega)}$$
$$= \left\{ f \in L^p(\partial\Omega); \int\limits_{\partial\Omega} \int\limits_{\partial\Omega} \left(\frac{|f(x) - f(y)|}{|x - y|} \right)^p \frac{dx \, dy}{|x - y|^{n-2}} < \infty \right\}.$$

Note that Gagliardo [Ga] has proven that the traces of $W^{1,1}(\Omega)$ are exactly $L^1(\partial\Omega)$; however, the trace operator does not possess such fine properties as when $p > 1$.

The technique used relies on the Hardy inequality or alternatively on the maximal inequality in $L^p(\mathbb{R}^1_+)$, $p > 1$, which in $L \log^+ L(\mathbb{R}^1_+)$ turns into the well known inequality

$$\int\limits_0^a M^- u(x) \, dx \leq C \left(a + \int\limits_0^a |u(x)| \log^+ |u(x)| \, dx \right), \qquad a > 0, \qquad (1)$$

where

$$M^- f(x) = \sup_{0<y<x} \frac{1}{x-y} \int_y^x |f(z)| \, dz$$

is the left maximal function. Thus supposing $u \in W^1 L \log^+ L(\Omega)$, one arrives after some effort at the conditions $\text{Tr} \, u \in L \log^+ L(\partial\Omega)$ and

$$\int_{\partial\Omega} \int_{\partial\Omega} \frac{|u(x) - u(y)|}{|x-y|} \frac{dx \, dy}{|x-y|^{n-2}} < \infty. \tag{2}$$

Conversely, given $u \in L \log^+ L(\partial\Omega)$, satisfying (2), standard extension operators map u into a space bigger than $W^1 L \log^+ L(\Omega)$; in other words, the loss of the smoothness is recovered, but this is not the case as to the integrability.

Using (1), however, we can at least characterize the kernel of Tr; that is the closure of $C_0^\infty(\Omega)$ in $W^1 L \log^+ L(\Omega)$. We have

$$\text{Ker} \, \text{Tr} = \{ f \in W^1 L \log^+ L(\Omega); \ f \cdot (\text{dist}(., \partial\Omega))^{-1} \in L^1 \}.$$

Indeed, after reducing Ω to a unit square, each $u \in \text{Ker} \, \text{Tr}$ can be considered as an element of $C_0^\infty((0, 1)^n)$, and the integrability of the mapping $x \mapsto u(x)(\text{dist}(x, \partial\Omega))^{-1}$ follows immediately from (1). On the other hand, if $u \notin \text{Ker} \, \text{Tr}$, then without any loss of generality we can suppose that the set

$$E = \{ x = (x' x_n) \in \partial((0, 1)^n); \ x_n = 0, \ u(x) \neq 0 \}$$

has a positive measure. We can also assume that u is absolutely continuous as a function of x_n for almost all $x' \in (0, 1)^n$. Then for such x' we have

$$u(x', x_n) = u(x', 0) + \int_0^{x_n} \frac{\partial u}{\partial x_n}(x', \eta) \, d\eta,$$

and it is clear that $u(x', x_n) x_n^{-1}$ cannot be integrable. Thus $\text{Ker} \, \text{Tr}$ has been characterized by an integral condition, the general problem of the characterization of traces remains open, at least according to the author's knowledge.

The above described phenomena of a loss of integrability, having sometimes rather unpleasant consequences as we just saw, concerns further operators: If T is a sublinear operator of weak type $(1, 1)$ and (p, p), then an inequality analogous to (1) holds (the Zygmund inequality; see e.g., Sadosky [S, page 192]). This applies particularly to singular operators. Let us also note that the inequality (1) and its weak counterpart have been studied in a nonweighted setting in $\mathbb{R}^n$ for maximal operators, with respect to various bases of intervals, including the strong maximal function

$$M^s f(x) = \sup_{I \ni x} \frac{1}{|I|} \int_I |f(y)| \, dy$$

(sup taken over intervals with sides parallel to coordinate axes). See de Guzmán [Gu], particularly, Appendix II by Cordóba and Fefferman. Necessary and sufficient conditions for weights such that

$$\int_Q Mf(x)\varrho(x)\,dx \le c_1 \left(\varrho(Q) + \int_Q (|f(x)|\log^+ |f(x)|)\varrho(x)\,dx \right)$$

or

$$\varrho(\{x \in \mathbb{R}^n;\ Mf(x) > \lambda\}) \le \int_{\mathbb{R}^n} \frac{|f(x)|}{\lambda} \left(1 + \log^+ \frac{|f(x)|}{\lambda}\right)\varrho(x)\,dx$$

have been found by Carbery, Chang, and Garnett [C], the latter inequality with two weights appeared in Krbec [Krb], the strong maximal function also in Bagby and Kurtz [BKu]. General theorems on the strong maximal function are due to Gogatishvili [Go1] and [Go2].

The generalized Orlicz spaces turn out to be very suitable for the study of boundedness with respect to various classical operators, and conditions on the growth of generating Young functions can be given both necessary and sufficient, guaranteeing the boundedness of an operator in question. We will restrict ourselves to nonweighted inequalities. Analogous results for weighted inequalities can reasonably be conjectured; however, only the case of the maximal operator has been successfully solved quite recently by Bloom and Kerman [BKe]. What follows is part of a joint research carried out with Gogatishvili and Kokilashvili.

14.2   Maximal functions and Riesz transforms

Let us recall several concepts and fix our terminology. We will consider the (off-centered) *maximal operator*

$$Mf(x) = \sup_{Q \ni x} \frac{1}{|Q|} \int_Q |f(y)|\,dy.$$

where Q are cubes with sides parallel to coordinate axes and the *Riesz transforms* R_j, $j = 1, \ldots, n$, defined by

$$R_j f(x) = c_n \int_{\mathbb{R}^n} \frac{x_j - y_j}{|x - y|^{n+1}} f(y)\,dy,$$

where

$$c_n = \frac{\Gamma\left(\frac{n+1}{2}\right)}{\pi^{(n+1)/2}}$$

(the exact value of c_n will not be of importance for us).

For $1 \leq \theta \leq \infty$ and a sequence $a = (a_1, a_2, \ldots) \subset R^1$, the symbol $\|a\|_\theta$ will denote the norm in the space ℓ_θ. If $f = (f_1, f_2, \ldots)$ is a sequence of real functions we will write $f \in L^1_{loc}$ if each component f_k belongs to L^1_{loc}. For such a vector $f = (f_1, f_2, \ldots) \in L^1_{loc}$ we define $Mf = (Mf_1, Mf_2, \ldots)$.

By Φ we will denote the set of all functions $\Phi : R^1 \to R^1$, nonnegative, even, increasing on $[0, \infty)$, and such that $\Phi(0+) = 0$, $\lim_{t \to \infty} \Phi(t) = \infty$. For $\Phi \in \Phi$, the $\Phi(L)$ class consists of all measurable functions $f : R^n \to R^n$ such that

$$\int_{R^n} \Phi(f(x)) \, dx < \infty.$$

The *complementary function* to $\Phi \in \Phi$ is defined by

$$\Psi(s) = \sup\{st - \Phi(t); \, t > 0\}, \qquad s \geq 0,$$

$$\Psi(s) = \Psi(-s), \qquad s < 0.$$

The *classical Young function* is a $\Phi \in \Phi$, which is convex and such that $\lim_{t \to 0+} \frac{\Phi(t)}{t} = \lim_{t \to \infty} \frac{t}{\Phi(t)} = 0$.

A function $\Phi \in \Phi$ satisfies the *(global)* Δ_2 *condition* if there exists $c > 0$ such that $\Phi(2t) \leq c\,\Phi(t), t > 0$.

If $\Phi(2t) \leq c\,\Phi(t)$ for small t (for large t), then the function Φ is said to satisfy the Δ_2 condition near 0 (near ∞).

The condition Δ_2 for a $\Phi \in \Phi$ puts restrictions on the growth of Φ from above: such a function can be majorized by a power function. This does not hold conversely (see [KR]). On the other hand, the growth of Φ from below (globally) can be gauged by the condition

$$\Phi(t) \leq \frac{1}{2a} \Phi(at) \text{ for some } a > 1 \text{ and all } t > 0 \tag{3}$$

and this can be shown to be equivalent to Δ_2 condition for the function complementary to Φ.

Let $\Phi \in \Phi$ satisfy the Δ_2 condition. Put

$$h_\Phi(\lambda) = \sup_{t>0} \frac{\Phi(\lambda t)}{\Phi(t)}, \qquad \lambda > 0.$$

Then we define the *lower index of* Φ by

$$i(\Phi) = \lim_{\lambda \to 0+} \frac{\log h_\Phi(\lambda)}{\log \lambda} = \sup_{0 < \lambda < 1} \frac{\log h_\Phi(\lambda)}{\log \lambda}$$

and the *upper index of* Φ by

$$I(\Phi) = \lim_{\lambda \to \infty} \frac{\log h_\Phi(\lambda)}{\log \lambda} = \inf_{1 < \lambda < \infty} \frac{\log h_\Phi(\lambda)}{\log \lambda}.$$

These numbers are reciprocals of the Boyd indices (cf., Matuszewska and Orlicz [MO], Boyd [B]).

A function $\Phi : [0, \infty) \to R^1$ is said to be *quasiconvex* on $[0, \infty)$ if there exists a convex function ω and a constant $c > 0$, such that

$$\omega(t) \leq \Phi(t) \leq c\omega(ct), \qquad t \in [0, \infty).$$

It is easy to show that $\Phi \in \Phi$ is quasiconcave iff

$$\frac{\Phi(x_1)}{x_1} \leq \frac{a\,\Phi(ax_2)}{x_2}$$

for all $0 < x_1 < x_2$, and this is in turn equivalent to the quasi-Jensen inequality

$$\Phi\left(\frac{1}{|I|}\int_I f(x)\,dx\right) \leq \frac{c}{|I|}\int_I \Phi(cf(x))\,dx$$

with a constant c independent of f (see [KK]).

Given $\Phi \in \Phi$, consider first the inequality of weak type

$$\Phi(\lambda)|\{x \in \mathbb{R}^n;\ Mf(x) > \lambda\}| \leq c\int_{\mathbb{R}^n} \Phi(cf(x))\,dx. \qquad (4)$$

THEOREM 1

A constant $c > 0$ exists such that the inequality (4) holds for all $f \in L^1_{\mathrm{loc}}$ and all $\lambda > 0$ iff the function Φ is quasiconvex.

PROOF   If Φ is quasiconvex, then in virtue of the Jensen inequality, we have for a suitable convex function ω,

$$\Phi(Mf(x)) \leq c\,\omega(cMf(x)) \leq cM(\omega(cf(x))) \leq cM(\Phi(cf(x))).$$

The function ω is increasing, therefore

$$\Phi(\lambda)\,|\{x \in \mathbb{R}^n;\ Mf(x) > \lambda\}| = \Phi(\lambda)\,|\{x \in \mathbb{R}^n;\ \Phi(Mf(x)) > \Phi(\lambda)\}|$$
$$\leq \frac{c\,\Phi(\lambda)}{c}\left|\left\{x \in \mathbb{R}^n;\ M(\Phi(cf(x))) > \frac{\Phi(\lambda)}{c}\right\}\right|$$
$$\leq c\int_{\mathbb{R}^n} \Phi(cf(x))\,dx,$$

where the last inequality follows from the weak type $(1, 1)$ of the operator M.

If (4) holds, then Φ is quasiconvex. Let $0 < t_1 < t_2$,

$$I = \{x = (x_1, \ldots, x_n) \in \mathbb{R}^n;\ 0 < x_i < \left(\frac{t_1}{t_2}\right)^{1/n},\qquad i = 1, \ldots, n\}$$

and put $f(x) = t_2 \chi_I(x)$. Then, for all $x \in (0, 1)^n$, we have $Mf(x) \geq t_2|I| = t_1$ and $|\{x \in \mathbb{R}^n; \ Mf(x) > t_1\}| \geq 1$ so that (4) yields

$$\Phi(t_1) \leq c \int_I \Phi(cf(x)) \, dx = c|I| \Phi(ct_2) = c \frac{t_1}{t_2} \Phi(ct_2),$$

and we see that Φ is quasiconvex. ∎

Now let us turn our attention to the *modular inequality*

$$\int_{\mathbb{R}^n} \Phi(Mf(x)) \, dx \leq c \int_{\mathbb{R}^n} \Phi(cf(x)) \, dx, \qquad f \in L^1_{loc} \qquad (5)$$

with a constant c independent of f. The question concerns a characterization of Φ from Φ such that (5) holds. In the case of classical Young functions, the necessary condition was proven by various authors. It reads that the complementary function to Φ satisfies the Δ_2 condition. This means that the lower index of Φ is bigger than 1, the sufficient part follows by applying the classical Jensen inequality after proving that Φ^α is convex for some $0 < \alpha < 1$. The general case of $\Phi \in \Phi$ is treated in [KK], and we will briefly discuss it here.

LEMMA 1
Let (3) hold. Then constants $\alpha \in (0, 1)$ and $a_1 > 1$ exist such that

$$\Phi^\alpha(t) < \frac{1}{2a_1} \Phi^\alpha(a_1 t), \qquad t \in (0, \infty).$$

PROOF For any $\alpha \in (0, 1)$,

$$\Phi^\alpha(t) \leq \frac{1}{(2a)^\alpha} \Phi^\alpha(at).$$

If

$$\log_{2a} \frac{3a}{2} < \alpha < 1,$$

then

$$\Phi^\alpha(t) \leq \frac{2}{3a} \Phi^\alpha(at).$$

Consequently,

$$\Phi^\alpha(t) \leq \frac{2}{3a} \frac{2}{3a} \Phi^\alpha(a^2 t) \leq \frac{1}{2a^2} \Phi^\alpha(a^2 t),$$

which is (3) with $a_1 = a^2$. ∎

LEMMA 2

If (3) *holds, then the function* Φ *is quasiconvex.*

PROOF   It suffices to find a constant c such that

$$\frac{\Phi(t_1)}{t_1} \leq \frac{c\,\Phi(ct_2)}{t_2}, \qquad \text{whenever } 0 < t_1 < t_2. \tag{6}$$

Let $0 < t_1 < t_2$ and $t_2 \leq at_1$. As Φ is increasing in $[0, \infty)$ it is

$$\frac{\Phi(t_2)}{t_2} \geq \frac{\Phi(t_1)}{at_1},$$

therefore

$$\frac{\Phi(t_1)}{t_1} \leq \frac{a\,\Phi(t_2)}{t_2} \leq \frac{a\,\Phi(at_2)}{t_2}.$$

Now let $0 < t_1 < t_2$, $t_2 > at_1$. Then ($[\,\cdot\,]$ denotes the integral part)

$$\begin{aligned}
\Phi(t_2) &= \Phi\left(\frac{t_2}{t_1} \cdot t_1\right) \\
&= \Phi\left(a^{\log_a t_2/t_1}t_1\right) \\
&\geq \Phi\left(a^{[\log_a (t_2/t_1)]}t_1\right) \\
&\geq (2a)^{-1+\log_a (t_2/t_1)}\,\Phi(t_1) \\
&\geq 2^{-1+\log_a (t_2/t_1)}a^{-1+\log_a (t_2/t_1)}\,\Phi(t_1) \\
&\geq \frac{t_2}{t_1}a^{-1}\,\Phi(t_1),
\end{aligned}$$

whence

$$\frac{\Phi(t_1)}{t_1} \leq \frac{a\,\Phi(t_2)}{t_2} < \frac{a\,\Phi(at_2)}{t_2}.$$

We conclude that (6) holds whenever $0 < t_1 < t_2$. ∎

THEOREM 2

Let $\Phi \in \Phi$. *Then the following statements are equivalent:*

(i)   there exists a positive constant c such that the inequality (4) *holds for all*
      $f \in L^1_{loc}$;

(ii) *the function* Φ^α *is quasiconvex for some* $\alpha \in (0, 1)$;

(iii) *there exists a positive constant* c *such that*

$$\int_0^\sigma \frac{\Phi(s)}{s^2} \, ds \leq \frac{c \, \Phi(c\sigma)}{\sigma}, \qquad 0 < \sigma < \infty;$$

(iv) *there exists a positive constant* c *such that for all* $t > 0$

$$\int_0^t \frac{d \, \Phi(u)}{u} \leq \frac{c \, \Phi(ct)}{t};$$

(v)   *there exists a constant* $a > 1$ *such that*

$$\Phi(t) < \frac{1}{2a} \Phi(at), \qquad t \geq 0.$$

PROOF   We only give a sketch. Suppose (i) holds. Let $B(x, r)$ denote the ball of radius r centered at x. Put $f_t(x) = \chi_{B(0,1)}(x) \cdot t, t > 0$. If $|x| > 1$, then

$$Mf_t(x) \geq \frac{1}{|B(x, 2|x|)|} \int_{B(x,2|x|)} \chi_{B(0,1)}(y)t \, dy \geq \frac{t}{2^n |x|^n},$$

which yields

$$\int_{|x|>1} \Phi\left(\frac{t}{2^n |x|^n}\right) dx \leq c_1 \, \Phi(c_1 t).$$

Whence

$$\int_0^{2^{-n}t} \frac{\Phi(s)}{s^2} \, ds \leq c \frac{\Phi(ct)}{t}. \tag{7}$$

This is the condition from (iii). Now supposing (iii) holds, it is easy to see that Φ is quasiconvex. Indeed, if $0 < s_1 < s_2$, then

$$\frac{\Phi(s_1)}{s_1} \leq 2 \int_{s_1}^{2s_1} \frac{\Phi(t)}{t^2} \, dt \leq 2 \int_0^{2s_2} \frac{\Phi(t)}{t^2} \, dt \leq \frac{c_2 \, \Phi(2c_2 s_2)}{s_2}.$$

Let $d > 2^n$ be a constant (whose value will be determined later). The estimate (7) implies

$$\int_{td^{-1}}^{t2^{-n}} \frac{\Phi(s)}{s^2} \, ds \leq \frac{c \, \Phi(ct)}{t},$$

therefore

$$d\,\Phi\left(\frac{t}{db}\right)\log\frac{d}{2^n} \le bc\,\Phi(ct).$$

If we put $t\,(db)^{-1} = \tau$ and choose d in such a manner that

$$(bc)^2\left(\log\frac{d}{2^n}\right)^{-1} < \frac{1}{2},$$

then

$$\Phi(\tau) \le \frac{1}{2bcd}\,\Phi(bcd\tau)$$

which is (3) with $a = bcd$, that is, (v) holds. Lemmas 1 and 2 imply (ii). As noticed earlier, the quasi-Jensen inequality and the L^p inequality (applied with some $p \in \left(1, \frac{1}{\alpha}\right)$ yield (i). ∎

In the case of the Riesz transforms, additional restriction on the growth of the function $\Phi \in \Phi$ is already necessary when the inequality of weak type is considered. The symbol $\Phi^e(L)$ denotes the set of those $f \in \Phi(L)$ for which R_j exists, $j = 1, \ldots, n$. We will not discuss conditions for the convergence of integrals considered.

THEOREM 3
Let $\Phi \in \Phi$. Then

$$\Phi(\lambda)|\{x \in \mathbb{R}^n;\ |R_j f(x)| > \lambda\}| \le c \int_{\mathbb{R}^n} \Phi(f(x))\,dx, \qquad j = 1, \ldots, n, \quad (8)$$

for some $c > 0$, all $\lambda > 0$ and all $f \in \Phi^e(L)$ iff Φ is quasiconvex and satisfies the Δ_2 condition.

PROOF   As to the sufficiency, define for $\lambda > 0$,

$$_\lambda f(x) = \begin{cases} f(x) & \text{if } |f(x)| \le \lambda, \\ 0 & \text{if } |f(x)| > \lambda, \end{cases}$$

and

$$^\lambda f(x) = \begin{cases} f(x) & \text{if } |f(x)| > \lambda, \\ 0 & \text{if } |f(x)| \le \lambda. \end{cases}$$

We have

$$\Phi(\lambda)|\{x \in \mathbb{R}^n;\ |R_j f(x)| > \lambda\}| \le \Phi(\lambda)\left|\left\{x \in \mathbb{R}^n;\ |R_j\,_\lambda f(x)| > \frac{\lambda}{2}\right\}\right|$$

$$+ \Phi(\lambda)\left|\left\{x \in \mathbb{R}^n;\ |R_j\,^\lambda f(x)| > \frac{\lambda}{2}\right\}\right|.$$

Since Φ is quasiconvex and satisfies the Δ_2 condition, we have, using the L^p result ($p \geq 1$),

$$\Phi(\lambda)\left|\left\{x \in \mathbb{R}^n;\ |R_j{}^\lambda f(x)| > \frac{\lambda}{2}\right\}\right| \leq \frac{c\,\Phi(\lambda)}{\lambda} \int_{\mathbb{R}^n} |{}^\lambda f(x)|\,dx$$

$$= \frac{c\,\Phi(\lambda)}{\lambda} \int_{\{|f(x)|>\lambda\}} |f(x)|\,dx$$

$$\leq c_1 \int_{\mathbb{R}^n} \Phi(af(x))\,dx$$

$$\leq c_2 \int_{\mathbb{R}^n} \Phi(f(x))\,dx.$$

On the other hand, if $\Phi \in \Delta_2$, one can show that $b > 0$ and $p > 1$ such that

$$\frac{\Phi(t_2)}{t_2^p} \leq \frac{b\,\Phi(t_1)}{t_1^p}, \qquad 0 < t_1 < t_2$$

(see [KK]). Once again employing the L^p inequality, we obtain

$$\Phi(\lambda)|\{x \in \mathbb{R}^n;\ |R_j f(x)| > \lambda\}| \leq \frac{c\,\Phi(\lambda)}{\lambda^p} \int_{\mathbb{R}^n} |{}_\lambda f(x)|^p\,dx$$

$$= \frac{c\,\Phi(\lambda)}{\lambda^p} \int_{\{|f(x)|\leq\lambda\}} |f(x)|^p\,dx$$

$$\leq c \int_{\mathbb{R}^n} |f(x)|^p \frac{\Phi(f(x))}{|f(x)|^p}\,dx$$

$$\leq c \int_{\mathbb{R}^n} \Phi(f(x))\,dx.$$

Let (8) hold, we prove that Φ is quasiconvex. Let $0 < t_1 < t_2$ and

$$I = \{x = (x_1, \ldots, x_n) \in \mathbb{R}^n;\ 0 < x_i < \left(\frac{t_1}{t_2}\right)^{1/n}, \quad i = 1, \ldots, n\},$$

$$f_1(x) = dt_2\chi_I(x),$$

where d is a positive constant. If $x = (x_1, \ldots, x_n) \in (2, 3)^n$, then

$$R_j f(x) = c_n dt_2 \int_I \frac{x_j - y_j}{|x - y|^{n+1}} f(y)\,dy \geq \frac{c_n d}{3^{n+1} n^{(n+1)/2}} t_2 \frac{t_1}{t_2}.$$

Putting $d = 3^{n+1} n^{(n+1)/2} c_n^{-1}$, we see that

$$|\{x \in \mathbb{R}^n; \ |R_j f(x)| > t_1\}| \geq 1.$$

Consequently,

$$\Phi(t_1) \leq c \int_I \Phi(f(x)) \, dx = \frac{ct_1}{t_2} \Phi(dt_2) \leq \max(c, d) \frac{t_1}{t_2} \Phi(\max(c, d)t_2),$$

and thus the function Φ is quasiconvex. Finally, Φ must satisfy the Δ_2 condition. Indeed, let $g(x) = \left(\frac{\lambda}{2}\right) \chi_I(x)$, $x \in \mathbb{R}^n$, where $I = (0, 1)^n$. Inserting g into (8), we obtain

$$\Phi(\lambda) |\{x \in \mathbb{R}^n; \ |R_j \chi_I(x)| > 2\}| \leq c \, \Phi\left(\frac{\lambda}{2}\right).$$

The quantity $|\{x \in \mathbb{R}^n; \ |R_j \chi_I(x)| > 2\}|$ is positive and independent of λ, whence the last inequality turns into the Δ_2 condition for Φ. ∎

Validity of the modular inequality requires more discussion.

THEOREM 4
Let $\Phi \in \Phi$. Then the following statements are equivalent:

(i)   *Φ satisfies the Δ_2 condition and Φ^α is quasiconvex for some $\alpha \in (0, 1)$;*

(ii)   *a constant $c > 0$ exists such that*

$$\int_{\mathbb{R}^n} \Phi(R_j f(x)) \, dx \leq c \int_{\mathbb{R}^n} \Phi(f(x)) \, dx, \qquad j = 1, \ldots, n,$$

for all $f \in \Phi^e(L)$.

PROOF   We restrict ourselves to the necessary part. Details of the rest can be found in [KK]. Set $f(x) = t \chi_{B(0,1)}(x)$ where $B(0, 1)$ is the unit ball centered at the origin. Then

$$\int_{|x| > 2\sqrt{n}} \Phi\left(\sum_{j=1}^{n} |R_j f(x)|\right) \leq c \, \Phi(t).$$

Let j_k, $k = 1, \ldots, m$, be those indices for which $|x_{j_k}| > 2$, then

$$|R_{j_k} f(x)| = c_n t \int_B \frac{|x_{j_k} - y_{j_k}|}{|x - y|^{n+1}} \, dy.$$

On the other hand, $|x_{j_k} - y_{j_k}| \geq |x_{j_k}| - |y_{j_k}| \geq 2^{-1}|x_{j_k}|$, and at the same time $|x - y| \leq |x| + |y| \leq 1 + |x| < c|x|$. Therefore

$$|R_{j_k} f(x)| \geq ct \frac{|x_{j_k}|}{|x|^{n+1}}.$$

Furthermore,

$$\sum_{j=1}^{n} |R_j f(x)| \geq \sum_{k=1}^{m} |R_{j_k} f(x)| \geq ct |x|^{-(n+1)} \sum_{k=1}^{m} |x_{j_k}| |x|^{n+1},$$

where $|x_{j_k}| > 2$. The absolute value of all the remaining coordinates is smaller than 1 so that

$$\sum_{j=1}^{n} |R_j f(x)| \geq ct |x|^{-(n+1)} \sum_{j=1}^{n} |x_i| \geq ct |x|^{-n}.$$

This is an estimate analogous to that for the maximal function from the beginning of the proof of Theorem 1; we can proceed from there and see that Φ^α is quasiconvex for some $\alpha \in (0, 1)$. ∎

Considering vector-valued inequalities, the characterization is a bit surprising as restrictions on the growth of the function Φ from above turn out to be necessary.

THEOREM 5
Let $\Phi \in \Phi$, $1 < \theta < \infty$. Then there exists a constant $c > 0$ such that the inequality

$$\Phi(\lambda) |\{x \in \mathbb{R}^n; \|Mf(x)\|_\theta > \lambda\}| \leq c \int_{\mathbb{R}^n} \Phi(\|f(x)\|_\theta) \, dx \qquad (9)$$

holds for every $\lambda > 0$ and every vector-valued function $f = (f_1, f_2, \ldots)$ from L^1_{loc} iff Φ is quasiconvex and satisfies the Δ_2 condition.

PROOF Again, we only sketch the necessary part. The sufficiency can actually be proven for any subadditive operator of weak type $(1, 1)$ and (p, p) with $p > 1$ large enough (for details see [KK]), using the Fefferman–Stein L^p theorem on vector-valued maximal inequalities.

The quasiconvexity of Φ is clear from the scalar case. We show that Φ satisfies the Δ_2 condition. Set $f_k(x) = \lambda \chi_{I_k}(x)$, $k = 1, 2, \ldots$, where

$$I_k = \{x \in \mathbb{R}^n; \ 2^{-k-2} < x_i < 2^{-k-1}, \quad i = 1, 2, \ldots, n\}, \quad k = 1, \ldots .$$

It is

$$\|f(x)\|_\theta = \left(\sum_{k=1}^{\infty} |f_k(x)|^\theta \right)^{1/\theta} = \lambda \chi_{\cup_k I_k}(x), \qquad x \in \mathbb{R}^n,$$

so that the right side of (9) is estimated by $c \Phi(\lambda)$.

Let now $1 \leq k \leq m \leq j$ where m will be fixed later and denote

$$I_{jk} = \{x \in \mathbb{R}^n; \ 2^{-j-2} < x_i < 2^{-k-1} \quad i = 1, \ldots, n \}.$$

If $x \in I_j$, then

$$Mf_k(x) \geq \frac{\lambda |I_k|}{|I_{jk}|} \geq \frac{\lambda}{2^n}$$

so that

$$\|Mf(x)\|_\theta \geq \left(\sum_{k=1}^m |Mf_k(x)|^\theta \right)^{1/\theta} \geq \frac{\lambda m^{1/\theta}}{2^n},$$

whence, in accordance with the hypothesis,

$$\Phi\left(\frac{\lambda m^{1/\theta}}{2^n} \right) \left| \bigcup_{j \geq m} I_j \right| \leq c\,\Phi(\lambda).$$

If m is now such that $m^{1/\theta} 2^{-n} \geq 2$, the Δ_2 condition for Φ follows. ∎

Now it is not surprising that the necessary and sufficient condition on Φ, in the case of strong type, reads as follows.

THEOREM 6
Let $\Phi \in \Phi$, $1 < \theta < \infty$. Then the following statements are equivalent:

(i)   *$\Phi \in \Delta_2$ and Φ^α is quasiconvex for some $\alpha \in (0, 1)$;*
(ii) *a constant c exists such that*

$$\int_{\mathbb{R}^n} \Phi(\|Mf(x)\|_\theta)\,dx \leq c \int_{\mathbb{R}^n} \Phi(\|f(x)\|_\theta)\,dx$$

for all $f = (f_1, f_2, \ldots) \in L^1_{\text{loc}}$.

PROOF   The implication (ii)⟹(i) follows, invoking Theorem 2 and Theorem 5. The converse is rather tedious, and we refer to [KK]. ∎

Also proofs of the following two theorems will be omitted. They go along the lines of those of Theorems 3 and 4.

THEOREM 7
Let $\Phi \in \Phi$, $1 < \theta < \infty$. Then the following statements are equivalent:

(i)   *Φ is quasiconvex and satisfies the Δ_2 condition;*
(ii) *a constant $c > 0$ exists such that*

$$\Phi(\lambda)|\{x \in \mathbb{R}^n;\ \|R_j f(x)\|_\theta > \lambda\}| \leq c \int_{\mathbb{R}^n} \Phi(\|f(x)\|_\theta)\,dx, \quad j = 1, \ldots, n,$$

holds for all $\lambda > 0$ and $f = (f_1, f_2, \ldots)$ such that $\|f(.)\|_\theta \in \Phi(L)$ and $R_j f_k$ exists for all $j = 1, \ldots, n$, $k = 1, 2, \ldots$.

THEOREM 8

Let $1 < \theta < \infty$. *Then the following statements are equivalent:*

(i) Φ *satisfies the* Δ_2 *condition and* Φ^α *is quasiconvex for some* $\alpha \in (0, 1)$;
(ii)   *there exists a constant* $c > 0$ *such that*

$$\int_{R^n} \Phi(\|R_j f(x)\|_\theta)\, dx \le c \int_{R^n} \Phi(\|f(x)\|_\theta)\, dx$$

for all $f = (f_1, f_2, \ldots)$ *with* $\|f(.)\|_\theta \in \Phi(L)$ *and such that* $R_j f_k$ *exists for all* $j = 1, \ldots, n$, $k = 1, 2, \ldots$.

References

[BKu] R. J. Bagby and D. S. Kurtz, *L*(log *L*) *spaces and weights for the strong maximal function,* J. Analyse Math. 44 (1984/85), 21–31.

[BS]   C. Bennett and R. Sharpley, *Interpolation of operators,* Academic Press, 1988.

[BKe] S. Bloom and R. Kerman, *Weighted Orlicz space integral inequalities for the Hardy–Littlewood maximal operator,* preprint, 1992.

[B]   D. W. Boyd, *Indices of function spaces and their relationship to interpolation,* Canad. J. Math. 21 (1969), 1245–1254.

[C]   A. Carbery, S.-Y. Chang, and J. Garnett, A_p *weights and* $L \log L$, Pacific J. Math. 120 (1985), 33–45.

[Ga]   E. Gagliardo, *Caratterizzazioni delle trace sulla frontiera relative ad alcune classi di funzioni in n variabili,* Rend. Sem. Mat. Padova 27 (1957), 284–305.

[Go1] A. Gogatishvili, *Weak type inequality for strong maximal functions (Russian),* In: Abstracts of Symposium on Continuum Mechanics and Related Problems of Analysis, Tbilisi 5–12.6.1991.

[Go2] A. Gogatishvili, *Weak type weighted inequalities for maximal functions with respect to a general basis,* to appear in Bull. Acad. Sci. Georgia.

[Gos] J.-P. Gossez, *Nonlinear elliptic boundary value problems for equations with rapidly (or slowly) increasing coefficients,* Trans. Amer. Math. Soc. 190 (1974), 163–205.

[Gu]   M. de Guzmán, *Differentiation of integrals in* R^n, Lecture Notes in Math. Vol. 481, Springer-Verlag, Berlin (1975).

[KT]   R. Kerman and A. Torchinsky, *Integral inequalities with weights for the Hardy maximal function,* Studia Math. 71 (1982), 277–284.

[KK]   V. Kokilashvili and M. Krbec, *Weighted inequalities in Lorentz and Orlicz spaces,* World Scientific, Singapore (1991).

[KR]   M. A. Krasnoselskii and J. B. Rutitskii, *Convex functions and Orlicz spaces,* Noordhof, Groningen, (1961). (Original Russian edition: Gos. Izd. Fiz. Mat. Lit., Moskva 1958.)

[Krb] M. Krbec, *Two weight weak type inequalities for the maximal function in the Zygmund class,* In: Function Spaces and Applications. Proceedings of the US-Swedish Seminar held in Lund, June 1986. Lecture Notes in Mathematics., Vol. 1302, pp. 317–320, Springer-Verlag, Berlin (1988).

[Kre] S. G. Krein, J. I. Petunin, and E. M. Semenov, *Interpolation of linear operators (Russian),* Nauka, Moskva, (1978). (English transl.: Translations of Mathematical Monographs, Amer. Math. Soc., Providence, RI 1982.)

[L]   M.-T. Lacroix, *Espaces de traces des espaces de Sobolev–Orlicz,* J. Math. Pures. Appl. 53 (1974), 439–458.

[MP]   L. Maligranda and L. E. Persson, *Bibliography on interpolation spaces,* University of Umeå.

[MO]   W. Matuszewska and W. Orlicz, *On certain properties of φ functions,* Bull. Acad. Polon. Sci. 8 (1960), 439–443.

[Mu]   J. Musielak, *Orlicz spaces and modular spaces,* Springer-Verlag, Lecture Notes in Math., Vol. 1034, Berlin (1983).

[P]   G. Palmieri, *Un approccio alla teoria degli spazi di traccia relativi agli spazi di Orlicz-Sobolev,* Bolletino U. M. I. 5, 16–B (1979), 100–119.

[RR]   M. M. Rao and Z. D. Ren, *Theory of Orlicz spaces,* M. Dekker, Inc., New York (1991).

[S]   C. Sadosky, *Interpolation of operators and singular integrals,* M. Dekker Inc., New York (1979).

[T]   A. Torchinsky, *Interpolation of operators and Orlicz classes,* Studia Math. 59 (1976), 177–207.

15

On the Herz Spaces with Power Weights

Shanzhen Lu
Fernando Soria[1]

ABSTRACT   *In this paper we develop some ideas about power weights due to Weiss and the second author, which we apply to the study of boundedness of certain operators on Herz spaces. To that end, we introduce a class of weighted Herz spaces, $K_q(\mathbb{R}^n; |x|^{-\alpha})$, $0 \le \alpha < n$, for which we find a characterization in terms of a decomposition with appropriate q-blocks in the sense of Taibleson and Weiss. As an application, we prove that many classical operators map $Kq(\mathbb{R}^n, |x|^{-\alpha})$ into itself, provided $0 < \alpha < n$, but the conclusion fails for $\alpha = 0$.*

15.1   Introduction

Herz spaces were introduced in connection with the study of absolutely convergent Fourier transforms. They are defined as follows:

Let $B_k = \{x \in \mathbb{R}^n : |x| \le 2^k\}$, $C_k = B_k - B_{k-1}$ and $\chi_k = \chi_{C_k}$, for $k \in \mathbb{Z}$. Given $1 < q < \infty$ and $q' = \frac{q}{(q-1)}$, we consider the Herz space $K_q(\mathbb{R}^n)$ as the class of those functions $f \in L_{\text{loc}}^q(\mathbb{R}^n \setminus \{0\})$ so that

$$\|f\|_{K_q} := \sum_{k \in \mathbb{Z}} |B_k|^{1/q'} \|f \chi_k\|_{L^q} < \infty.$$

(See [H] or [LY].) It is well known that $K_q(\mathbb{R}^n)$ represents a homogeneous version of the Beurling algebra $A^q(\mathbb{R}^n)$.

In this section, we will show that K_q has a useful decomposition characterization, and the same holds for the Herz spaces with power weights defined below. By means of this characterization, we shall prove that many classical operators are bounded on these spaces with respect to the power weights $w_\alpha(x) = |x|^{-\alpha}$, $0 < \alpha < n$, but this is not true for $\alpha = 0$.

Let us begin with some definitions. Let $w_\alpha(x) = |x|^{-\alpha}$, $0 \le \alpha < n$, and $1 < q < \infty$. Set q' as above. Then, the Herz space with respect to the weight

---

[1] Both authors were supported by DGICYT, the first one under a sabbatical fellowship.

w_α, $K_q(\mathbb{R}^n, w_\alpha)$, consists of those functions $f \in L_{loc}^q(\mathbb{R}^n \setminus \{0\}, w_\alpha)$ satisfying

$$\|f\|_{K_q(w_\alpha)} := \sum_{k \in \mathbb{Z}} [w_\alpha(B_k)]^{1/q'} \|f \chi_k\|_{L^q(w_\alpha)} < \infty,$$

where $w_\alpha(E) = \int_E w_\alpha(x) \, dx$.

To establish the block decomposition of $K_q(\mathbb{R}^n, w_\alpha)$, we need to extend the notion of q-block of Taibleson and Weiss [TW] into the weighted case.

With the same notation, we state that a measurable function $b(x)$, is a central (q, w_α)-block if the support of b is contained in a ball $B = B(0, r)$, with center at the origin, and so that

$$\|b\|_{q, w_\alpha} \leq [w_\alpha(B)]^{-1/q'}.$$

We now have the following theorem.

THEOREM 1

Suppose $1 < q < \infty$ and $w_\alpha(x) = |x|^{-\alpha}, 0 \leq \alpha < n$. Then, the following are equivalent:

  (i) $f \in K_q(\mathbb{R}^n, w_\alpha)$
  (ii) f can be represented as $f(x) = \sum_{k \in \mathbb{Z}} \lambda_k b_k(x)$, where each b_k is a central (q, w_α)-block and $\sum |\lambda_k| < \infty$. Moreover, the infimum of $\sum |\lambda_k|$ over all such representations, gives an equivalent norm on $K_q(\mathbb{R}^n, w_\alpha)$.

Before we begin the proof, let us note that any central (q, w_α)-block lies in the unit ball of $L^1(\mathbb{R}^n, w_\alpha)$. Therefore, the above series is convergent in the (strong) L^1 sense.

PROOF   The proof is straightforward. Let us assume first that $f \in K_q(\mathbb{R}^n, w_\alpha)$. Write

$$f(x) = \sum_k f(x) \chi_k(x) = \sum_k \lambda_k b_k(x),$$

where $\lambda_k = [w_\alpha(B_k)]^{1/q'} \|f \chi_k\|_{q, w_\alpha}$ and $b_k(x) = \frac{1}{\lambda_k} f(x) \chi_k(x)$, if $\lambda_k \neq 0$ and $b_k(x) = 0$ if $\lambda_k = 0$. Since by hypothesis, $f \in L_{loc}^q(\mathbb{R}^n \setminus \{0\}, w_\alpha)$, λ_k is finite, and hence b_k is a central (q, w_α)-block with supporting ball at B_k. Moreover,

$$\sum_{k \in \mathbb{Z}} |\lambda_k| = \sum_{k \in \mathbb{Z}} [w_\alpha(B_k)]^{1/q'} \|f \chi_k\|_{q, w_\alpha} = \|f\|_{K_q(w_\alpha)}.$$

Reciprocally, to prove (ii) $\Rightarrow$ (i), it suffices to show that if b is a central (q, w_α)-block, then $b \in K_q(\mathbb{R}^n, w_\alpha)$ (uniformly). Let $B = B(0, r)$ be the supporting ball of b. Take $j \in \mathbb{Z}$ with $2^{j-1} \leq r < 2^j$. Then

$$\sum_{k \in \mathbb{Z}} [w_\alpha(B_k)]^{1/q'} \|b\chi_k\|_{q.w_\alpha} \leq \sum_{k \leq j} [w_\alpha(B_k)]^{1/q'} [w_\alpha(B)]^{-1/q'}$$

$$\sim \left(\sum_{k \leq j} 2^{k(n-\alpha)/q'} \right) \left(2^{-j(n-\alpha)/q'} \right) \sim C_\alpha. \quad \blacksquare$$

As an immediate consequence of Theorem 1, the following proposition shows how $K_q(\mathbb{R}^n, w_\alpha)$ is close to $L^1(\mathbb{R}^n, w_\alpha)$.

PROPOSITION 1

Let $1 < q < \infty$, $w_\alpha(x) = |x|^{-\alpha}$, $0 \leq \alpha < n$. If f is positive, radial, and decreasing, then $f \in K_q(\mathbb{R}^n, w_\alpha)$, if and only if, $f \in L^1(\mathbb{R}^n, w_\alpha)$.

PROOF It suffices to show that if f is positive, radial, and decreasing, then $f \in L^1(\mathbb{R}^n, w_\alpha)$ implies $f \in K_q(\mathbb{R}^n, w_\alpha)$. For such f, let us define $E_k = \{x \in \mathbb{R}^n : f(x) \geq 2^k\}$, $k \in \mathbb{Z}$. We write

$$f(x) = \sum_{k \in \mathbb{Z}} f(x) \chi_{\{E_k \setminus E_{k+1}\}}(x) = \sum_{k \in \mathbb{Z}} \lambda_k b_k(x),$$

where

$$b_k(x) = \frac{1}{2^{k+1} w_\alpha(E_k)} f(x) \chi_{E_k \setminus E_{k+1}}(x).$$

Since f is radial and decreasing, E_k is a ball with center at the origin, sup $b_k \subset E_k$ and $\|b_k\|_{q.w_\alpha} \leq [w_\alpha(E_k)]^{-1/q'}$. Thus each b_k is a central (q, w_α)-block. Moreover,

$$\sum_k |\lambda_k| = \sum_k 2^{k+1} w_\alpha(E_k) \sim \|f\|_{L^1(w_\alpha)}.$$

Therefore, it follows from Theorem 1 that $\sum \lambda_k b_k(x)$ is in $K_q(\mathbb{R}^n, w_\alpha)$, and hence $f \in K_q(\mathbb{R}^n, w_\alpha)$. $\blacksquare$

15.2 Applications to boundedness of operators

As an application of Theorem 1, we shall prove that many classical operators map $K_q(\mathbb{R}^n, |x|^{-\alpha})$ into itself provided $0 < \alpha < n$. Among them, we can consider Calderón–Zygmund operators, Carleson's maximal operator in one dimension, strong singular multipliers of Fefferman, maximal Bochner–Riesz means at critical index, oscillatory integral operators of Phong–Ricci–Stein, etc., and, in general, any sublinear operator bounded on $L^q(\mathbb{R}^n)$ and satisfying

$$|Tf(x)| \le C \int \frac{|f(y)|}{|x-y|^n} \, dy$$

for x "far" from the support of f.

We shall state our result in the following general form.

THEOREM 2

Let $w_\alpha(x) = |x|^{-\alpha}$, $0 < \alpha < n$. If the sublinear operator T satisfies

$$|Tf(x)| \le C \int \frac{|f(y)|}{|x-y|^n} \, dy,$$

and T maps $L^q(\mathbb{R}^n)$ into $L^q(\mathbb{R}^n)$, $1 < q < \infty$, then T maps $K_q(\mathbb{R}^n, w_\alpha)$ into $K_q(\mathbb{R}^n, w_\alpha)$.

PROOF   By Theorem 1, it suffices to show that the inequality

$$\|Tb\|_{K_q(\mathbb{R}^n, w_\alpha)} \le C$$

holds for any central (q, w_α)-block $b(x)$, where C is independent of b.

Let $B = B(0, r)$ be the supporting ball of $b(x)$. There exists a $j \in \mathbb{Z}$ such that $2^{j-2} < r \le 2^{j-1}$.

Now we write

$$\|Tb\|_{K_q(\mathbb{R}^n, w_\alpha)} = \sum_{k \in \mathbb{Z}} [w_\alpha(B_k)]^{1/q'} \|(Tb)\chi_k\|_{q, w_\alpha}$$

$$= \left(\sum_{k \le j} + \sum_{k > j} \right) \cdots \overset{\text{def}}{=} \sum_1 + \sum_2.$$

Under the hypothesis of our theorem, it follows (see [SW]) that T maps $L^q(\mathbb{R}^n, w_\alpha)$ into $L^q(\mathbb{R}^n, w_\alpha)$. Thus

$$\|(Tb)\chi_k\|_{q, w_\alpha} \le C \left(\int_B |b(x)|^q w_\alpha(x) \, dx \right)^{1/q}$$

$$\le C[w_\alpha(B)]^{(1/q)-1} \le C[w_\alpha(B_j)]^{(1/q)-1}.$$

Hence

$$\sum_1 \le C \sum_{k \le j} \left[\frac{w_\alpha(B_k)}{w_\alpha(B_j)} \right]^{1-(1/q)}$$

$$= C \sum_{k \le j} \frac{1}{2^{(j-k)(n-\alpha)(1-(1/q))}} \le C \sum_{\ell=0}^{\infty} \frac{1}{2^{\ell(n-\alpha)(1-(1/q))}} < \infty.$$

To estimate $\sum_2$, we use the size condition of T, and obtain

$$\|(Tb)\chi_k\|_{q,w_\alpha}^q = \int_{C_k} |Tb(x)|^q w_\alpha(x)\, dx$$

$$\leq C \int_{C_k} \left(\int_B \frac{|b(y)|}{|x-y|^n}\, dy \right)^q w_\alpha(x)\, dx.$$

Note that if $x \in C_k$, $y \in B$ and $j < k$, then $|x - y| \sim |x|$. Therefore, we have

$$\|(Tb)\chi_k\|_{q,w_\alpha}^q \leq C \int_{C_k} \frac{w_\alpha(x)}{|x|^{nq}} \left(\int_B |b(y)|\, dy \right)^q$$

$$\leq C \int_{C_k} \frac{dx}{|x|^{nq+\alpha}} \left(\int_B |b(y)|^q\, dy \right) |B|^{q/q'}$$

$$\leq C \int_{C_k} \frac{dx}{|x|^{nq+\alpha}} \left(\frac{1}{ess\, \inf_B w_\alpha} \right) \left(\int_B |b(y)|^q w_\alpha(y)\, dy \right) |B|^{q/q'}.$$

Since $w_\alpha \in A_1$, (Muckenhoupt class), we have

$$\frac{w_\alpha(B)}{|B|} \leq C\, ess\, \inf_{y \in B} [w_\alpha(y)].$$

From this inequality, it follows

$$\|(Tb)\chi_k\|_{q,w_\alpha}^q \leq C \int_{C_k} \frac{dx}{|x|^{nq+\alpha}} \frac{|B|}{w_\alpha(B)} [w_\alpha(B)]^{1-q} |B|^{q-1}$$

$$\leq C \left(\int_{C_k} \frac{dx}{|x|^{nq+\alpha}} \right) \left(\frac{|B|}{w_\alpha(B)} \right)^q.$$

Thus

$$\sum_2 \leq C \sum_{k>j} [w_\alpha(B_k)]^{1/q'} \left(\int_{C_k} \frac{dx}{|x|^{nq+\alpha}} \right)^{1/q} \frac{|B|}{w_\alpha(B)}$$

$$\leq C \sum_{k>j} \frac{1}{2^{(k-j)\alpha}} \leq C \sum_{\ell=0}^{\infty} \frac{1}{2^{\ell\alpha}} < \infty.$$

Observe that we have used here the assumption $\alpha > 0$. This completes the proof of Theorem 2. ∎

REMARK 1 We shall point out that the conclusion of Theorem 2 is false for $\alpha = 0$, even in the 1-dimensional case. ∎

Let H be the Hilbert transform. It is easy to see

$$|(H\chi_{(-1,1)})(x)| \sim \frac{1}{|x|}, \qquad \text{if } |x| > 2.$$

It follows from this that $H\chi_{(-1,1)} \notin L^1(\mathbb{R})$, and $H\chi_{(-1,1)} \notin K_q(\mathbb{R})$, while $\chi_{(-1,1)}$ is a multiple of a central (q, w_0)-block. In Section 15.3 of this chapter, we shall find a substitute result in this case.

As in [SW], Theorem 2 can be extended in several ways. We present here only one of such possible generalizations.

THEOREM 3
Let $1 < \rho_0 \leq q < \infty$ and $w_\alpha(x) = |x|^{-\alpha}, n - \frac{n}{\rho_0} < \alpha < n$. If the sublinear operator T satisfies

$$|Tf(x)| \leq C \left(\int_{\mathbb{R}^n} \frac{|f(y)|^{\rho_0}}{|x - y|^n} \, dy \right)^{1/\rho_0}$$

and T maps $L^q(\mathbb{R}^n)$ into $L^q(\mathbb{R}^n)$, then T maps $K_q(\mathbb{R}^n, |x|^{-\alpha})$ into $K_q(\mathbb{R}^n, |x|^{-\alpha})$.

As an example, let us consider the smooth square function of Rubio de Francia defined by

$$(Sf)(x) := \left(\sum_{k,j} |(\varphi_{k,j} \widehat{f})^\vee (x)|^2 \right)^{1/2},$$

where φ is a function in the Schwarz class and $\varphi_{k,j}(y) = \varphi(2^k y - n_{k,j})$ with $n_{k,j} \neq n_{k,j'}$ (integers) if $j \neq j'$. Since S satisfies (see [SW])

$$(Sf)(x) \leq C \left(\int_{\mathbb{R}} \frac{|f(y)|^2}{|x - y|} \, dy \right)^{1/2},$$

and S maps $L^q(\mathbb{R})$ into $L^q(\mathbb{R})$ with $q \geq 2$ (see [R]), Theorem 3 can be applied to this operator to yield that S maps $K_q(\mathbb{R}, |x|^{-\alpha})$ into $K_q(\mathbb{R}, |x|^{-\alpha})$, provided $2 \leq q < \infty$ and $\frac{1}{2} < \alpha < 1$.

The proof of Theorem 3 requires only some technical modifications from the proof of the previous theorem in the spirit of the arguments given in [SW]. The details are left to the reader.

15.3 Hardy spaces associated with $K_q(\mathbb{R}^n, |x|^{-\alpha})$ and its characterization

We have pointed out in Section 15.2 that the conclusion of Theorem 2 fails for $\alpha = 0$, but that there is, however, a substitute result for this case. To see this, we introduce the Hardy space associated with $K_q(\mathbb{R}^n, w_\alpha)$, denoted by $HK_q(\mathbb{R}^n, w_\alpha)$, as the class of all functions $f \in L^q_{loc}(\mathbb{R}^n \setminus \{0\}, w_\alpha)$, which satisfy

$$Gf \in K_q(\mathbb{R}^n, w_\alpha),$$

where Gf is the grand maximal function of f, introduced by Feffermam and Stein [FS]. That is,

$$Gf(x) = \sup_{\varphi \in A_N} \sup_{|x-y|<t} |(f * \varphi_t)(y)|,$$

where A_N is a class of appropriate smooth functions satisfying the uniform estimate

$$\sup_{|\alpha|,|\beta| \leq N} |x^\alpha D^\beta \varphi(x)| \leq C,$$

for $N \in \mathbb{N}$ sufficiently large.

The result is then the following.

THEOREM 4
If the operator T, defined by

$$Tf(x) = P.V. \int_{\mathbb{R}^n} K(x, y) f(y) \, dy,$$

is bounded on $L^q(\mathbb{R}^n)$, $1 < q < \infty$, and $K(x, y)$ satisfies

$$|K(x, y) - K(x, 0)| \leq \frac{C|y|^\delta}{|x - y|^{n+\delta}}, \text{ if } |y| < \frac{|x - y|}{2},$$

where $0 < \delta \leq 1$, then T maps $HK_q(\mathbb{R}^n, w_\alpha)$ into $K_q(\mathbb{R}^n, w_\alpha)$, $0 \leq \alpha < n$.

REMARK 2   It is easy to see that the conditions in Theorem 4 are satisfied by δ-Calderón–Zygmund operators. (See [T] for the definition.)   ∎

Also, the case $0 < \alpha < n$ in the theorem follows from Theorem 2. Therefore we are only interested in the proof when $\alpha = 0$, which is the endpoint result mentioned in Section 15.2.

To prove Theorem 4, let us characterize $HK_q(\mathbb{R}^n, w_0) = HK_q$ in terms of an atomic decomposition. To do that, we state that the L^q function $a(x)$ is a central $(1, q)$-atom if it satisfies the following:

(1)   Sup $a \subset B$, for some ball centered at the origin.
(2) $\|a\|_q \leq |B|^{-1/q'}$.
(3) $\int a(x) dx = 0$.

THEOREM 5
Let $1 < q < \infty$. Then the following are equivalent:

(i) $f \in HK_q$;
(ii) f can be represented as $f(x) = \sum_k \lambda_k a_k(x)$ where each a_k is a central $(1, q)$-atom and $\sum |\lambda_k| < \infty$. Moreover, the infimum of $\sum |\lambda_k|$ over all such representations gives an equivalent norm on HK_q.

PROOF OF THEOREM 4   As we stated above, we will only consider the case $\alpha = 0$. From Theorem 5, it suffices to show $\|Ta\|_{K_q} \leq C$, for all central $(1, q)$-atoms a, with C independent of a.

Let $B = B(0, r)$ be the supporting ball of such an atom and let j be so that $2^{j-1} \leq r < 2^j$. Write

$$\|Ta\|_{K_q} = \sum_k |B_k|^{1/q'} \|(Ta)\chi_k\|_q = \left(\sum_{k \leq j+1} + \sum_{k > j+1} \right) \cdots = \sum_1 + \sum_2.$$

By the L^q-boundedness hypothesis of T, we obtain

$$\|(Ta)\chi_k\|_q \leq C\|a\|_q \leq C|B|^{-1/q'}.$$

Thus

$$\sum_1 \leq C \sum_{k \leq j+1} |B_k|^{1/q'} |B|^{-1/q'} \sim C \sum_{k \leq j+1} 2^{n(k-j)/q'} \sim C.$$

To estimate Σ_2, we make use of the size condition on $K(x, y)$. Since

$$Ta(x) = \int_B a(y) \left(K(x, y) - K(x, 0) \right) dy,$$

due to the cancellation property of a, and $|x - y| > 2|y|$ if $x \in C_k$, $y \in B$ and $k > j + 1$, that size condition gives, for $x \in C_k$,

$$|(Ta)(x)| \leq C \int_B |a(y)| \frac{|y|^\delta}{|x - y|^{n+\delta}} \, dy \leq C \frac{2^{j\delta}}{|x|^{n+\delta}}.$$

Hence

$$\sum_2 \leq C \sum_{k>j+1} 2^{nk/q'} 2^{j\delta} \left(\int_{C_k} \frac{dx}{|x|^{(n+\delta)q}} \right)^{1/q} \sim C \sum_{k>j+1} 2^{-(k-j)\delta} \sim C_\delta,$$

where we have used that $\delta > 0$. This completes the proof of Theorem 4. ∎

PROOF OF THEOREM 5   Let us assume first that $f \in HK_q$. Take ψ a radial smooth function with $\sup \psi \subset \{x : \frac{1}{4} < |x| < 2\}$ and $\psi(x) = 1$ for $\frac{1}{2} \leq |x| \leq 1$. Set $\psi_k(x) = \psi(2^{-k}x)$ and define

$$\Psi_k(x) = \frac{\psi_k(x)}{\sum_{j \in \mathbb{Z}} \psi_j(x)}, \qquad x \neq 0.$$

Observe that $\sup \Psi_k \subset \tilde{C}_k = \{x; 2^{k-2} < |x| < 2^{k+1}\}$. Let $\tilde{\chi}_k = \frac{1}{|\tilde{C}_k|}\chi_{\tilde{C}_k}$ and write

$$f(x) = \sum_{k \in \mathbb{Z}} \left(f(x)\Psi_k(x) - \left(\int f\Psi_k \right) \tilde{\chi}_k(x) \right) + \sum_{k \in \mathbb{Z}} \left(\int f\Psi_k \right) \tilde{\chi}_k(x)$$

$$= \sum_1 + \sum_2.$$

Set

$$g_k(x) = f(x)\Psi_k(x) - \left(\int f\Psi_k \right) \tilde{\chi}_k(x) = \lambda_k a_k(x)$$

with

$$\lambda_k = |B_k|^{1/q'} \|g_k\|_q \text{ and } a_k(x) = \frac{1}{\lambda_k} g_k(x), \text{ if } \lambda_k \neq 0.$$

Since $|g_k(x)| \leq Gf(x)\chi_{\tilde{C}_k}(x) \in L^q$, each a_k is a $(1, q)$-atom and

$$\sum_k |\lambda_k| \leq \sum_k |B_k|^{1/q'} \sum_{j=k-1}^{k+1} \|(Gf)\chi_j\|_q < \infty.$$

This takes care of Σ_1. For Σ_2 we write

$$\sum_2 = \sum_k \left(\sum_{j=-\infty}^{k} \left(\int f\Psi_j \right) \right) (\tilde{\chi}_k(x) - \tilde{\chi}_{k+1}(x))$$

$$= \sum_k h_k(x) = \sum_k \mu_k A_k(x),$$

with $\mu_k = |B_k|^{1/q'} \|h_k\|_q$.
Now,

$$\int \left(\sum_{j=-\infty}^{k} \Psi_j \right) \sim |B_{k+1}| \sim 2^{kn},$$

and a well known result in [FS] gives us

$$\left| \int f(x) \left(\sum_{j=-\infty}^{k} \Psi_j \right) \right| \leq 2^{kn} Gf(x), \text{ for } x \in B_{k+1}.$$

Notice also that

$$|\tilde{\chi}_k(x) - \tilde{\chi}_{k+1}(x)| \leq \frac{C}{2^{kn}} \sum_{j=k-1}^{k+1} \chi_j(x).$$

From the above inequalities, it follows

$$\sum_k \mu_k \leq C \sum_k \sum_{j=k-1}^{k+2} \|(Gf)\chi_j\|_q |B_k|^{1/q'} < \infty.$$

To prove now the other implication, it suffices to show that if a is $(1, q)$-atom then $Ga \in K_q$ with norm uniformly bounded. The proof of this is just a duplicate of the proof presented in Theorem 4, taking into account the following observation:
There exists a constant C such that

$$|\nabla\varphi(x)| \leq \frac{C}{(1 + |y|)^{n+1}}$$

for every $\varphi \in A_N$. Therefore, we can deduce the pointwise estimate (with the same arguments as for the case $\delta = 1$ in Theorem 4)

$$|Ga(x)| \leq C \frac{r}{|x|^{n+1}} \text{ if } |x| \geq 2r,$$

for an atom, a, supported in the ball $B(0, r)$. The rest follows the same pattern as in Theorem 4. ∎

References

[FS]   C. Fefferman and E. M. Stein, H^p *spaces of several variables*, Acta Math., 129 (1972), 137–193.

[H]   C. Herz, *Lipschitz spaces and Bernstein's theorem on absolutely convergent Fourier transform*, J. Math. Mech. 18 (1968), 283–324.

[LY]   S. Z. Lu and D. C. Yang, *The Littlewood–Paley function and φ-transform characterizations of a new Hardy space HK_2 associated with the Herz space*, Studia Math. 101 (1992), 3 285–298.

[R]   J. L. Rubio de Francia, *A Littlewood–Paley inequality for arbitrary intervals*, Rev. Mat. Iberoamericana 1 (1985), 1–14.

[SW]   F. Soria and G. Weiss, *A remark on singular integrals and power weights*, Indiana Univ. Math. J. 43 (1994), 187–204.

[TW]   M. H. Taibleson and G. Weiss, *Certain function spaces associated with a.e. convergence of Fourier series*, Univ. of Chicago Conf. in honor of Zygmund, Woodsworth, Vol. 1, 95–113, (1983).

[T]   A. Torchinsky, *Real-Variable Methods in Harmonic Analysis*, V.123 Pure and Applied Math., Academic Press, Inc., 1986.

16

On the One-Sided Hardy–Littlewood Maximal Function in the Real Line and in Dimensions Greater than One

F. J. Martín-Reyes[1]

16.1 Introduction

The goal of this paper is to conduct a survey about weights for one-sided Hardy–Littlewood maximal functions in $\mathbb{R}$ and some applications in the study of the one-sided fractional integrals. We also include some development of the theory in dimensions greater than one.

The one-sided Hardy–Littlewood maximal functions are defined by

$$M^+ f(x) = \sup_{h>0} \frac{1}{h} \int_x^{x+h} |f| \text{ and } M^- f(x) = \sup_{h>0} \frac{1}{h} \int_{x-h}^x |f|.$$

Sawyer characterized in [S] the good weights for these operators. The results for M^+ are the following.

THEOREM 1
(See [S].*) Assume that u and v are measurable functions such that $0 < u, v < \infty$ a.e. and $1 \leq p < \infty$. The operator M^+ applies $L^p(v)$ into weak-$L^p(u)$, if and only if, the pair (u, v) satisfies the A_p^+ condition, i.e., there exists a constant $C > 0$ such that*

$$M^- u \leq Cv \text{ a.e. if } p = 1$$

[1] This research has been partially supported by DGICYT grant (PB88-0324) and Junta de Andalucía.

1980 *Mathematics Subject Classification* (1985 *Revision*), 42B25

and

$$\int_a^b u \left(\int_b^c v^{-1/(p-1)} \right)^{p-1} \leq C(b-a)^p \text{ for all numbers } a < b < c, \text{ if } p > 1.$$

THEOREM 2

(See [S].) Let u and v remain as in Theorem 1 . Let $1 < p < \infty$. The operator M^+ applies $L^p(v)$ into $L^p(u)$, if and only if, the pair (u, v) satisfies the S_p^+ condition, i.e., there exists a constant $C > 0$ such that for all numbers $a < b$

$$\int_a^b |M^+(v^{-1/(p-1)}\chi_{(a,b)})|^p u \leq C \int_a^b v^{-1/(p-1)} < \infty.$$

THEOREM 3

(See [S].) Let w be a measurable function such that $0 < w < \infty$ a.e., and let $1 < p < \infty$. The operator M^+ applies $L^p(w)$ into $L^p(w)$, if and only if, w satisfies A_p^+, i.e., the pair (w, w) satisfies A_p^+.

Observe that A_p^+ resembles the A_p condition of Muckenhoupt ([Mu] and [CF]). The difference is that in the A_p condition the integrals of u and $v^{-1/(p-1)}$ are taken over the same interval, while in the A_p^+ condition, the function u is integrated on the left part of the interval and $v^{-1/(p-1)}$ on the right part.

Analogous results hold for M^- changing A_p^+ and S_p^+ by the conditions A_p^- and S_p^- in the obvious way.

We can also consider the one-sided maximal operator associated to a Borel measure finite on compact sets,

$$M_\mu^+ f(x) = \sup_{h>0} \frac{1}{\mu([x, x+h))} \int_{[x,x+h)} |f| \, d\mu$$

(the average is taken to be 0 if $\mu([x, x+h)) = 0$). This operator applies $L^1(d\mu)$ into weak-$L^1(d\mu)$ ([Be]). The good weights for this operator were studied in [MOT] for measures equivalent to the Lebesgue measure and in [A] for general measures.

Throughout this chapter, the letter C will always mean a positive constant not necessarily the same at each occurrence. Also, if $1 \leq p < \infty$, then p' will be the conjugate exponent, i.e., the number such that $p + p' = pp'$ if $p > 1$ and ∞ if $p = 1$.

16.2   Theorem 1 and the one-sided fractional maximal functions

That A_p^+ is a necessary condition is proven ([S], [MOT] and [A]) in the same way that the Muckenhoupt's case is, while the proofs of the other implication in the three mentioned papers are based on the following well known fact: if f is a nice function then $\{M^+ f > \lambda\} = \cup_i (a_i, b_i)$ where (a_i, b_i) are the connected components and therefore

$$\lambda \leq \frac{1}{b_i - x} \int_x^{b_i} |f(t)| \, dt \text{ for all } x \in (a_i, b_i). \tag{1}$$

This decomposition reduces the study of M^+ to the study of weighted Hardy inequalities on each connected component.

Now consider the one-sided fractional maximal operator

$$M_\alpha^+ f(x) = \sup_{h>0} \frac{1}{h^{1-\alpha}} \int_x^{x+h} |f(t)| \, dt, \qquad 0 \leq \alpha < 1.$$

The problem with this operator is the following: if we want to study the weighted weak type inequalities, we do not have a property like (1) for M_α^+. So, we would like another approach to study the weighted weak type inequalities. The following lemma makes it possible.

LEMMA 1

Let $0 \leq \alpha < 1$, $1 \leq p < \frac{1}{\alpha}$, $\frac{1}{q} = \frac{1}{p} - \alpha$. Let u and v be positive, measurable functions. If the pair (u^q, v^q) satisfies A_r^+ where $r = 1 + \frac{q}{p'}$ then

$$\left(M_\alpha^+ f\right)^q \leq C \|f\|_{L^p(v^p)}^{q-p} M_\mu^+ (f^p v^p u^{-q})$$

where $d\mu = u^q(x) \, dx$, for all measurable functions.

Once this lemma is established, we can prove the characterization of the weighted weak type inequalities for M_α^+.

THEOREM 4

Let $0 \leq \alpha < 1$, $1 \leq p < \frac{1}{\alpha}$, $\frac{1}{q} = \frac{1}{p} - \alpha$. Let u and v be positive measurable functions. Then there exists $C > 0$ such that for every $\lambda > 0$ and all measurable functions f

$$\left(\int_{\{x : M_\alpha^+ f(x) > \lambda\}} u^q\right)^{1/q} \leq \frac{C}{\lambda} \left(\int_{-\infty}^\infty |f|^p v^p\right)^{1/p},$$

if and only if, the pair (u^q, v^q) satisfies A_r^+ where $r = 1 + \frac{q}{p'}$.

REMARK 1   Lemma 1 and Theorem 4 were obtained jointly with de la Torre and they are related to some results that will appear in [MPT] and [MT2].   ∎

PROOF OF THEOREM 4   That A_r^+ is necessary is proven in the usual way, i.e., for fixed $a < b < c$, we test the inequality by functions $f = w^{-p'} \chi_E$ where

$E \subset (b, c)$ if $p > 1$, and by functions $f = \chi_E$ if $p = 1$. For the converse, we may assume $f \geq 0$. Then by Lemma 1, if $d\mu = u^q(x)\, dx$,

$$\int_{\{x: M_\alpha^+ f(x) > \lambda\}} u^q \leq \int_{\{x: M_\mu^+ (f^p v^p u^{-q})(x) > C\lambda^q \|f\|_{L^p(v^p)}^{p-q}\}} u^q,$$

and because M_μ^+ is of weak type $(1,1)$ (see [Be]), with respect to the measure μ, we obtain

$$\int_{\{x: M_\alpha^+ f(x) > \lambda\}} u^q \leq \frac{C}{\lambda^q \|f\|_{L^p(v^p)}^{p-q}} \|f\|_{L^p(v^p)}^p = \frac{C}{\lambda^q} \|f\|_{L^p(v^p)}^q. \qquad \blacksquare$$

But how is the lemma proven? In the Muckenhoupt's case ([MW]), we mean without pluses, this property is obvious. However, we think it is not in our case (but perhaps it is).

PROOF OF LEMMA 1 For x and $h > 0$ fixed, we choose a decreasing sequence $\{x_k\}$ such that

$$x_0 = x + h \quad \text{and} \quad \int_{x_{k+1}}^{x_k} u^q = \int_x^{x_{k+1}} u^q.$$

Observe that $\int_{x_{k+2}}^{x_{k+1}} u^q$ is comparable to $\int_x^{x_k} u^q$. More precisely

$$\int_x^{x_k} u^q = 4 \int_{x_{k+2}}^{x_{k+1}} u^q.$$

Assume $p > 1$. Then from the Hölder inequality and the assumption in the lemma,

$$\int_x^{x+h} |f| = \sum_{k=0}^\infty \int_{x_{k+1}}^{x_k} |f| \leq \sum_{k=0}^\infty \left(\int_{x_{k+1}}^{x_k} |f|^p v^p\right)^{1/p} \left(\int_{x_{k+1}}^{x_k} v^{-p'}\right)^{1/p'}$$

$$\leq C \sum_{k=0}^\infty \frac{\left(\int_{x_{k+1}}^{x_k} |f|^p v^p\right)^{1/p}}{\left(\int_{x_{k+2}}^{x_{k+1}} u^q\right)^{1/q}} (x_k - x_{k+2})^{1-\alpha}$$

$$\leq C \sum_{k=0}^\infty \left(\frac{\int_x^{x_k} |f|^p v^p}{\int_x^{x_k} u^q}\right)^{1/q} \left(\int_{x_{k+1}}^{x_k} |f|^p v^p\right)^{(1/p)-(1/q)} (x_k - x_{k+2})^{1-\alpha}.$$

Now by the definition of M_μ^+, with $d\mu = u^q(x)\, dx$, and the Hölder inequality applied to the sum, we obtain

$$\int_x^{x+h} |f| \leq C \left(M_\mu^+(|f|^p v^p u^{-q})\right)^{1/q}(x) h^{1-\alpha} \|f\|_{L^p(v^p)}^{\alpha p},$$

which proves the case $p > 1$ of the lemma, taking into account the relation between α, p, and q.

Now assume $p = 1$. Then the pair (u^q, v^q) satisfies A_1^+ and therefore

$$\int_x^{x+h} |f| = \sum_{k=0}^{\infty} \int_{x_{k+1}}^{x_k} |f| v v^{-1} \le C \sum_{k=0}^{\infty} \int_{x_{k+1}}^{x_k} |f| v \left(\frac{x_k - x_{k+2}}{\int_{x_{k+2}}^{x_{k+1}} u^q} \right)^{1/q}$$

$$\le C \sum_{k=0}^{\infty} \left(\frac{\int_x^{x_k} |f| v}{\int_x^{x_k} u^q} \right)^{1/q} \left(\int_{x_{k+1}}^{x_k} |f| v \right)^{1-(1/q)} (x_k - x_{k+2})^{1-\alpha}.$$

As before, this inequality proves the case $p = 1$ of the lemma. ∎

Therefore, we have seen in this section that the weighted weak type inequalities for M^+, and more generally for M_α^+, can be studied without reducing it to Hardy inequalities but reducing it to the weak type $(1, 1)$ of M_μ^+ for certain Borel measures μ.

16.3 The case of equal weights

The three papers we mentioned before ([S], [MOT], and [A]) follow the same pattern to prove (Theorem 3) that if w satisfies A_p^+, $1 < p$, then $\int |M^+ f|^p w \le C \int |f|^p w$. These proofs depend on the characterization of the pairs of weights for which M^+ is of strong type (p, p) (in this sense, they are not direct proofs). More precisely, it is proven that $w \in A_p^+ \Rightarrow w \in S_p^+$, and then by Theorem 2

$$\int_{-\infty}^{\infty} |M^+ f|^p w \le C \int_{-\infty}^{\infty} |f|^p w.$$

As a consequence of this result and a factorization theorem, it is possible to obtain $w \in A_p^+ \Rightarrow w \in A_{p-\varepsilon}^+$, $1 < p < \infty$.

It seems interesting to have a proof of the strong type for equal weights independent of the problem for different weights. Of course, there will be a direct proof if we are able to give a direct proof of $w \in A_p^+ \Rightarrow w \in A_{p-\varepsilon}^+$, $1 < p < \infty$. We succeeded in this problem ([M]). The idea of the proof is the following: if we remember Muckenhoupt's case, we recall that the usual step is to establish the reverse Hölder inequality. But for weights $w \in A_p^+$, the reverse Hölder inequality does not hold. However, there is a substitute: if $w \in A_p^+$ then there exist positive numbers δ and C, such that

$$\int_a^b w^{1+\delta} \le C \int_a^b w \left(M^-(w \chi_{(a,b)})(b) \right)^{\delta} \tag{2}$$

for every bounded interval (a, b). This inequality, together with the fact that $w \in A_p^+$, implies $w \in A_{p-\varepsilon}^+$. The proof uses tricks similar to the ones in the proof of Lemma 1 and the ideas in [CF].

Perhaps, an important consequence of this direct proof is that the method is useful to extend the study of weights for M^+ to $L_{p,q}$ and Orlicz spaces. These extensions have been done by Ortega and Pick ([O1], [O2], and [OP]). But in this paper we will not follow this direction.

Let's return to the one-sided fractional maximal operator. Assume that $0 < \alpha < 1$, $1 < p < \frac{1}{\alpha}$, $\frac{1}{q} = \frac{1}{p} - \alpha$. We know by Theorem 4 that if w^q satisfies A_r^+, $r = 1 + \frac{q}{p'}$, then the weak type inequality holds. Now, since $w^q \in A_r^+$ implies $w^q \in A_{r-\varepsilon}^+$, an interpolation argument (see [MW]) gives the strong type inequality.

THEOREM 5

(See [AS].) Assume that $0 < \alpha < 1$, $1 < p < \frac{1}{\alpha}$, $\frac{1}{q} = \frac{1}{p} - \alpha$. The following statements are equivalent.

(a) w^q *satisfies* A_r^+, $r = 1 + \frac{q}{p'}$.

(b) *There exists a constant $C > 0$ such that for all measurable functions f*

$$\left(\int |M_\alpha^+ f|^q w^q \right)^{1/q} \leq C \left(\int |f|^p w^p \right)^{1/p}.$$

This theorem was proved in [AS] by using complex interpolation arguments. Our proof uses only real variable methods, and it is a direct proof in the sense that we do not use the result of strong type about pairs of weights (see [MT2]).

16.4 A_∞^+ weights and one-sided sharp functions

Now we pay attention to the inequality (2). Is it a good analogue of the reverse Hölder inequality? Is there a concept of one-sided A_∞ weights? These questions have been answered in [MPT] in the following way, although in a more general setting.

THEOREM 6

Let w be a positive, locally integrable function on the real line. The following are equivalent.

(a) *There exist positive numbers δ and C, such that*

$$\int_a^b w^{1+\delta} \leq C \int_a^b w \left(M^-(w\chi_{(a,b)})(b) \right)^\delta \tag{2}$$

for every bounded interval (a, b).

(b) *There exists $p > 1$, such that $w \in A_p^+$.*

(c) $w \in A_{\infty}^+$, which means that there exist positive numbers δ and C, such that

$$\frac{\int_E w}{\int_a^c w} \leq C \left(\frac{|E|}{c-b} \right)^{\delta}$$

for all numbers $a < b < c$ and all subsets $E \subset (a, b)$.

(d)   There exist positive numbers δ and C, such that

$$\frac{|E|}{c-a} \leq C \left(\frac{\int_E w}{\int_a^b w} \right)^{\delta}$$

for all numbers $a < b < c$ and all subsets $E \subset (b, c)$.

REMARK 2   Observe that the equivalence between *(c)* and *(d)* is nothing but the one-sided version of the comparability relation of measures introduced in [CF]. However, observe that in our case, we do not need to assume that the weight satisfies the doubling condition.   ∎

The A_{∞}^+ weights can be used to obtain weighted distribution function inequalities, for example (see [MPT]) between M_{α}^+ and the one-sided fractional integral (Weyl operator) defined by

$$I_{\alpha}^+ f(x) = \int_x^{\infty} \frac{f(y)}{(y-x)^{1-\alpha}} \, dy.$$

We study this operator by introducing a new concept that we have called one-sided sharp maximal function (see [MT3]). This sharp function is defined by

$$f_+^{\sharp}(x) = \sup_{h>0} \frac{1}{h} \int_x^{x+h} \left(f(y) - \frac{1}{h} \int_{x+h}^{x+2h} f \right)^+ dy$$

where $z^+ = \max(z, 0)$. It is easy to see that

$$f_+^{\sharp}(x) \leq 3M^+ f(x).$$

There is not an opposite pointwise inequality, but by means of a weighted distribution function inequality between M^+ and $f_+^{\sharp}$, it is possible to prove the following theorem.

THEOREM 7

(See [MT3].*) Assume* $w \in A_{\infty}^+$, $f \geq 0$ *and* $M^+ f \in L^{p_0}(w)$ *for some* p_0, $0 < p_0 < \infty$. *Then for every* p, $p_0 \leq p < \infty$,

$$\int_{-\infty}^{\infty} (M^+ f)^p w \leq C \int_{-\infty}^{\infty} f_+^{\sharp p} w.$$

Furthermore, with $f_+^\sharp$ we can define a one-sided BMO space that is related to A_p^+ weights in the same way that BMO is related to Muckenhoupt's classes ([MT]).

Now we use this sharp function to study the weights for the one-sided fractional integrals.

The weights for which I_α^+ is of strong type, with respect to $w(x)\,dx$, are characterized in [AS]. The result is collected in the following theorem.

THEOREM 8

(See [AS].) Let $0 < \alpha < 1$, $1 < p < \frac{1}{\alpha}$ and $\frac{1}{q} = \frac{1}{p} - \alpha$. Let w be a positive, locally integrable function. The following are equivalent:

(a)   *There exists C such that for all functions $f \in L^p(w^p)$*

$$\left(\int_{-\infty}^\infty |I_\alpha^+ f|^q w^q \right)^{1/q} \le C \left(\int_{-\infty}^\infty |f|^p w^p \right)^{1/p}.$$

(b)   *The function w^q satisfies A_r^+ where $r = 1 + \frac{q}{p'}$.*

Andersen and Sawyer use a method of Welland to prove that $(b) \Rightarrow (a)$. We can give an alternative proof by using the one-sided sharp maximal function ([MT]). In what follows, we sketch the proof.

Consider $f \ge 0$. Then it is easy to see that

$$(I_\alpha^+ f)_+^\sharp(x) \le C M_\alpha^+ f(x).$$

Then by Theorem 7, the above inequality, and Theorem 5, we obtain

$$\int_{-\infty}^\infty |I_\alpha^+ f|^q w^q \le \int_{-\infty}^\infty |M^+(I_\alpha^+ f)|^q w^q \le C \int_{-\infty}^\infty \left((I_\alpha^+ f)_+^\sharp \right)^q w^q$$

$$\le C \int_{-\infty}^\infty |M_\alpha^+ f|^q w^q$$

$$\le C \left(\int_{-\infty}^\infty |f|^p w^p \right)^{q/p}.$$

The other implication is clear.

16.5   Weights for one-sided maximal functions in $\mathbb{R}^n$, $n > 1$

One of the applications of the study of the weights for the one-sided Hardy–Littlewood maximal functions is in ergodic theory ([MT1]). We can use transference arguments to dominate the ergodic maximal function

$$f^*(x) = \sup_{k \geq 0} \frac{1}{k+1} \sum_{i=0}^{k} |T^i f(x)|$$

associated to an invertible positive operator with positive inverse. In the same way, we would like to study the ergodic maximal operator

$$f^*(x) = \sup_{k \geq 0} \frac{1}{(k+1)^n} \sum_{i_1=0}^{k} \cdots \sum_{i_n=0}^{k} |T_1^{i_1} \cdots T_n^{i_n} f(x)|$$

associated to n positive invertible operators $T_i, i = 1, \ldots, n$, with positive inverse. To do this, we must study the one-sided Hardy–Littlewood maximal function in dimension $n > 1$. We dedicate the last two sections of this chapter to this generalization.

We will work in $\mathbb{R}^2$ although all the results hold in $\mathbb{R}^n$, $n > 2$. We may choose at least four possible one-sided operators in $\mathbb{R}^2$. Their definitions are

$$M^{\pm\pm} f(x, y) = \sup_{h > 0} \frac{1}{h^2} \left| \int_x^{x \pm h} \int_y^{y \pm h} |f(s, t)| \, ds \, dt \right|.$$

By symmetry it suffices to pay attention to one of them. We will work with M^{++}, i.e., with

$$M^{++} f(x, y) = \sup_{h > 0} \frac{1}{h^2} \int_x^{x+h} \int_y^{y+h} |f(s, t)| \, ds \, dt.$$

To deal with this operator, we introduce some notation. If Q is the square $[x - h, x] \times [y - h, y]$ we denote by Q^* the square $[x, x + h] \times [y, y + h]$.

Our question is: which are the good weights for M^{++}? We do not know, but a necessary condition is collected in the following proposition, where $|Q|$ will denote the Lebesgue measure of Q.

PROPOSITION 1

Let u and v be positive measurable functions, and let $1 \leq p < \infty$. Assume that M^{++} applies $L^p(v)$ into weak-$L^p(u)$. Then the pair (u, v) satisfies the A_p^{++} condition, which means that there exists a constant C such that

$$\int_Q u \left(\int_{Q^*} v^{-1/(p-1)} \right)^{p-1} \leq C |Q|^p \text{ for all squares } Q \text{ if } p > 1$$

and

$$M^{--} u \leq C v \text{ a.e. if } p = 1$$

(the word square will always mean a square with sides parallel to the axis).

The proof follows as usual, testing the weak-type inequality by functions $f = v^{-1/(p-1)}\chi_E$ in the case $p > 1$, and by functions $f = \chi_E$ in the case $p = 1$, where $E \subset Q^*$.

Now, let's take $u = v = w$ in the condition A_p^{++}. If $1 < p < \infty$, A_p^{++} and Hölder's inequality give

$$\int_Q w \left(\int_{Q^*} w^{-1/(p-1)} \right)^{p-1} \le C|Q|^p = C|Q^*|^p \le C \int_{Q^*} w \left(\int_{Q^*} w^{-1/(p-1)} \right)^{p-1}.$$

Thus

$$\int_Q w \le C \int_{Q^*} w \quad \text{for all squares } Q, \tag{3}$$

which is a certain doubling condition. Since $A_1^{++} \Rightarrow A_p^{++}$, $p > 1$, we have proven the following proposition.

PROPOSITION 2
If w satisfies A_p^{++}, then (3) holds.

Analogous results hold for M^{--}, changing the roles of Q and Q^*.

Assume now that M^{++} and M^{--} apply $L^p(w)$ into weak-$L^p(w)$. If $p > 1$ we have by Proposition 1 and the analogous one of Proposition 2 for M^{--} that

$$\int_Q w \left(\int_{Q^*} w^{-1/(p-1)} \right)^{p-1} \le C|Q|^p$$

and

$$\int_{Q^*} w \le C \int_Q w,$$

for all squares Q. Thus

$$\int_{Q^*} w \left(\int_{Q^*} w^{-1/(p-1)} \right)^{p-1} \le C \int_Q w \left(\int_{Q^*} w^{-1/(p-1)} \right)^{p-1} \le C|Q|^p = C|Q^*|^p.$$

Therefore, we have determined that w satisfies the condition A_p of Muckenhoupt ([Mu], [CF]), which is the one that characterizes the weak type (p, p) (and the strong type (p, p) if $p > 1$) of the Hardy–Littlewood maximal operator defined by

$$Mf(x, y) = \sup_{(x,y)\in Q} \frac{1}{|Q|} \int_Q |f|$$

where the supremum is taken over all the squares, such that $(x, y) \in Q$. Since it is clear that $M^{++} \le M$ and $M^{--} \le M$, we have obtained the following proposition.

PROPOSITION 3
The operators M^{++} and M^{--} apply $L^p(w)$ into weak-$L^p(w)$, if and only if, M applies $L^p(w)$ into weak-$L^p(w)$.

The proof has been completed for $p > 1$. The result is also valid if $p = 1$ (the proof is similar to the one for $p > 1$).

16.6 The one-sided maximal function associated to Borel measures in $\mathbb{R}^n, n > 1$

As in the previous section, we will work on $\mathbb{R}^2$, but the results can be easily generalized to $\mathbb{R}^n$ with $n > 2$.

The operator M_μ^+, the one-sided operator associated to a Borel measure μ in $\mathbb{R}$, has played an important role in Section 16.2. Therefore, to completely solve the problem of the previous section, it seems appropriate to try to understand what happens in $\mathbb{R}^2$ with the operator

$$M_\mu^{++} f(x, y) = \sup_{h>0} \frac{1}{\mu([x, x+h) \times [y, y+h))} \int_{[x,x+h)\times[y,y+h)} |f| d\mu$$

associated to a Borel measure finite on compact sets. We know that the usual maximal operator associated to μ

$$M_\mu f(x, y) = \sup_{(x,y)\in Q} \frac{1}{\mu(Q)} \int_Q |f| d\mu$$

is not always of weak type $(1, 1)$ with respect to the measure μ (see [Sj]) (the supremum is taken over all the squares such that $(x, y) \in Q$). However the centered maximal operator

$$M_\mu^c f(x, y) = \sup_{(x,y)=\text{center of } Q} \frac{1}{\mu(Q)} \int_Q |f| d\mu$$

is always of weak type $(1, 1)$ (now the supremum is taken over all the squares such that (x, y) is the center of Q). We know also [Sj] that if μ satisfies the doubling condition

$$\mu(2Q) \le C\mu(Q) \text{ for all squares } Q,$$

where $2Q$ is the square with the same center as Q and $|2Q| = 4|Q|$, then M_μ is of weak type $(1,1)$ with respect to the measure μ. Our operator M_μ^{++} has a behavior similar to M_μ (see Theorem 9).

From now on, the word square will mean squares of the type $Q = [a, a+h) \times [b, b+h)$ and Q^* will be $[a+h, a+2h) \times [b+h, b+2h)$.

THEOREM 9

Let μ be a Borel measure on $\mathbb{R}^2$ finite on compact sets. If there exists a constant $C > 0$ such that

$$\mu(Q) \le C\mu(Q^*) \tag{4}$$

for all squares Q then M_μ^{++} is of weak type $(1,1)$ with respect to the measure μ.

PROOF   Let us fix a point $(x, y) \in \mathbb{R}^2$ and a square $Q = [x, x+h) \times [y, y+h)$. Let F be the convex hull of the set

$$\left[x - \frac{h}{4}, x\right) \times \left[y - \frac{h}{4}, y\right) \cup [x, x+h) \times [y, y+h), \qquad h > 0.$$

The assumption in the theorem, which is a certain doubling condition, implies

$$\mu(F) \le C\mu(Q) \tag{5}$$

where C is a constant, depending only on the constant in the theorem and the dimension (in this case, 2). Once this inequality is established, we have

$$\frac{1}{\mu(Q)} \int_Q |f| \, d\mu \le \frac{C}{\mu(F)} \int_F |f| \, d\mu. \tag{6}$$

Thus if $\mathcal{F}(x, y)$ is the family of the convex hulls of the sets

$$\left[x - \frac{h}{4}, x\right) \times \left[y - \frac{h}{4}, y\right) \cup [x, x+h) \times [y, y+h), \qquad h > 0,$$

we have obtained

$$M_\mu^{++} \le C M_{\mu.\mathcal{F}}$$

where

$$M_{\mu.\mathcal{F}} f(x, y) = \sup_{F \in \mathcal{F}(x,y)} \frac{1}{\mu(F)} \int_F |f| \, d\mu.$$

Now, by a Besicovitch recovering argument (see [G]), we find that $M_{\mu.\mathcal{F}}$ is of weak type $(1,1)$ and therefore our operator M_μ^{++} is also of weak type $(1,1)$ with respect to μ.

To prove the inequality (5), we observe, first, that if

$$S = \left[x - \frac{h}{4}, x\right) \times \left[y - \frac{h}{4}, y\right),$$

then we see from the assumption of the theorem that

$$\mu(S \cup Q) \le C\mu(Q).$$

Now $F - (S \cup Q)$ is the union of two triangles $T_1 \cup T_2$. By symmetry, it is enough to see that the triangle T_1 in the second quadrant satisfies

$$\mu(T_1) \le C\mu(Q).$$

Observe that

$$T_1 \subset \cup_{i=1}^{\infty} Q_i$$

where Q_i is the square

$$\left[x - \frac{h}{2^{i+1}}, x + \frac{h}{2^{i+1}} \right) \times \left[y + h - \frac{h}{2^{i-1}}, y + h - \frac{h}{2^i} \right).$$

It is clear that the squares $Q_i{}^*$ are included in Q and

$$\sum_{i=1}^{\infty} \chi_{Q_i{}^*} \le C$$

for a certain constant. Then by the assumption of the theorem and these properties

$$\mu(T_1) \le \sum_{i=1}^{\infty} \mu(Q_i) \le C \sum_{i=1}^{\infty} \mu(Q_i{}^*) = C\mu(Q),$$

as we intended to prove. ∎

Example The following example shows that M_μ^{++} is not always of weak type $(1,1)$.

Consider points $z_k = (x_k, y_k)$, $k \ge 1$, such that $y_k = \frac{1}{x_k}$ and $x_k \uparrow 0$. Let $z_0 = (0, 0)$ and consider $\mu = \sum_{k=0}^{\infty} \delta_k$, where δ_k is the Dirac measure concentrated at the point z_k. Let B be a ball with center, the origin, and radius ε (ε small). If $f = \chi_B$ then

$$M_\mu^{++} f(z_k) \ge \frac{1}{2}, \quad \text{for all } k \ge 1.$$

Therefore, $\mu\left(\left\{ M_\mu^{++} f \ge \frac{1}{2} \right\}\right) = \infty$, and, consequently, we do not have weak-type $(1, 1)$ inequality.

References

[A]   K. F. Andersen, *Weighted inequalities for maximal functions associated with general measures*, Trans. Amer. Math. Soc. 326 (1991), 907–920.

[AS]   K. F. Andersen and E. T. Sawyer, *Weighted norm inequalities for the Riemann–Liouville and Weyl fractional integral operators*, Trans. Amer. Math. Soc. 308 (1988), 547–557.

[Be]   A. Bernal, *A note on the one-dimensional maximal function*, Proc. Royal Soc. Edinburgh 111A (1989), 1989.

[CF]   R. R. Coifman and C. Fefferman, *Weighted norm inequalities for maximal functions and singular integrals*, Studia Math. 51 (1974), 241–250.

[G]   M. de Guzmán, *Differentiation of Integrals in* $\mathbb{R}^n$, Springer–Verlag, Berlin, Heidelberg, New York, 1975.

[M]   F. J. Martín-Reyes, *New proofs of weighted inequalities for the one sided Hardy–Littlewood maximal functions,* Proc. Amer. Math. Soc. 117 (1993), 691–698.

[MOT]   F. J. Martín-Reyes, P. Ortega Salvador and A. de la Torre, *Weighted inequalities for one-sided maximal functions,* Trans. Amer. Math. Soc. 319-2 (1990), 517–534.

[MPT]   F. J. Martín-Reyes, L. Pick, and A. de la Torre, A_∞^+ *condition,* Canadian Journal of Mathematics. 45 (1993), 1231–1244.

[MT1]   F. J. Martín-Reyes and A. de la Torre, *The dominated ergodic estimate for mean bounded, invertible, positive operators,* Proc. Amer. Math. Soc. 104 (1988), 69–76.

[MT2]   F. J. Martín-Reyes and A. de la Torre, *Two weight norm inequalities for fractional one-sided maximal operators,* Proc. Amer. Math. Soc. 117 (1993), 483–489.

[MT3]   F. J. Martín-Reyes and A. de la Torre, *One-sided BMO spaces,* Journal of The London Mathematical Society. 49 (1994), 529–542.

[Mu]   B. Muckenhoupt, *Weighted norm inequalities for the Hardy maximal function,* Trans. Amer. Math. Soc. 165 (1972), 207–226.

[MW]   B. Muckenhoupt and R. L. Wheeden, *Weighted norm inequalities for fractional integrals,* Trans. Amer. Math. Soc. 192 (1974), 261–274.

[O1]   P. Ortega, *Weighted inequalities for one sided maximal functions in Orlicz spaces,* to appear in Studia Math.

[O2]   P. Ortega, *Pesos para operadores maximales y teoremas ergódicos en espacios L_p, $L_{p.q}$ y de Orlicz,* Doctoral thesis, Universidad de Málaga, 1991.

[OP]   P. Ortega and L. Pick, *Two weight weak and extra-weak type inequalities for the one-sided maximal operator,* Proc. Royal Soc. Edinburgh. 123A (1993), 1109–1118.

[S]   E. Sawyer, Weighted inequalities for the one sided Hardy-Litlewood maximal functions, Trans. Amer. Math. Soc. 297 (1986), 53–61.

[Sj]   J. O. Sjögren, *A remark on the maximal function for measures in* $\mathbb{R}^n$, Amer. J. Math. 105 (1983), 1231–1233.

17

Characterization of the Besov Spaces

M. Paluszyński
M. Taibleson[1]

Homogeneous Lipschitz spaces were first defined by Herz (see [Her]). He used the same norm as (1)–(2) and defined his Λ_{pq}^{β} space as the closure in this norm of the space $\widehat{O}$, the Schwartz class functions with the Fourier transforms supported compactly, away from the origin. Subsequently, Johnson (see [Jo]) defined the homogeneous Lipschitz spaces as subspaces of the space of temperatures in the upper half space. For technical reasons, Johnson's Lipschitz spaces consist of temperatures with better than polynomial decay in the "vertical" variable (see [Jo], page 304). Johnson's definition agrees with Herz's except in the case of p or q being infinite. The modern definition of the Besov spaces can be found in [Fr]. The definition that we use is equivalent to this latter definition, and the equivalence can be shown using Calderón's reproducing formula in a similar fashion as in this paper. The main result of this paper is the characterization of the Besov spaces, $\dot{B}_p^{\beta,q}$, when $1 \leq p, q \leq \infty$ and $\beta > 0$.

The characterization established here follows the idea of controlling the mean oscillation of the function. In the case where $\beta \geq 1$ it has been shown that replacing the mean of a function over a cube by the so-called Gramm–Schmidt polynomial, yields a characterization of $\dot{B}_p^{\beta,q}$ [JTW]. The characterization described in this chapter arises naturally in the study of the commutator operator introduced in [CRW]. The study of the smoothing properties of this operator and its generalization leads to the function spaces described by conditions (b)–(e) of Theorem 1. These conditions involve the integral means of the difference operator applied to the function. These means, as well as the Gramm–Schmidt approach, reduce to the ordinary mean oscillation in the case when $0 < \beta < 1$. The results for the smoothing properties of the generalized commutator operator [P] will appear elsewhere.

The proof of Theorem 1 employs the version of Calderón's reproducing formula due to Janson and Taibleson [JT].

---

[1] The authors would like to thank Guido Weiss for his help in this project and his valuable suggestions.

Let Δ_h^k denote the k-th difference operator. That is,

$$\Delta_h^1 f(x) = \Delta_h f(x) = f(x+h) - f(x),$$
$$\Delta_h^{k+1} f(x) = \Delta_h^k f(x+h) - \Delta_h^k f(x), \qquad k \geq 1.$$

Suppose $0 < \beta < k$, with k an integer. For $f \in L_{\text{loc}}^1$, $x_0, z \in \mathbb{R}^n$, $t, d > 0$, $1 \leq r \leq \infty$ let

$$\Omega_f(x_0, z, t, d) = \frac{1}{|Q|} \frac{1}{|Q^z|} \int_Q \int_{Q^z} \left[\Delta_{y-x}^k f(x) \right] dy \, dx,$$

$$\Omega_f'(x_0, z, t, d) = \frac{1}{|Q|} \frac{1}{|Q^z|} \int_Q \int_{Q^z} \left[\Delta_{(y-x)/k}^k f(x) \right] dy \, dx,$$

$$\Omega_{f,r}(x_0, z, t, d) = \frac{1}{|Q|^{1/r}} \frac{1}{|Q^z|} \left(\int_Q \left| \int_{Q^z} \left[\Delta_{y-x}^k f(x) \right] dy \right|^r dx \right)^{1/r},$$

$$\Omega_{f,r}'(x_0, z, t, d) = \frac{1}{|Q|^{1/r}} \frac{1}{|Q^z|} \left(\int_Q \left| \int_{Q^z} \left[\Delta_{(y-x)/k}^k f(x) \right] dy \right|^r dx \right)^{1/r},$$

where $Q = Q(x_0, td)$, $Q^z = Q(x_0 + zt, t)$, and $Q(x, t)$ denotes a cube with sides parallel to the coordinate axes, centered at x, and with sidelength t. When d is known and fixed, we will suppress the dependence on d. If, in addition, $r = 1$, we omit r. If $r = \infty$, the last two definitions are interpreted in the usual way:

$$\Omega_{f,\infty}(x_0, z, t, d) = \sup_{x \in Q} \left| \frac{1}{|Q^z|} \int_{Q^z} \left[\Delta_{y-x}^k f(x) \right] dy \right|,$$

$$\Omega_{f,\infty}'(x_0, z, t, d) = \sup_{x \in Q} \left| \frac{1}{|Q^z|} \int_{Q^z} \left[\Delta_{(y-x)/k}^k f(x) \right] dy \right|.$$

Theorem 1, which is the main result, establishes the equivalence of four characterizations of homogeneous Besov spaces, $\dot{B}_p^{\beta,q}$.

We use the following definition of $\dot{B}_p^{\beta,q}$.

DEFINITION 1   *If* $1 \leq p, q \leq \infty$, $0 < \beta$, $k_0 = [\beta] + 1$, $f \in S' \cap L_{\text{loc}}^1$, *then* $f \in \dot{B}_p^{\beta,q}$, *if*

$$\left(\int_{\mathbb{R}^n} \left\| \frac{\Delta_h^{k_0} f(\cdot)}{|h|^\beta} \right\|_p^q \frac{dh}{|h|^n} \right)^{1/q} < \infty. \tag{1}$$

If $q = \infty$, *this condition is interpreted as*

$$\sup_{0 \neq h \in \mathbb{R}^n} \left\| \frac{\Delta_h^{k_0} f(\cdot)}{|h|^\beta} \right\|_p < \infty. \tag{2}$$

The $\dot{B}_p^{\beta,q}$ *norm is represented by the left side of (1) or (2), respectively.*

THEOREM 1

Let $1 \leq p, q \leq \infty$, $1 \leq r \leq p$, and $k > \beta > 0$ as above. Suppose $f \in S' \cap L^1_{loc}$. The following are equivalent:

(a) $f = f_1 + P$, where $f_1 \in \dot{B}_p^{\beta,q}$ and P is a polynomial of degree less than k.

(b)   For every $d > 0$ and $U \subset \mathbb{R}^n$, U open and bounded

$$\left(\int_0^\infty \left(t^{-\beta} \left\| \Omega_f(\cdot, z, t, d) \right\|_p \right)^q \frac{dt}{t} \right)^{1/q} \leq C < \infty, \quad \forall z \in U, \qquad (3)$$

with C independent of $z \in U$. If $q = \infty$, this should be replaced by

$$\sup_{t>0} t^{-\beta} \left\| \Omega_f(\cdot, z, t, d) \right\|_p \leq C < \infty, \quad \forall z \in U. \qquad (4)$$

(c)   The same condition as (b), with Ω_f replaced by Ω_f'.

(d)   For every $d > 0$, $z_0 \in \mathbb{R}^n$

$$\left(\int_0^\infty \left(t^{-\beta} \left\| \Omega_{f,r}(\cdot, z_0, t, d) \right\|_p \right)^q \frac{dt}{t} \right)^{1/q} \leq C < \infty, \qquad (5)$$

if $q = \infty$, (5) should be replaced by

$$\sup_{t>0} t^{-\beta} \left\| \Omega_{f,r}(\cdot, z_0, t, d) \right\|_p \leq C < \infty; \qquad (6)$$

(e)   The same condition as d) with $\Omega_{f,r}$ replaced by $\Omega_{f,r}'$.

If these conditions hold, then $\| f_1 \|_{\dot{B}_p^{\beta,q}}$ is comparable with the best possible C in (3)–(6). The constants of comparability depend only on $|z_0|$, d, k, and n in (d), (e) and on the diameter of U, d, k, and n in (b), (c).

The most interesting case for (d) and (e) is when $r = 1$. Then these conditions involve a generalization (if $k > 1$) of the mean oscillation of f.

The proof of this theorem will proceed in the following way: (a) implies (d) and (e); (d) implies (b); (e) implies (c); (b) implies (a); (c) implies (a).

REMARK 1   The equivalence remains valid, if we replace "for every" by "there exist" in (b)–(e), as can be observed by examining the proof.   ∎

PROOF   (a) $\Rightarrow$ (d), (e): Fix $d > 0$ and $z_0 \in \mathbb{R}^n$.

$$\Delta_h^k f(x) = \sum_{l=0}^k c_l^k f(x + lh),$$

where

$$c_l^k = (-1)^{k-l} \binom{k}{l}. \qquad (7)$$

Observe that $\Delta_h^{m+l} f(x) = \Delta_h^m \Delta_h^l f(x)$. Recall that $r \leq p \leq \infty$ and $k_0 = [\beta] + 1$. Let l be any integer. Using Minkowski's integral inequality and substitution, we obtain

$$
\left\| \int_Q \left| \int_{Q^{z_0}} \Delta_{y-x}^{k_0} f(x + l(y-x)) \, dy \right|^r dx \right\|_{p/r}^{1/r}
$$

$$
\leq (dt)^{n/r} \int_{B(0,\sigma t)} \left\| \Delta_h^{k_0} f \right\|_p dh,
$$

with $\sigma = \sqrt{n} \left(|z_0| + \frac{(1+d)}{2} \right)$ dependent only on $|z_0|$, d, and n. $B(x,t)$ denotes a ball centered at x with the radius t. Thus

$$
\left\| \Omega_{f,r}(\cdot, z_0, t) \right\|_p = \left\| \left(\frac{1}{|Q|} \int_Q \frac{1}{|Q^{z_0}|} \int_{Q^{z_0}} \left| \Delta_{y-x}^k f(x) \, dy \right|^r dx \right)^{1/r} \right\|_p
$$

$$
= d^{-n/r} t^{-n(1+1/r)} \left\| \int_Q \left| \int_{Q^{z_0}} \Delta_{y-x}^{k-k_0} \Delta_{y-x}^{k_0} f(x) \, dy \right|^r dx \right\|_{p/r}^{1/r}
$$

$$
\leq d^{-n/r} t^{-n(1+1/r)} \sum_{l=0}^{k-k_0} \left| c_l^{k-k_0} \right|
$$

$$
\cdot \left\| \int_Q \left| \int_{Q^{z_0}} \Delta_{y-x}^{k_0} f(x + l(y-x)) \, dy \right|^r dx \right\|_{p/r}^{1/r}
$$

$$
\leq C t^{-n} \int_{B(0,\sigma t)} \left\| \Delta_h^{k_0} f \right\|_p dh. \qquad \blacksquare
$$

The above computations require the obvious modifications in the case $r \leq p = \infty$. The constant C is dependent only on k. Similarly, for $1 \leq r \leq p \leq \infty$

$$
\left\| \Omega'_{f,r}(\cdot, z_0, t) \right\|_p \leq C t^{-n} \int_{B(0,\sigma t)} \left\| \Delta_h^{k_0} f \right\|_p dh.
$$

Thus we have the following estimate

$$
\left(\int_0^\infty \left(t^{-\beta} \left\| \Omega_{f,r}(\cdot, z_0, t) \right\|_p \right)^q \frac{dt}{t} \right)^{1/q}
$$

$$
\leq C \left(\int_0^\infty \left(t^{-\beta-n} \int_{B(0,\sigma t)} \left\| \Delta_h^{k_0} f \right\|_p dh \right)^q \frac{dt}{t} \right)^{1/q}
$$

$$
\leq C \left(\int_0^\infty \int_{B(0,\sigma t)} \frac{1}{t^n} \left\| \frac{\Delta_h^{k_0} f}{|h|^\beta} \right\|_p^q dh \frac{dt}{t} \right)^{1/q}
$$

$$= C \left(\int_0^\infty \int_{r/\sigma}^\infty \frac{dt}{t^{n+1}} \int_{r\Sigma_{n-1}} \left\| \frac{\Delta_{r\omega}^{k_0} f}{r^\beta} \right\|_p^q d\omega \, dr \right)^{1/q}$$

$$= C \left(\int_0^\infty \int_{r\Sigma_{n-1}} \left\| \frac{\Delta_{r\omega}^{k_0} f}{r^\beta} \right\|_p^q d\omega \frac{dr}{r^n} \right)^{1/q}$$

$$= C \left(\int_{\mathbb{R}^n} \left\| \frac{\Delta_h^{k_0} f}{|h|^\beta} \right\|_p^q \frac{dh}{|h|^n} \right)^{1/q}$$

$$= C \|f\|_{\dot{B}_p^{\beta,q}}, \tag{8}$$

with C only dependent on k, $|z_0|$, d, and n. (8) follows from Jensen's inequality, and Σ_{n-1} denotes the unit sphere in $\mathbb{R}^n$. Similarly,

$$\left(\int_0^\infty \left(t^{-\beta} \left\| \Omega'_{f,r}(\cdot, z_0, t) \right\|_p \right)^q \frac{dt}{t} \right)^{1/q} \leq C \|f\|_{\dot{B}_p^{\beta,q}}.$$

If $q = \infty$, the argument requires the obvious modifications.

(d) $\Rightarrow$ (b), (e) $\Rightarrow$ (c): Fix $d > 0$, $U \subset \mathbb{R}^n$ open and bounded, and let S be the diameter of U. Choose $z_0 \in U$. Let $d' = d + S$, $z \in B(0, S)$.

$$\left| \Omega_f(x_0, z + z_0, t, d) \right|$$

$$= \frac{1}{|Q|} \frac{1}{|Q^{z+z_0}|} \left| \int_{Q(x_0, td)} \int_{Q(x_0 + z_0 t + zt, t)} \left[\Delta_{y-x}^k f(x) \right] dy \, dx \right|$$

$$\leq \frac{1}{|Q|} \frac{1}{|Q^{z+z_0}|} \int_{Q(x_0, td)} \left| \int_{Q((x_0+zt)+z_0 t, t)} \left[\Delta_{y-x}^k f(x) \right] dy \right| dx$$

$$\leq \frac{1}{|Q|} \frac{1}{|Q^{z+z_0}|} \int_{Q((x_0+zt), td')} \left| \int_{Q((x_0+zt)+z_0 t, t)} \left[\Delta_{y-x}^k f(x) \right] dy \right| dx$$

$$= C\Omega_f(x_0 + zt, z_0, t, d')$$

$$\leq C\Omega_{f,r}(x_0 + zt, z_0, t, d'),$$

where the second inequality follows, since $Q(x_0, td) \subset Q(x_0 + zt, td')$ for $z \in B(0, S)$, and the last follows by Jensen's inequality. An identical argument shows that

$$\left| \Omega'_f(x_0, z + z_0, t, d) \right| \leq C\Omega'_{f,r}(x_0 + zt, z_0, t, d').$$

In both cases, C depends only on d, S, and n. Thus we obtain

$$\left\| \Omega_f(\cdot, z + z_0, t, d) \right\|_p \leq C \left\| \Omega_{f,r}(\cdot, z_0, t, d') \right\|_p,$$

$$\left\| \Omega'_f(\cdot, z + z_0, t, d) \right\|_p \leq C \left\| \Omega'_{f,r}(\cdot, z_0, t, d') \right\|_p.$$

Any element of U can be written as $z_0 + z$ with $z \in B(0, S)$, so (b) follows from (d) and (c) follows from (e).

(b) $\Rightarrow$ (a), (c) $\Rightarrow$ (a): Throughout this part we have d fixed, and we suppress its appearance. For $z \in \mathbb{R}^n$, let

$$\mu^z(x) = c_0^k d^{-n} X_{Q(0,d)}(x) + c_1^k X_{Q(-z,1)}(x)$$

$$+ \sum_{l=2}^{k} \frac{c_l^k}{l^n(l-1)^n d^n} X_{lQ(-z,1)} * X_{(l-1)Q(0,d)}(x),$$

$$v^z(x) = c_0^k d^{-n} X_{Q(0,d)}(x) + \sum_{l=1}^{k-1} \frac{c_l^k k^{2n}}{l^n(k-l)^n d^n} X_{(l/k)Q(-z,1)}$$

$$* X_{((k-l)/k)Q(0,d)}(x) + c_k^k X_{Q(-z,1)}(x),$$

where c_l^k is as in (7). We need the following lemma.

LEMMA 1
For every $t > 0$, $x_0, z \in \mathbb{R}^n$ the following formulas hold:

$$\Omega_f(x_0, z, t) = \left(\mu_t^z * f\right)(x_0),$$

$$\Omega_f'(x_0, z, t) = \left(v_t^z * f\right)(x_0),$$

where $\mu_t^z(x) = \left(\frac{1}{t^n}\right)\mu^z\left(\frac{x}{t}\right)$, and $v_t^z(x) = \left(\frac{1}{t^n}\right)v^z\left(\frac{x}{t}\right)$.

PROOF   In view of the equality immediately above (7), it is sufficient to show that for $0 \leq l \leq k$

$$\frac{1}{|Q|}\frac{1}{|Q^z|}\int_Q \int_{Q^z} f(x + l(y - x))\,dy\,dx$$

$$= \begin{cases} d^{-n}\left(X_{Q(0,d)}\right)_t * f(x_0); & l = 0 \\ \left(X_{Q(-z,1)}\right)_t * f(x_0); & l = 1 \\ \frac{1}{l^n(l-1)^n d^n}\left(X_{lQ(-z,1)} * X_{(l-1)Q(0,d)}\right)_t * f(x_0); & l = 2,\ldots,k, \end{cases} \quad (9)$$

and

$$\frac{1}{|Q|}\frac{1}{|Q^z|}\int_Q \int_{Q^z} f\left(x + \left(\frac{l}{k}\right)(y - x)\right)\,dy\,dx$$

$$= \begin{cases} d^{-n}\left(X_{Q(0,d)}\right)_t * f(x_0); & l = 0, \\ \frac{k^{2n}}{l^n(l-1)^n d^n}\left(X_{(l/k)Q(-z,1)} * \right. \\ \qquad \left. X_{((k-l)/k)Q(0,d)}\right)_t * f(x_0); & l = 1,\ldots,k-1, \\ \left(X_{Q(-z,1)}\right)_t * f(x_0); & l = k. \end{cases} \quad (10)$$

The equalities (9) and (10) follow by the change of variables, and thus Lemma 1 is proven.   ∎

We will need the following claim.

Claim 1 *Recall the definition of $c_l = c_l^k$ given by (7), and let u be an integer. We have*

(a) $\displaystyle\sum_{l=0}^{k} c_l l^u = 0 \text{ for } 0 \leq u < k,$

(b) $\displaystyle\sum_{l=0}^{k} c_l l^u > 0 \text{ for } 0 < k \leq u,$

(c) $\displaystyle\sum_{l=0}^{k} c_l l^k = k!.$

PROOF The proof is immediate, once we observe that $\sum_{l=0}^{k} c_l^k l^u$ is the number of ways in which we can place u balls in k cells so that no cell is empty (see [Fe], page 59, Exercise 7). ∎

Let $z_1, \ldots, z_n$ be n linearly independent vectors in $\mathbb{R}^n$. Let

$$\mu(x) = (\mu^{z_1}(x), \ldots, \mu^{z_n}(x)),$$

and,

$$\nu(x) = (\nu^{z_1}(x), \ldots, \nu^{z_n}(x)).$$

Thus μ and ν are two vector-valued measures, depending (implicitly) on $z_1, \ldots, z_n$, k and d. Define $\widehat{\mu}$ and $\widehat{\nu}$ component-wise:

$$\widehat{\mu}(\xi) = (\widehat{\mu^{z_1}}(\xi), \ldots, \widehat{\mu^{z_n}}(\xi))$$

$$\widehat{\nu}(\xi) = (\widehat{\nu^{z_1}}(\xi), \ldots, \widehat{\nu^{z_n}}(\xi)).$$

We have our next lemma.

LEMMA 2
For every choice of linearly independent $z_1, \ldots, z_n$, $\widehat{\mu}$ and $\widehat{\nu}$ satisfy a Tauberian condition; that is, for every $\xi \in \mathbb{R}^n$, $\xi \neq 0$ there exists $t_1 > 0$ and $t_2 > 0$ such, that $|\widehat{\mu}(t_1 \xi)| \neq 0$ and $|\widehat{\nu}(t_2 \xi)| \neq 0$.

PROOF Let $P = Q(0, 1)$, $P^z = Q(-z, 1)$.

$$\widehat{\mu^z}(\xi) = c_0^k \widehat{\chi}_P(\xi d) + c_1^k \widehat{\chi}_{P^z}(\xi) + \sum_{l=2}^{k} \frac{c_l^k}{l^n (l-1)^n} \widehat{\chi}_{lP^z}(\xi) \widehat{\chi}_{(l-1)P}(\xi d)$$

$$= \sum_{l=0}^{k} c_l^k e^{2\pi i \langle z, \xi \rangle l} \widehat{\chi}_P(l\xi) \widehat{\chi}_P((l-1)\xi d).$$

$$\widehat{\chi}_P(\eta) = \prod_{j=1}^{n} \text{sinc}(\eta_j); \text{ where } \eta = (\eta_1, \ldots, \eta_n), \text{ and } \text{sinc}(x) = \frac{\sin(\pi x)}{\pi x}. \text{ So,}$$

$$\widehat{\mu}^z(\xi) = \sum_{l=0}^{k} c_l^k e^{2\pi i \langle z, \xi \rangle l} \prod_{j=1}^{n} \text{sinc}(l\xi_j) \prod_{j=1}^{n} \text{sinc}((l-1)\xi_j d).$$

In a similar way, we obtain

$$\widehat{\nu}^z(\xi) = \sum_{l=0}^{k} c_l^k e^{2\pi i \langle z, \xi \rangle (l/k)} \prod_{j=1}^{n} \text{sinc}\left(\frac{k-l}{k}\xi_j d\right) \prod_{j=1}^{n} \text{sinc}\left(\frac{l}{k}\xi_j\right).$$

Fix ξ_0 such that $|\xi_0| = 1$. We are supposed to show that $t \mapsto \widehat{\mu}(t\xi_0)$ and $t \mapsto \widehat{\nu}(t\xi_0)$ are not identically 0 for $t > 0$. We will do that by showing that there exists a j, $j = 1, \ldots, n$, such that $\widehat{\mu^{z_j}}(t\xi_0)$ and $\widehat{\nu^{z_j}}(t\xi_0)$ have a nontrivial derivative of order k at 0. Define

$$F_{\xi_0}(x) = \prod_{j=1}^{n} \text{sinc}(\xi_{0j}x). \tag{11}$$

Thus

$$\widehat{\mu}^z(t\xi_0) = \sum_{l=0}^{k} c_l^k e^{2\pi i \langle z, \xi_0 \rangle lt} F_{\xi_0}(lt) F_{\xi_0}((l-1)td),$$

$$\widehat{\nu}^z(t\xi_0) = \sum_{l=0}^{k} c_l^k e^{2\pi i \langle z, \xi_0 \rangle (l/k)t} F_{\xi_0}\left(\frac{k-l}{k}td\right) F_{\xi_0}\left(\frac{l}{k}t\right).$$

From among $z_1, \ldots, z_n$ we choose z_0, such that $\langle z_0, \xi_0 \rangle \neq 0$. Compute

$$\left(\frac{d}{dt}\right)^k \widehat{\mu}^{z_0}(t\xi_0)|_{t=0} = \sum_{l=0}^{k} c_l^k \sum_{m \leq k} \binom{k}{m} (2\pi i \langle z_0, \xi_0 \rangle l)^m \sum_{p \leq k-m} \binom{k-m}{p}$$

$$\cdot l^p F_{\xi_0}^{(p)}(0)((l-1)d)^{k-m-p} F_{\xi_0}^{(k-m-p)}(0)$$

$$= \sum_{m \leq k} \binom{k}{m} \sum_{p \leq k-m} \binom{k-m}{p} (2\pi i \langle z_0, \xi_0 \rangle)^m F_{\xi_0}^{(p)}(0)$$

$$\cdot F_{\xi_0}^{(k-m-p)}(0) d^{k-m-p} \sum_{l=0}^{k} c_l^k l^{m+p}(l-1)^{k-m-p}$$

$$= k! \sum_{m \leq k} \sum_{p \leq k-m} \binom{k}{m} \binom{k-m}{p} (2\pi i \langle z_0, \xi_0 \rangle)^m$$

$$\cdot F_{\xi_0}^{(p)}(0) \cdot F_{\xi_0}^{(k-m-p)}(0) d^{k-m-p}$$

$$= k! \left(e^{2\pi i \langle z_0, \xi_0 \rangle \cdot} F_{\xi_0}(d \cdot) F_{\xi_0}(\cdot)\right)^{(k)}(0). \tag{12}$$

$$\left(\frac{d}{dt}\right)^k \widehat{v^{z_0}}(t\xi_0)|_{t=0} = \sum_{l=0}^k c_l^k \sum_{m\le k} \binom{k}{m} \left(2\pi i\langle z_0,\xi_0\rangle \frac{l}{k}\right)^m$$

$$\cdot \sum_{p\le k-m} \binom{k-m}{p}\left(\frac{k-l}{k}\right)^p d^p F_{\xi_0}^{(p)}(0)$$

$$\cdot \left(\frac{l}{k}\right)^{k-m-p} F_{\xi_0}^{(k-m-p)}(0)$$

$$= k^{-k} \sum_{m\le k}\binom{k}{m}\sum_{p\le k-m}\binom{k-m}{p}(2\pi i\langle z_0,\xi_0\rangle)^m\, F_{\xi_0}^{(p)}(0)d^p$$

$$\cdot F_{\xi_0}^{(k-m-p)}(0)\sum_{l=0}^k c_l^k l^{k-p}(k-l)^p$$

$$= \frac{k!}{k^k}\sum_{m\le k}\sum_{p\le k-m}\binom{k}{m}\binom{k-m}{p}$$

$$\cdot (2\pi i\langle z_0,\xi_0\rangle)^m\, F_{\xi_0}^{(p)}(0)d^p \cdot F_{\xi_0}^{(k-m-p)}(0)(-1)^p$$

$$= \frac{k!}{k^k}\left(e^{2\pi i\langle z_0,\xi_0\rangle\cdot}F_{\xi_0}(-d\,\cdot)F_{\xi_0}(\cdot)\right)^{(k)}(0)$$

$$= \frac{k!}{k^k}\left(e^{2\pi i\langle z_0,\xi_0\rangle\cdot}F_{\xi_0}(d\,\cdot)F_{\xi_0}(\cdot)\right)^{(k)}(0). \tag{13}$$

We have used the Leibniz rule, and (12) and (13) follow from Claim 1*(a)* and *(c)*. We now need the following. ∎

Claim 2 *Let* F_{ξ_0} *be given by (11). Write*

$$e^{2\pi i\langle z_0,\xi_0\rangle x}F_{\xi_0}(dx)F_{\xi_0}(x) = \sum_{m=0}^\infty a_m x^m$$

(We can do this, since the function in question is analytic.) If $|\xi_0| = 1$, *then*

$$a_m = \begin{cases} i^m e_m; & e_m > 0, & \text{if } m \text{ is even,} \\ i^m\langle z_0,\xi_0\rangle e_m; & e_m > 0, & \text{if } m \text{ is odd.} \end{cases}$$

Observe, that kth derivative at 0 of $t\mapsto \widehat{\mu^{z_0}}(t\xi_0)$ and $t\mapsto \widehat{v^{z_0}}(t\xi_0)$ is thus nonzero (recall that $\langle z_0,\xi_0\rangle \ne 0$).

PROOF

$$\operatorname{sinc}(\xi_{0j}x) = \sum_{m=0}^\infty A_m\xi_{0j}^{2m}x^{2m}; \qquad A_m = (-1)^m\frac{\pi^{2m}}{(2m+1)!},$$

so,

$$F_{\xi_0}(x) = \prod_{j=1}^{n} \text{sinc}\,(\xi_{0j}x) = \sum_{m=0}^{\infty} x^{2m} \left(\sum_{\substack{i_1,\dots,i_n \geq 0 \\ i_1+\cdots+i_n=m}} A_{i_1} \xi_{01}^{2i_1} \cdots A_{i_n} \xi_{0n}^{2i_n} \right)$$

$$= \sum_{m=0}^{\infty} x^{2m} (-1)^m \pi^{2m} \left(\sum_{\substack{i_1,\dots,i_n \geq 0 \\ i_1+\cdots+i_n=m}} \frac{\xi_{01}^{2i_1}}{(2i_1+1)!} \cdots \frac{\xi_{0n}^{2i_n}}{(2i_n+1)!} \right)$$

$$= \sum_{m=0}^{\infty} x^{2m} (-1)^m B_m,$$

with

$$B_m = \pi^{2m} \sum_{\substack{i_1,\dots,i_n \geq 0 \\ i_1+\cdots+i_n=m}} \frac{\xi_{01}^{2i_1}}{(2i_1+1)!} \cdots \frac{\xi_{0n}^{2i_n}}{(2i_n+1)!} > 0.$$

The last inequality follows, since all terms of the sum are nonnegative, and at least one is strictly positive. To find the strictly positive term, observe that since $|\xi_0| = 1$, at least one $\xi_{0j} \neq 0$. The term corresponding to some i_j equal to m is then strictly positive. So

$$F_{\xi_0}(dx)F_{\xi_0}(x) = \sum_{m=0}^{\infty} x^{2m}(-1)^m \left[\sum_{j=0}^{m} B_j d^{2j} B_{m-j} \right]$$

$$= \sum_{m=0}^{\infty} x^{2m}(-1)^m C_m,$$

with

$$C_m = \sum_{j=0}^{m} B_j d^{2j} B_{m-j} > 0.$$

Finally, let $D_m = C_{m/2}$ if m is even, $D_m = 0$ if m is odd. Thus

$$F_{\xi_0}(dx)F_{\xi_0}(x) = \sum_{m=0}^{\infty} x^m i^m D_m.$$

$$e^{2\pi i \langle z_0, \xi_0 \rangle x} F_{\xi_0}(dx)F_{\xi_0}(x) = \left(\sum_{m=0}^{\infty} \frac{(2\pi i \langle z_0, \xi_0 \rangle)^m x^m}{m!} \right) \left(\sum_{m=0}^{\infty} x^m i^m D_m \right)$$

$$= \sum_{m=0}^{\infty} x^m \left(\sum_{j=0}^{m} \frac{(2\pi i \langle z_0, \xi_0 \rangle)^j}{j!} i^{m-j} D_{m-j} \right)$$

$$= \sum_{m=0}^{\infty} x^m a_m.$$

If m is even, then

$$a_m = i^m \sum_{\substack{j=0 \\ j-\text{even}}}^{m} \frac{(2\pi \langle z_0, \xi_0 \rangle)^j}{j!} D_{m-j} = i^m e_m,$$

where $e_m > 0$, since it is a sum of nonnegative numbers, and one term in this sum (D_m) must be strictly positive. If m is odd, then

$$a_m = i^m \sum_{\substack{j=1 \\ j-\text{odd}}}^{m} \frac{(2\pi \langle z_0, \xi_0 \rangle)^j}{j!} D_{m-j}$$

$$= i^m (2\pi \langle z_0, \xi_0 \rangle) \sum_{\substack{j=1 \\ j-\text{odd}}}^{m} \frac{(2\pi \langle z_0, \xi_0 \rangle)^{j-1}}{j!} D_{m-j}$$

$$= i^m (\langle z_0, \xi_0 \rangle) e_m.$$

As above, $e_m > 0$ since it is a sum of nonnegative numbers, and, again, one term (D_{m-1}) must be strictly positive. Claim 2 is proven. ∎

Thus, for arbitrary $\xi_0 \in \mathbb{R}^n$, $|\xi_0| = 1$ we found $z_0 \in \{z_1, \ldots, z_n\}$, such that both $t \mapsto \widehat{\mu}^{z_0}(t\xi_0)$ and $t \mapsto \widehat{v}^{z_0}(t\xi_0)$ are not identically 0. Lemma 2 is proven.

REMARK 2 (a) Observe that Lemma 2 remains true if we replace the cubes in the definition of the Ω_f functions with balls having the same centers as the respective cubes and diameters equal to the respective side lengths. In fact, to prove such a modified Lemma 2 we need only an obvious modification of Claim 2.

(b) An examination of the proof of Lemma 2 shows that

$$\left(\frac{d}{dt}\right)^u \widehat{\mu}^{z_0}(t\xi_0)|_{t=0} = 0,$$

$$\left(\frac{d}{dt}\right)^u \widehat{v}^{z_0}(t\xi_0)|_{t=0} = 0,$$

for $u < k$. Since the direction ξ_0 is arbitrary, it follows that all partial derivatives of $\widehat{\mu}$ and $\widehat{v}$, of order less than k, vanish at 0. Thus the methods of Janson and Taibleson [JT] which we exploit later in this part of the proof of Theorem 1, can be used to show directly that (a) implies (b) and (c) in Theorem 1. Thus μ and v are compactly supported, finite vector-valued Borel measures, such that $\widehat{\mu}$ and $\widehat{v}$ satisfy a Tauberian condition. We now invoke a version of Calderón's reproducing formula, due to Janson and Taibleson ([JT], Theorem 2 and the preceding lemma). ∎

THEOREM 2

(See Janson–Taibleson [JT].) There exist vector-valued functions $\eta = (\eta^1, \ldots, \eta^n)$ and $\omega = (\omega^1, \ldots, \omega^n)$, such that η^i, $\omega^i \in S$; $i = 1, \ldots, n$; associated with μ and v respectively, and an integer N ($N - 1$ is the greater number of $[\beta]$ and the order of f as a tempered distribution) such, that

$$f = \int_0^\infty f * \mu_t * \eta_t \, \frac{dt}{t}, \quad \text{where } \mu_t * \eta_t = \sum_{i=1}^n \mu_t^{z_i} * \eta_t^i,$$

$$f = \int_0^\infty f * v_t * \omega_t \, \frac{dt}{t}, \quad \text{where } v_t * \omega_t = \sum_{i=1}^n v_t^{z_i} * \omega_t^i,$$

both integrals converging in $S_N' = S'/\mathbb{P}_{N-1}$ where $\mathbb{P}_{N-1}$ is the space of polynomials of degree at most $N - 1$.

The integrals are the Pettis integrals [R]; i. e.,

$$\left\langle \int_0^\infty g(t) \frac{dt}{t}, \phi \right\rangle = \lim_{\substack{\epsilon \to 0 \\ A \to \infty}} \int_\epsilon^A \langle g(t), \phi \rangle \frac{dt}{t}, \tag{14}$$

where $g : \mathbb{R}^+ \to S'$, $\phi \in S$. Convergence in S_N' means that the limit (14) exists for $\phi \in (S_N)' = \{\phi \in S : \int \phi(x) x^l \, dx = 0, \ |l| < N\}$, and defines an element in S_N'.

Let $e_j = (0, \ldots, 1, \ldots, 0)$ be the unit vector in $\mathbb{R}^n$ in the direction of the jth variable, $j = 1, \ldots, n$. Let $h \in \mathbb{R}^+$, and denote $\kappa^j = (\delta_{-e_j} - \delta_0)^{*N}$ (Nth convolution power). We have $\Delta_{he_j}^N f(x) = \kappa_h^j * f(x)$. The operator $f \mapsto \kappa_h^j * f$ is bounded on S_N', so we can write

$$f * \kappa_h^j = \int_0^\infty f * \kappa_h^j * \mu_t * \eta_t \frac{dt}{t},$$

$$f * \kappa_h^j = \int_0^\infty f * \kappa_h^j * v_t * \omega_t \frac{dt}{t}, \tag{15}$$

both integrals converging in S_N'. Observe that

$$\widehat{\kappa^j}^{(\gamma)}(0) = 0 \text{ for } |\gamma| < N,$$

so,

$$\|\kappa_h^j * \eta_t^i\|_{L^1} \le C_\eta \min\left\{1, \left(\frac{h}{t}\right)^N\right\}, \quad i, j = 1, \ldots, n, \tag{16}$$

and,

$$\|\kappa_h^j * \omega_t^i\|_{L^1} \le C_\omega \min\left\{1, \left(\frac{h}{t}\right)^N\right\}, \quad i, j = 1, \ldots, n, \tag{17}$$

(see [Hei]) with the constants independent of j, $j = 1, \ldots, n$. The conditions (b) and (c) in Theorem 1, together with (16) and (17), imply that the integrals in (15) actually converge into $L^p(\mathbb{R}^n)$. Using the Minkowski's inequality, the estimates (16)–(17), and Hardy's inequalities, we obtain

$$\left(\int_0^\infty \left(h^{-\beta} \| f * \kappa_h^j \|_p \right)^q \frac{dh}{h} \right)^{1/q}$$

$$\leq C \sum_{i=1}^n \left(\int_0^\infty \left(h^{-\beta} \int_0^h \| f * \mu_t^{z_i} \|_p \frac{dt}{t} \right)^q \frac{dh}{h} \right)^{1/q}$$

$$+ C \sum_{i=1}^n \left(\int_0^\infty \left(h^{N-\beta} \int_h^\infty \| f * \mu_t^{z_i} \|_p \frac{dt}{t^{1+N}} \right)^q \frac{dh}{h} \right)^{1/q}$$

$$\leq C \sum_{i=1}^n \left(\int_0^\infty \left(h^{-\beta} \| f * \mu_h^{z_i} \|_p \right)^q \frac{dh}{h} \right)^{1/q}.$$

Similarly,

$$\left(\int_0^\infty \left(h^{-\beta} \| f * \kappa_h^j \|_p \right)^q \frac{dh}{h} \right)^{1/q} \leq C \sum_{i=1}^n \left(\int_0^\infty \left(h^{-\beta} \| f * \nu_h^{z_i} \|_p \right)^q \frac{dh}{h} \right)^{1/q},$$

with the usual interpretation of the case of $q = \infty$ in both cases. Thus both (b) and (c) in Theorem 1 (independently) imply

$$\left(\int_0^\infty \left(h^{-\beta} \| f * \kappa_h^j \|_p \right)^q \frac{dh}{h} \right)^{1/q} \leq C < \infty, \tag{18}$$

with C independent of j, $j = 1, \ldots, n$.

Define $\kappa = (\kappa^1, \ldots, \kappa^n)$, the vector-valued measure, where on the jth coordinate we have the measure $\kappa^j = (\delta_{-e_j} - \delta_0)^{*N}$. Since $\widehat{\kappa^j}(\xi) = (e^{2\pi i \langle \xi, e_j \rangle} - 1)^N$, we see, that $\widehat{\kappa}$ satisfies the Tauberian condition. Hence, by the result of Janson and Taibleson, since $(N - 1)$ is at least equal to the order of f as the tempered distribution, there exists a vector-valued function $\rho = (\rho^1, \ldots, \rho^n)$ with $\rho^i \in S$, $i = 1, \ldots, n$, such, that

$$f = \int_0^\infty f * \kappa_t * \rho_t \frac{dt}{t} \text{ in } S_N'.$$

Observe, that (18) implies

$$\left(\int_0^\infty \left(h^{-\beta} \| f * \kappa_h \|_p \right)^q \frac{dh}{h} \right)^{1/q} < \infty. \tag{19}$$

We now follow the argument of Janson and Taibleson ([JT], Application 1). Recall that $k_0 = [\beta] + 1$, and let $\tilde{\rho}^i(x) = \rho^i(-x)$, $\tilde{\rho} = (\tilde{\rho}^1, \ldots, \tilde{\rho}^n)$. It follows from [Hei], that for $\phi \in S_{k_0}$ $\| \tilde{\rho}_t^i * \phi \|_1 \leq C \min\{1, t^{-k_0}\}$, where C only depends on ρ^i

and ϕ. Consider the functional

$$\phi \mapsto f_1(\phi) = \int_0^\infty \langle f * \kappa_t * \rho_t, \phi \rangle \frac{dt}{t}$$

on S_{k_0}. Fix $\phi \in S_{k_0}$, then

$$\int_0^\infty |\langle f * \kappa_t * \rho_t, \phi \rangle| \frac{dt}{t} \leq \sum_{i=1}^n \int_0^\infty |\langle f * \kappa_t^i, \tilde{\rho}_t^i * \phi \rangle| \frac{dt}{t}$$

$$\leq \sum_{i=1}^n \int_0^\infty \|f * \kappa_t^i\|_p \|\tilde{\rho}_t^i * \phi\|_{p'} \frac{dt}{t}, \qquad (20)$$

where $\frac{1}{p} - \frac{1}{p'} = 1$, $1 \leq p, p' \leq \infty$. It is easy to see that

$$\|\tilde{\rho}_t^i * \phi\|_{p'} \leq C \min\{1, t^{-k_0}\}.$$

Consider $0 < t \leq 1$, then $\|\tilde{\rho}_t^i * \phi\|_{p'} \leq \|\tilde{\rho}_t^i\|_1 \|\phi\|_{p'} = C = C \min\{1, t^{-k_0}\}$. Consider $t \geq 1$. From [Hei] we see that each $\tilde{\rho}^i$ is compactly supported, and thus there exists $\Phi \in S$, such that $\tilde{\rho}_t^i * \phi = \tilde{\rho}_t^i * \phi * \Phi_t$. Then $\|\tilde{\rho}_t^i * \phi\|_{p'} \leq \|\tilde{\rho}_t^i * \phi\|_1 \|\Phi_t\|_{p'} \leq C \min\{1, t^{-k_0}\} t^{n(1/p'-1)} \leq C t^{-k_0}$. So $\|\tilde{\rho}_t^i * \phi\|_{p'} \leq C \min\{1, t^{-k_0}\}$. So

$$(20) \leq C \int_0^1 \|f * \kappa_t\|_p \frac{dt}{t} + C \int_1^\infty \|f * \kappa_t\|_p \frac{dt}{t^{1+k_0}}$$

$$\leq C \left(\int_0^1 \left(t^{-\beta} \|f * \bar{\kappa}_t\|_p \right)^q \frac{dt}{t} \right)^{1/q} \left(\int_0^1 t^{\beta q'} \frac{dt}{t} \right)^{1/q'}$$

$$+ C \left(\int_1^\infty \left(t^{-\beta} \|f * \bar{\kappa}_t\|_p \right)^q \frac{dt}{t} \right)^{1/q} \left(\int_1^\infty t^{(\beta-k_0)q'} \frac{dt}{t} \right)^{1/q'}$$

$$\leq C \left(\frac{1}{\beta q'} \right)^{1/q'} + C \left(\frac{1}{(k_0 - \beta)q'} \right)^{1/q'} < \infty,$$

with the obvious modifications, if $q = \infty$. Thus f_1 is defined for all $\phi \in S_{k_0}$. Moreover, $f_1 = f$ acting on S_N. f is continuous on S_N, and $S_{k_0} = S_N + F$ (topologic sum with F finite dimensional). So, $f_1 \in S'_{k_0}$.

We now choose a representative of f_1 in S'_{k_0}, and again call it f_1. $f - f_1 = 0$ on S_N, and thus $f - f_1 \in \mathbb{P}_{N-1}$. Thus $f_1 \in L^1_{\text{loc}}$ and $f = f_1 + P$, with P a polynomial. Now we choose a direction $e_0 \in \mathbb{R}^n$, $|e_0| = 1$, and let $\kappa = (\delta_{-e_0} - \delta_0)^{*k_0}$. $\hat{\kappa}^{(\alpha)}(0) = 0$ provided $|\alpha| < k_0$, and hence the convolution with κ is bounded on S'_{k_0}. Thus

$$f_1 * \kappa_s = \int_0^\infty f * \kappa_t * \rho_t * \kappa_s \frac{dt}{t} \text{ in } S'_{k_0}.$$

Moreover,

$$\left(\int_0^\infty \left(\left\|s^{-\beta}\int_0^\infty f * \kappa_t * \rho_t * \kappa_s \frac{dt}{t}\right\|_p\right)^q \frac{ds}{s}\right)^{1/q}$$

$$\leq \sum_{i=1}^n \left(\int_0^\infty \left(s^{-\beta}\int_0^\infty \|f * \kappa_t^i\|_p \, \|\rho_t^i * \kappa_s\|_1 \frac{dt}{t}\right)^q \frac{ds}{s}\right)^{1/q}$$

$$\leq c\sum_{i=1}^n \left(\int_0^\infty (s^{-\beta}\|f * \kappa_s^i\|_p)^q \frac{ds}{s}\right)^{1/q},$$

where, as in the proof of (18), we used Hardy's inequalities and the inequality

$$\|\rho_t^i * \kappa_s\|_1 \leq C\min\left\{1, \left(\frac{s}{k}\right)^{k_0}\right\}, \qquad j = 1, \ldots, n,$$

due to Heideman [Hei]. The examination of the proof in [Hei] shows that the above constant can be chosen to be independent of e_0, provided that $|e_0| = 1$. Therefore, by (18), the integral

$$s^{-\beta}\int_0^\infty f * \kappa_t * \rho_t * \kappa_s \frac{dt}{t}$$

exists in the space $l^q\left(L^p(\mathbb{R}^n)\frac{ds}{s}\right)$, and, we obtain

$$\left(\int_0^\infty (s^{-\beta}\|f_1 * \kappa_s\|_p)^q \frac{ds}{s}\right)^{1/q} \leq C < \infty,$$

with C independent of the direction e_0 ($|e_0| = 1$). We have

$$\int_{\mathbb{R}^n}\left\|\frac{\Delta_h^{k_0}f_1(\cdot)}{|h|^\beta}\right\|_p^q \frac{dh}{|h|^n} = \int_{\Sigma_{n-1}}\int_0^\infty \left(s^{-\beta}\|\Delta_{se_0}^{k_0}f_1\|_p\right)^q \frac{ds}{s} de_0$$

$$= \int_{\Sigma_{n-1}}\int_0^\infty (s^{-\beta}\|f_1 * \kappa_s\|_p)^q \frac{ds}{s} de_0$$

$$\leq C < \infty,$$

with the usual interpretation, if $q = \infty$. By definition, we see that $f_1 \in \dot{B}_p^{\beta,q}$. From the proof of (a) $\Rightarrow$ (b) and (a) $\Rightarrow$ (c) we see that (b) (or (c)) holds with f replaced by f_1. Thus we deduce, that (b) (or (c)) holds with f replaced by the polynomial P.

REMARK 3 Observe that so far we have only used the fact, that (b) or (c) in Theorem 1 hold for some linearly independent collection $z_1, \ldots, z_n \in \mathbb{R}^n$. The fact that these conditions hold uniformly in U, some open set in $\mathbb{R}^n$, will be necessary for the final step of the proof. ∎

The proof of Theorem 1 will be concluded after we establish the following lemma:

LEMMA 3

Suppose $P(x) = \sum_{|\gamma| \leq N} a_\gamma x^\gamma$ is a polynomial. Suppose there exist $d > 0$, $U \subset \mathbb{R}^n$ an open set, such that, for every $x_0 \in \mathbb{R}^n$ and $z \in U$

$$\left(\int_0^\infty \left(t^{-\beta} \|\Omega_P(\cdot, z, t, d)\|_p \right)^q \frac{dt}{t} \right)^{1/q} \leq C < \infty, \tag{21}$$

or, for $q = \infty$

$$\sup_{t>0} t^{-\beta} \|\Omega_P(\cdot, zt, d)\|_p \leq C < \infty, \tag{22}$$

with C independent of $z \in U$. Alternately, we assume (21) or (22), respectively, hold with Ω_P replaced with Ω'_P. Then $N = $ degree of $P < k$, and the left side of (21) or (22), respectively, vanishes. (We recall that $k > \beta$ is used in the definition of Ω operators.)

PROOF   Observe

$$\Omega_P(tx_0, z, t, d)$$

$$= \sum_{j=0}^N t^j \sum_{|\gamma|=j} a_\gamma d^{-n} \int_{Q(x_0, d)} \int_{Q(x_0+z.1)}$$

$$\sum_{l=0}^k c_l^k (x + l(y-x))^\gamma \, dy \, dx. \tag{23}$$

Thus $\Omega_P(tx_0, z, t, d)$ is a polynomial in t, for x_0, z, d fixed. We are going to show that the degree of this polynomial is less that k, and, in fact, that this is a zero polynomial. Observe that for t, z, and d fixed $\Omega_P(tx_0, z, t, d)$ is a polynomial in x_0:

$$\Omega_P(tx_0, z, t, d) = \sum_{j=0}^N t^j \sum_{|\gamma|=j} a_\gamma d^{-n} \int_{Q(x_0, d)} \int_{Q(x_0+z.1)}$$

$$\cdot \sum_{l=0}^k c_l^k (x + l(y-x))^\gamma \, dy \, dx$$

$$= \sum_{j=0}^N t^j \sum_{|\gamma|=j} a_\gamma d^{-n} \int_{Q(0,d)} \int_{Q(z.1)}$$

$$\cdot \sum_{l=0}^k c_l^k (x_0 + x + l(y-x))^\gamma \, dy \, dx$$

$$= \sum_{j=0}^{N} t^j \sum_{|\gamma|=j} a_\gamma \sum_{\alpha \leq \gamma} \binom{\gamma}{\alpha} x_0^\alpha \sum_{l=0}^{k} c_l^k d^{-n}$$

$$\cdot \int_{Q(0,d)} \int_{Q(z,1)} (x + l(y - x))^{\gamma - \alpha} \, dy \, dx.$$

This polynomial is in $L^p(\mathbb{R}^n)$ for a. e. $t > 0$. Thus (except, possibly, in the case when $p = \infty$) it is a zero polynomial for a. e. $t > 0$, and, being continuous in t, it vanishes identically for all $t > 0$. If $p = \infty$, then

$$\|\Omega_P(t \cdot , z, t, d)\|_\infty$$

$$= \left| \sum_{j=0}^{N} t^j \sum_{|\gamma|=j} a_\gamma \sum_{l=0}^{k} c_l^k d^{-n} \int_{Q(0,d)} \int_{Q(z,1)} (x + l(y - x))^\gamma \, dy \, dx \right|$$

$$= \left| \sum_{j=0}^{N} u_j t^j \right|,$$

where

$$u_j = \sum_{|\gamma|=j} a_\gamma \sum_{l=0}^{k} c_l^k d^{-n} \int_{Q(0,d)} \int_{Q(z,1)} (x + l(y - x))^\gamma \, dy \, dx.$$

From (21) or (22) we obtain $u_j = 0$ for $j > \beta$ (looking at large t's). If $j \leq \beta < k$ and $|\gamma| = j$, then

$$\sum_{l=0}^{k} c_l^k (x + l(y - x))^\gamma = \sum_{l=0}^{k} c_l^k \sum_{\alpha \leq \gamma} \binom{\gamma}{\alpha} x^\alpha l^{|\gamma - \alpha|} (y - x)^{\gamma - \alpha}$$

$$= \sum_{\alpha \leq \gamma} \binom{\gamma}{\alpha} x^\alpha (y - x)^{\gamma - \alpha} \sum_{l=0}^{k} c_l^k l^{|\gamma - \alpha|}$$

$$= 0,$$

by Claim 1(a). Thus $u_j = 0$ for all $j \geq 0$. Consequently,

$$\Omega_P(tx_0, z, t, d) = 0 \qquad \forall x_0 \in \mathbb{R}^n, t > 0, \qquad 1 \leq p \leq \infty.$$

Hence, by (23), we obtain for every $j \geq 0$

$$0 = \sum_{|\gamma|=j} a_\gamma d^{-n} \int_{Q(x_0,d)} \int_{Q(x_0+z,1)} \sum_{l=0}^{k} c_l^k (x + l(y - x))^\gamma \, dy \, dx$$

$$= \sum_{|\gamma|=j} a_\gamma d^{-n} \int_{Q(x_0,d)} \int_{Q(x_0,1)} \sum_{l=0}^{k} c_l^k (x + l(y - x) + lz)^\gamma \, dy \, dx$$

$$= \sum_{|\gamma|=j} a_\gamma \sum_{\alpha \le \gamma} \binom{\gamma}{\alpha} z^\alpha d^{-n} \int_{Q(x_0,d)} \int_{Q(x_0,1)}$$

$$\cdot \sum_{l=0}^{k} c_l^k (x + l(y - x))^{\gamma - \alpha} l^{|\alpha|} \, dy \, dx.$$

This is a polynomial in z, vanishing identically on an open set U; thus it is a zero polynomial. Consider its homogeneous part of degree j, which is the highest degree:

$$0 = \sum_{|\gamma|=j} a_\gamma z^\gamma d^{-n} \int_{Q(x_0,d)} \int_{Q(x_0,1)} \sum_{l=0}^{k} c_l^k l^j \, dy \, dx$$

$$= \left(\sum_{l=0}^{k} c_l^k l^j \right) \sum_{|\gamma|=j} a_\gamma z^\gamma.$$

For $j \ge k$, $\sum c_l^k l^j > 0$ by Claim 1(b), and thus $a_\gamma = 0$ for $|\gamma| = j$. Thus $N < k$, where N is the degree of P. The identical argument works for Ω'_p. Lemma 3 is proven. ∎

Thus we have shown that $f = f_1 + P$, $f_1 \in \dot{B}_p^{\beta,q}$ and P is a polynomial of degree less than k. Theorem 1 is proven.

References

[CRW] R. R. Coifman, R. Rochberg and G. Weiss, *Factorization theorems for Hardy spaces in several variables,* Annals Math. 103 (1976), 611–635.

[Fe] W. Feller, *An introduction to probability theory and its applications,* vol. I, 3rd ed., John Wiley & Sons, New York, 1968.

[Fr] M. Frazier, B. Jawerth and G. Weiss, Littlewood-Paley Theory and the Study of Function Spaces, CBMS-AMS Regional Conf. Series 79 (1991).

[Hei] N. J. H. Heideman, *Duality and fractional integration in Lipschitz spaces,* Studia Math. 50 (1974), 65–85.

[Her] C. Herz, *Lipschitz spaces and Bernstein's theorem on absolutely convergent Fourier transforms,* J. Math. and Mech. 18 (1968), 283–323.

[JT] S. Janson and M. Taibleson, *I teoremi di rappresentazione di Calderón,* Rend. Sem. Mat. Univers. Politecn. Torino 39 (1981), 27–35.

[JTW] S. Janson, M. Taibleson and G. Weiss, *Elementary characterization of the Morrey-Campanato spaces,* Lect. Notes in Math. 992 (1983), 101–114.

[Jo] R. Johnson, *Temperatures, Riesz potentials, and the Lipschitz spaces of Herz*, Proc. London. Math. Soc. 27 (1973), 290–316.

[P] M. Paluszynski, *Characterization of Lipschitz Spaces Via the Commutator Operator of Coifman, Rochberg and Weiss; A Multiplier Theorem for the Semigroup of Contractions*, Ph. D. Thesis, Washington University, 1992.

[R] W. Rudin, *Functional analysis*, 2nd ed., McGraw-Hill, New York, 1991.

18

Oscillatory Singular Integrals on Hardy Spaces

Yibiao Pan

18.1 Introduction

We shall consider the boundedness properties of oscillatory singular integral operators on Hardy spaces. Let $K(x)$ be a Calderón-Zygmund kernel, which we will assume satisfies the following conditions:

(i) $K \in C^1(\mathbb{R}^n \setminus \{0\})$;

(ii)   There is a constant $A > 0$ such that

$$|K(x)| \leq A|x|^{-n}, \ |\nabla K(x)| \leq A|x|^{-n-1};$$

(iii) $\displaystyle\int_{a<|x|<b} K(x)dx = 0$ holds for $b > a > 0$.

We use $\|K\|_{cz}$ to denote the least value of A for which (ii) holds. Let $P(x, y)$ be a real-valued polynomial in $\mathbb{R}^n \times \mathbb{R}^n$. Define the operator T:

$$Tf(x) = \text{P.V.} \int_{\mathbb{R}^n} e^{iP(x,y)} K(x - y) f(y) \, dy. \tag{1}$$

T is called an oscillatory singular integral operator with polynomial phase. The following result on the L^p boundedness of T is due to Ricci and Stein [RS].

THEOREM 1
Let T be given as in (1). Then the operator, T, can be extended to be a bounded operator on $L^p(\mathbb{R}^n)$ to itself with $1 < p < \infty$. The bound of T (which of course depends on $\|K\|_{cz}$, p, and n) can be taken to depend only on the degree of P and is otherwise independent of the coefficients of P.

In [CC] Chanillo and Christ proved the weak $(1, 1)$ boundedness of T.

THEOREM 2
Let T be given as in (1). Then the operator, T, is of weak-type on $L^1(\mathbb{R}^n)$ with a bound depending only on $\|K\|_{cz}$, n, and the degree of P.

The question that arises naturally here is whether such operators are bounded on the Hardy space H^1 (or H^p with $p \leq 1$). The present chapter is devoted to the study of this problem, mainly in the translation invariant case (where $P(x, y) = P(x - y)$) and its extension to operators with general phase functions. Results for nonconvolutional operators are also discussed (see Section 18.5). Some of the results described here are in [HP] and [P1–P4]. Results that have not appeared before will be proven in detail.

18.2 Polynomial phase

Let $S(\mathbb{R}^n)$ denote the Schwartz class and assume that $\phi \in S(\mathbb{R}^n)$ and

$$\int_{\mathbb{R}^n} \phi(x)dx \neq 0.$$

For $f \in S'(\mathbb{R}^n)$, we set

$$f^*(x) = \sup_{t>0} |f * \phi_t(x)|,$$

where $x \in \mathbb{R}^n$, $\phi_t(x) = t^{-n}\phi\left(\frac{x}{t}\right)$. The Hardy space $H^p(\mathbb{R}^n)$, $0 < p < \infty$, is defined to be the space of all f, such that

$$\|f\|_{H^p} = \|f^*\|_p < \infty.$$

Let K be a Calderón-Zygmund kernel and $P(x)$ be a real-valued polynomial on $\mathbb{R}^n$. Let T be given by

$$Tf(x) = \text{P.V.} \int_{\mathbb{R}^n} e^{iP(x-y)} K(x - y)f(y)\, dy. \tag{2}$$

In [HP] we proved the following.

THEOREM 3
Suppose $\nabla P(0) = 0$, $deg(P) = d$. Then there exists a constant C, which depends on only $\|K\|_{cz}$, n, and d, such that

$$\|Tf\|_{H^1} \leq C\|f\|_{H^1}, \tag{3}$$

for all $f \in H^1(\mathbb{R}^n)$.

REMARK 1 (a) The above theorem becomes false if the condition $\nabla P(0) = 0$ is removed. To see this, we let

$$n = 1, \quad P(x) = kx, \quad Tf(x) = \int e^{ik(x-y)} \frac{f(y)}{x - y} \, dy.$$

Define f by

$$f(x) = \begin{cases} 1, & \text{for } 0 \le x \le \pi \\ -1, & \text{for } -\pi \le x < 0 \\ 0, & \text{otherwise,} \end{cases} \tag{4}$$

let $g_k(x) = e^{-ikx} f(x)$. Suppose (3) holds for T. Then one has $H(g_k) \in L^1$ where H denotes the Hilbert transform. This implies that $g_k \in H^1(\mathbb{R})$ and

$$\int_{-\pi}^{\pi} e^{-ikx} f(x) dx = 0$$

for all $k \in \mathbb{Z}$. Hence $f \equiv 0$ a.e., which contradicts (4).

(b) There is no analogue of Theorem 3 for H^p if $p < 1$. To show this, we first let

$$p \in \left(\frac{1}{2}, 1 \right), \quad P(x) = x^2, \quad Tf(x) = \int e^{i(x-y)^2} \frac{f(y)}{x - y} \, dy.$$

Let $a(x)$ be a p-atom defined by

$$a(x) = \begin{cases} \delta^{-1/p} e^{-ix^2}, & \text{if } \frac{\delta}{4} \le x \le \frac{\delta}{2} \\ -\delta^{-1/p} e^{-ix^2}, & \text{if } -\frac{\delta}{2} \le x - \frac{\delta}{4} \\ 0, & \text{otherwise,} \end{cases} \tag{5}$$

by direct computation, one has

$$\|Ta\|_{H^p} \ge C \delta^{2(p-1)} \|a\|_{H^p}. \tag{6}$$

By letting $\delta \to 0$, we see that T cannot be bounded on H^p.

For $p \in (0, \frac{1}{2}]$, the unboundedness of T on H^p follows from the above argument, the L^2 boundedness of T, and interpolation between L^2 and H^p.

(c) Let $\psi \in C^\infty(\mathbb{R}^n)$, and assume that $\psi(x) = 0$, $|x| \le 1$, and $\psi(x) = 1$, $|x| \ge 2$. Set $T_\alpha f = \dfrac{e^{i|x|^\alpha}}{|x|^n} \psi(x) * f$. Such operators have been studied by many authors, and it is known that T_α is bounded on $H^1(\mathbb{R}^n)$, if $\alpha > 0$ and $\alpha \ne 1$ (see [Sj]). When $\alpha = 2m$ is a positive, even integer, the boundedness of T_α on $H^1(\mathbb{R}^n)$ implies the boundedness of T on $H^1(\mathbb{R}^n)$, where T is given by (2) with $P(x) = |x|^{2m}$. See also [CKS].

(d) Theorem 3 remains true when $H^1(\mathbb{R}^n)$ is replaced by the weighted Hardy space $H_w^1(\mathbb{R}^n)$, where $w \in A_1$. ∎

A proof of the result in Theorem 3 in the one-dimensional case was given in [H], but the method used there cannot be extended to higher dimensions. The proof of Theorem 3, which will appear in [HP], has the following main ingredients: (1) L^2 theory, which is contained in Ricci-Stein's theorem; (2) the use of induction on the degree of P; (3) the reduction of the problem to the study of the L^2 norm of certain oscillatory integral operators (with smooth kernels) S_j. For the estimation of $\| S_j \|_{2,2}$, we are led to consider $S_j^* S_j$. Two important tools used there are (i) van der Corput's lemma and (ii) a certain inequality on polynomials due to Ricci and Stein. Further details can be found in [HP].

18.3   Generic phase

In this section we discuss extensions of the result in Section 18.2 to operators with general smooth phase functions. Let $\Phi \in C^\infty(\mathbb{R}^n)$, $\varphi \in C_0^\infty(\mathbb{R}^n)$. Also let K be a Calderón-Zygmund kernel. We consider the following oscillatory singular integral operators:

$$T_\lambda f(x) = \text{P.V.} \int_{\mathbb{R}^n} e^{i\lambda \Phi(x-y)} K(x-y)\varphi(x-y) f(y)\, dy, \qquad (7)$$

where $\lambda \in \mathbb{R}$. The boundedness of T_λ on H^1 for any fixed λ is trivial, but what is of interest is the uniform boundedness of T_λ on $H^1(\mathbb{R}^n)$ for $\lambda \in \mathbb{R}$. If Φ is a real-valued polynomial and $\nabla\Phi(0) = 0$, then by Theorem 3, T_λ are uniformly bounded on $H^1(\mathbb{R}^n)$. More generally, we have the following result.

THEOREM 4
(See [P3].) Suppose $\nabla\Phi(0) = 0$. If $\frac{\partial^\alpha \Phi(0)}{\partial x^\alpha} \neq 0$, for some multiindex α with $|\alpha| > 1$, then there exists a constant C, which is independent of λ such that

$$\| T_\lambda f \|_{H^1(\mathbb{R}^n)} \leq C \| f \|_{H^1(\mathbb{R}^n)}$$

for $f \in H^1(\mathbb{R}^n)$, $\lambda \in \mathbb{R}$.

For example, Theorem 4 implies that when Φ is real-analytic and satisfies $\nabla\Phi(0) = 0$, T_λ are uniformly bounded on H^1.

REMARK 2   The assumption $\Phi \in C^\infty$ in Theorem 4 can be somewhat weakened. Suppose $\frac{\partial^\alpha \Phi(0)}{\partial x^\alpha} \neq 0$ and $k = |\alpha| > 1$. Then one may replace the assumption $\Phi \in C^\infty$ by $\Phi \in C^{k(k-1)+1}$.   ∎

As usual, the proof of Theorem 4 requires the corresponding L^2 estimates for T_λ, which are contained in the following theorem.

THEOREM 5

If $\frac{\partial^\alpha \Phi(0)}{\partial x^\alpha} \neq 0$ *for some multiindex* α *with* $|\alpha| > 1$, *then for* $p \in (1, \infty)$ *there exists a constant* C_p, *which is independent of* λ *such that*

$$\|T_\lambda f\|_{L^p(\mathbb{R}^n)} \leq C_p \|f\|_{L^p(\mathbb{R}^n)} \tag{8}$$

for all $f \in L^p(\mathbb{R}^n)$.

Theorem 5 is a special case of Theorem 2 in [P1]. An important idea used in the proof of Theorem 4 is the following van der Corput-type lemma established in [P3].

LEMMA 1

Let $\Psi \in C^\infty(\mathbb{R}^n)$, $\varphi \in C_0^\infty(\mathbb{R}^n)$. *Let* k *be a positive integer and* $\alpha \in \mathbb{R}^N$ *with* $|\alpha| = k$. *Assume that* $\left|\frac{\partial^\alpha \Psi(x)}{\partial x^\alpha}\right| \leq B \leq M$ *for all* $x \in \text{supp}(\varphi)$. *Define that* $V = \{x \in \mathbb{R}^n | dist(x, \text{supp}(\varphi)) \leq B\}$. *We also assume that* $\left|\frac{\partial^\beta \Psi(x)}{\partial x^\beta}\right| \leq A$ *for all* $|\beta| = k+1$ *and* $x \in V$. *Then there exists a constant* C, *which depends only on* A, M, k, *and* φ, *such that*

$$\left|\int_{\mathbb{R}^n} e^{i\lambda \Psi(x)} \varphi(x)\, dx\right| \leq C\lambda^{-\varepsilon/k} \int_V \left|\frac{\partial^\alpha \Psi}{\partial x^\alpha}(x)\right|^{-\varepsilon[1+(1/k)]} dx \tag{9}$$

holds for $\varepsilon \in [0, 1]$.

One would like to know if the condition $\frac{\partial^\alpha \Phi(0)}{\partial x^\alpha} \neq 0$ in Theorem 4 and 5 is really necessary. We shall address this problem below.

Let $\Gamma = (t, \Phi(t))$, $t \in [-1, 1]$, be a curve on $\mathbb{R}^2$. Consider the Hilbert transform along Γ:

$$H_\Gamma g(x_1, x_2) = \int_{-1}^1 f(x_1 - t, x_2 - \Phi(t)) \frac{dt}{t}. \tag{10}$$

Then H_Γ is bounded on $L^2(\mathbb{R}^2)$, if and only if,

$$\sup_{\lambda \in \mathbb{R}} \|T_\lambda\|_{L^2(\mathbb{R}) \to L^2(\mathbb{R})} < \infty \tag{11}$$

where

$$T_\lambda f(x) = \int_{|x-y| \leq 1} e^{i\lambda \Phi(x-y)} \frac{f(y)}{x - y}\, dy. \tag{12}$$

In [NW] Nagel and Wainger found a C^∞ nonzero function Φ, which is odd, convex, and satisfies $\Phi^{(k)}(0) = 0$ for all $k \in \mathbb{N}$, such that H_Γ is unbounded on $L^2(\mathbb{R}^2)$. This implies that

$$\sup_\lambda \| T_\lambda \|_{L^2(\mathbb{R}) \to L^2(\mathbb{R})} = \infty.$$

For the same operators T_λ on H^1, our calculation in [P3] shows that

$$\sup_\lambda \| T_\lambda \|_{H^1(\mathbb{R}) \to H^1(\mathbb{R})} = \infty;$$

therefore, both Theorem 4 and Theorem 5 become false if the condition

$$\frac{\partial^\alpha \Phi(0)}{\partial x^\alpha} \neq 0$$

is removed.

Φ is said to be degenerate if all the derivatives of Φ at the origin are zero. It has been shown that the uniform boundedness of T_λ may still hold when Φ is degenerate. The result of Nagel, Vance, Wainger, and Weinberg [NVWW] on H_Γ implies the following.

THEOREM 6
Let T_λ be given as in (12). Assume that $\Phi \in C^2([-1, 1])$ and $\Phi''(t) > 0$ for $t > 0$.

(i)   Suppose Φ is even. Then T_λ are uniformly bounded on $L^2(\mathbb{R})$, if and only if, there are C and d, such that $\Phi'(Ct) \geq 2\Phi'(t)$ for $0 < t < d$.

(ii)   Suppose Φ is odd. Then T_λ are uniformly bounded on $L^2(\mathbb{R})$, if and only if, there are C and d, such that $h(Ct) \geq 2h(t)$ for $0 < t < d$ where $h(t) = t\Phi'(t) - \Phi(t)$.

The above theorem suggests that the uniform H^1 boundedness of T_λ may hold for certain degenerate functions Φ. Recently, we were able to prove the following result.

THEOREM 7
(See [P4].) Let T_λ be given as in (12). Assume $\Phi \in C^3([-1, 1])$ and $\Phi'(0) = 0$. Also, suppose that Φ is even or odd and there exists a $d > 0$, such that $\Phi'''(t) \geq 0$ for $t \in [0, d]$. Then there is a $C > 0$ independent of λ, such that

$$\| T_\lambda f \|_{H^1(\mathbb{R})} \leq C \| f \|_{H^1(\mathbb{R})}.$$

As an example we see that the operators T_λ with $\Phi(x) = e^{-1/x^2}$ are uniformly bounded on $H^1(\mathbb{R})$. In the next section we shall state and prove an n-dimensional version of Theorem 7.

18.4 Degenerate phase

THEOREM 8

Let $x \in \mathbb{R}^n$ and $\Phi(x) = \phi(|x|)$, where $\phi \in C^1([0, d]) \cap C^3((0, d])$ for some $d > 0$. Also, let Φ satisfy (i) $\phi'(0) = 0$ and (ii) $\phi''(t) \geq 0$, $\phi'''(t) \geq 0$ for $0 < t \leq d$. Let T_λ be defined by

$$T_\lambda f(x) = P.V. \int_{\mathbb{R}^n} e^{i\lambda \Phi(x-y)} K(x-y) \varphi(x-y) f(y) \, dy.$$

Then T_λ are uniformly bounded on $H^1(\mathbb{R}^n)$ for $\lambda \in \mathbb{R}$.

Examples of the phase functions described in Theorem 8 include $\Phi(x) = e^{-1/|x|^2}$ and $\Phi(x) = e^{-1/|x|}$, where $|x| = (x_1^2 + \cdots + x_n^2)^{1/2}$.

LEMMA 2

(van der Corput, see [St1] and [Z].) Suppose ψ and u are smooth in $[a, b]$ and ψ is real valued. If $|\psi^{(k)}(t)| \geq 1$ for $t \in [a, b]$, then

$$\left| \int_a^b e^{i\lambda \psi(t)} u(t) \, dt \right| \leq c_k |\lambda|^{-1/k} (\|u\|_\infty + \|u'\|_1),$$

when (i) $k \geq 2$ or when (ii) $k = 1$, if in addition, it is assumed that $\psi'(t)$ is monotonic.

We will need the following L^2 estimate for T_λ.

LEMMA 3

Let Φ and T_λ be given as in Theorem 8. Then there is a constant C, which is independent of λ, such that

$$\|T_\lambda f\|_{L^2(\mathbb{R}^n)} \leq C \|f\|_{L^2(\mathbb{R}^n)}.$$

PROOF Without loss of generality, one may assume that $\phi(0) = 0$ and supp $(\varphi) \subset \{|x| \leq d\}$. By taking the Fourier transform, we find

$$\widehat{T_\lambda f}(\xi) = m_\lambda(\xi) \hat{f}(\xi), \tag{13}$$

where $\xi \in \mathbb{R}^n$ and

$$m_\lambda(\xi) = \int_{\mathbb{R}^n} e^{i(\xi \cdot x + \lambda \Phi(x))} K(x) \varphi(x) \, dx. \tag{14}$$

Let

$$m_{\lambda,0}(\xi) = \int_{|x| \leq \eta} e^{i(\xi \cdot x + \lambda \Phi(x))} K(x) \varphi(x) \, dx, \tag{15}$$

$$m_{\lambda,j}(\xi) = \int_{2^j \eta \leq |x| \leq 2^{j+1} \eta} e^{i(\xi \cdot x + \lambda \Phi(x))} K(x) \varphi(x) \, dx, \tag{16}$$

where $j \geq 0$, and $\eta = \phi^{-1}\left(\frac{1}{|\lambda|}\right)$. We obtain

$$|m_{\lambda,0}(\xi)| \leq \left| \int_{\mathbb{R}^n} e^{i\xi \cdot x} (e^{i\lambda \Phi(x)} - 1) K(x) \varphi(x) \, dx \right| + \left| \int_{\mathbb{R}^n} e^{i\xi \cdot x} K(x) \varphi(x) \, dx \right|$$

$$\leq C + C \int_0^\eta \left| \frac{\lambda \phi(r)}{r} \right| dr \leq C. \tag{17}$$

By van der Corput's lemma, we obtain

$$|m_{\lambda,j}(\xi)| = \left| \int_{S^{n-1}} \int_{2^j \eta}^{2^{j+1} \eta} e^{i(r\xi \cdot x' + \lambda \phi(r))} r^{n-1} K(rx') \varphi(rx') \, dr \, d\sigma(x') \right|$$

$$\leq C |\lambda \phi''(2^j \eta)|^{-1/2} (2^j \eta)^{-1}. \tag{18}$$

Hence

$$|m_\lambda(\xi)| \leq C + C|\lambda|^{-1/2} \sum_{j \geq 0} ((2^j \eta)^2 \phi''(2^j \eta))^{-1/2}$$

$$\leq C + C|\lambda|^{-1/2} \sum_{j \geq 0} (\phi(2^j \eta))^{-1/2} \leq C, \tag{19}$$

where we used the fact $\phi(2t) \geq 2\phi(t)$. It follows from Plancherel's theorem that $T_\lambda's$ are uniformly bounded on $L^2(\mathbb{R}^n)$.

Let $\delta > 0$ and $v \in S^{n-1}$. Choose $h_1 \in C_0^\infty(\mathbb{R})$ and $h_2 \in C_0^\infty(\mathbb{R}^n)$, such that the following hold:

(i) $h_1(t) \equiv 1$, for $1 \leq t \leq 2$ and supp $(h_1) \subset \left[\frac{3}{4}, 3\right]$;

(ii) $h_2(x) \equiv 1$ for $|x| \leq 1$ and supp $(h_2) \subset \{|x| \leq 2\}$.

For $k \in Z$ define the operator $S_j = S_{v,j}^{\lambda,\delta} : L^2(\mathbb{R}^n) \to L^2(\mathbb{R})$ by

$$(S_j f)(t) = h_1(2^{-j}t) \int_{\mathbb{R}^n} e^{i\lambda \Phi(tv - y)} \varphi(tv - y) h_2\left(\frac{y}{\delta}\right) f(y) \, dy. \tag{20}$$

∎

LEMMA 4

Suppose Φ is given as in Theorem 8 and $2^j \geq 12\delta$. Then there exists a constant C, which is independent of λ, δ, v, and j, such that

$$\|S_j f\|_{L^2(\mathbb{R})} \leq C(2^j \delta^{2n-1})^{1/4} |\lambda \phi''(2^{j-1})|^{-1/4} \|f\|_{L^2(\mathbb{R}^n)}. \tag{21}$$

PROOF It suffices to prove that

$$\|S_j^* g\|_{L^2(\mathbb{R}^n)} \leq C(2^j \delta^{2n-1})^{1/4} |\lambda \phi''(2^{j-1})|^{-1/4} \|g\|_{L^2(\mathbb{R})} \tag{22}$$

where

$$(S_j^* g)(x) = h_2\left(\frac{x}{\delta}\right) \int_{\mathbb{R}} e^{-i\lambda \Phi(tv-x)} h_1(2^{-j} t) \varphi(tv - x) g(t) \, dt. \tag{23}$$

We write

$$\|S_j^* g\|_{L^2(\mathbb{R}^n)}^2 = \int_{\mathbb{R}} \int_{\mathbb{R}} L_j(t, s) g(s) \overline{g(t)} \, ds \, dt \tag{24}$$

where

$$L_j(t, s) = h_1(2^{-j} t) h_1(2^{-j} s) \int_{\mathbb{R}^n} e^{i(\lambda(\Phi(tv-x)-\Phi(sv-x)))}$$

$$\cdot \varphi(tv - x)\varphi(sv - x)\left(h_2\left(\frac{x}{\delta}\right)\right)^2 dx. \tag{25}$$

It suffices therefore to prove that $L_j(t, s)$ is the kernel of a bounded operator on $L^2(\mathbb{R})$ with norm not exceeding $C(2^j \delta^{2n-1})^{1/2} |\lambda \phi''(2^{j-1})|^{-1/2}$.

Let P be an $n \times n$ orthogonal matrix whose first row is given by the vector v. By letting $x = yP$, we find

$$L_j(t, s) = h_1(2^{-j} t) h_1(2^{-j} s) \int_{\mathbb{R}^n} e^{i\lambda(\Phi(tv-yP)-\Phi(sv-yP))}$$

$$\cdot \varphi(tv - yP)\varphi(sv - yP)\left(h_2\left(\frac{yP}{\delta}\right)\right)^2 dy. \tag{26}$$

Let

$$F(t, y) = \Phi(tv - yP) = \phi(|tv - yP|);$$

$$W = t - v \cdot (yP);$$

$$Z = |tv - yP|.$$

Then for $t \in \left[\frac{(3 \cdot 2^j)}{4}, (3 \cdot 2^j)\right]$, we obtain

$$\frac{\partial F(t, y)}{\partial y_1} = \phi'(Z)\left(-\frac{W}{Z}\right);$$

$$\left|\frac{\partial^2 F(t, y)}{\partial t \partial y_1}\right| = \left|\phi''(Z)\left(-\frac{W^2}{Z^2}\right) + \phi'(Z)\left(\frac{W^2 - Z^2}{Z^3}\right)\right| \geq \frac{1}{2}\phi''(2^{j-1});$$

$$\frac{\partial^3 F(t, y)}{\partial t \partial^2 y_1} = \phi'''(Z)\frac{W^3}{Z^3} + 3(Z\phi''(Z) - \phi'(Z))\left(\frac{W(Z^2 - W^2)}{Z^5}\right) \geq 0,$$

where we used $0 \leq W \leq Z$ and $Z\phi''(Z) \geq \phi'(Z)$. By van der Corput's lemma, we have

$$|L_j(t, s)| \leq C\delta^{n-1}(|\lambda|\phi''(2^{j-1})|t - s|)^{-1}|h_1(2^{-j}t)h_1(2^{-j}s)|. \tag{27}$$

On the other hand, we have

$$|L_j(t, s)| \leq C\delta^n|h_1(2^{-j}t)h_1(2^{-j}s)|. \tag{28}$$

Hence

$$|L_j(t, s)| \leq C\delta^{n-1/2}(|\lambda|\phi''(2^{j-1})|t - s|)^{-1/2}|h_1(2^{-j}t)h_1(2^{-j}s)|. \tag{29}$$

By (29), we find

$$\operatorname*{Supp}_t \int_{\mathbb{R}} |L_j(t, s)|\, ds \leq C2^{j/2}\delta^{n-1/2}(|\lambda|\phi''(2^{j-1}))^{-1/2}, \tag{30}$$

$$\operatorname*{Supp}_s \int_{\mathbb{R}} |L_j(t, s)|\, dt \leq C2^{j/2}\delta^{n-1/2}(|\lambda|\phi''(2^{j-1}))^{-1/2}. \tag{31}$$

(30) and (31) imply that L_j is the kernel of a bounded operator on $L^2(\mathbb{R})$ with norm not exceeding $C(2^j\delta^{2n-1})^{1/2}|\lambda\phi''(2^{j-1})|^{-1/2}$, which concludes the proof of the lemma. ∎

PROOF OF THEOREM 8 By the atomic decomposition of $H^1(\mathbb{R}^n)$, it suffices to show that there is a constant $C > 0$, such that

$$\|T_\lambda a\|_1 \leq C, \tag{32}$$

where a is an atom, i.e., a is a function that satisfies (i) Supp $(a) \subset Q$ for some cube $Q \subset \mathbb{R}^n$; (ii) $\|a\|_\infty \leq |Q|^{-1}$; (iii) $\int_Q a(x)\, dx = 0$. We may also assume that Q is centered at the origin with sidelength δ. By Lemma 3 we find

$$\int_{|x| \leq A\delta} |T_\lambda a(x)|\, dx \leq \left(\int_{|x| \leq A\delta} 1\, dx \right)^{1/2} \|T_\lambda a\|_2 \leq CA^{n/2}. \tag{33}$$

In view of (33), we may assume that δ is very small. Next we prove that

$$\int_{|x| > 6\delta} |T_\lambda a(x)|\, dx \leq C. \tag{34}$$

First, we observe that for $R \geq 6\delta$,

$$\int_{6\delta \leq |x| \leq R} |T_\lambda a(x)|\, dx$$

$$\leq \int_{6\delta \leq |x| \leq R} \left| \int_{\mathbb{R}^n} (e^{i\lambda\Phi(x-y)} - e^{i\lambda\Phi(x)}) K(x - y)\varphi(x - y)a(y)\, dy \right| dx$$

$$+ \int_{6\delta \leq |x| \leq R} \left| \int_{\mathbb{R}^n} K(x - y)a(y)\, dy \right| dx$$

$$\leq C + C|\lambda\delta| \int_{6\delta}^{R} \frac{\phi'(2r)}{r} \, dr$$

$$\leq C|\lambda\delta|\phi'(2R). \tag{35}$$

If $|\lambda\delta|$ is small, one may choose R, such that (32) follows from (33) and (35). Now we assume that $|\lambda\delta|$ is large. Let $R = \max\left\{\left(\frac{1}{2}\right)(\phi')^{-1}\left(\frac{1}{|\lambda\delta|}\right), 6\delta\right\}$, by (35) we obtain

$$\int_{6\delta\leq|x|\leq R} |T_\lambda a(x)| dx \leq C. \tag{36}$$

On the other hand, we find

$$\int_{|x|\geq 8R} |T_\lambda a(x)| \, dx$$

$$\leq \int_{|x|\geq 8R} \left| K(x) \int_{\mathbb{R}^n} e^{i\lambda\Phi(x-y)}\varphi(x-y)a(y) \, dy \right| dx$$

$$+ C\int_{|x|>8R} \int_{\mathbb{R}^n} |K(x-y) - K(x)| \, |a(y)| \, dy \, dx$$

$$\leq C + \int_{S^{n-1}} d\sigma(x') \sum_{2^j\geq 4R} \int_{2^j}^{2^{j+1}}$$

$$\cdot \left| \int_{\mathbb{R}^n} e^{i\lambda\Phi(rx'-y)}\varphi(rx'-y)a(y) \, dy \right| \frac{dr}{r}$$

$$\leq C + C\int_{S^{n-1}} \left(\sum_{2^j\geq 4R} \left(\int_{2^j}^{2^{j+1}} \frac{dr}{r^2} \right)^{1/2} \|S_{x',j}^{\lambda,\delta}a\|_{L^2(\mathbb{R})} \right) d\sigma(x')$$

$$\leq C + \sum_{2^j\geq 4R} 2^{-j/2}(2^j\delta^{2n-1})^{1/4}|\lambda\phi''(2^{j-1})|^{-1/4}\|a\|_{L^2(\mathbb{R}^n)}$$

$$\leq C + |\lambda\delta|^{-1/4} \sum_{2^j\geq 2R} (\phi'(2^j))^{-1/4} \leq C, \tag{37}$$

where we used $\phi'(2t) \geq 2\phi'(t)$. In addition, we obtain

$$\int_{R\leq|x|\leq 8R} |T_\lambda a(x)| dx \leq C \left(\int_{R\leq|x|\leq 8R} \frac{dx}{|x|^n} \right) \|a\|_{L^1(\mathbb{R}^n)} \leq C. \tag{38}$$

By combining (33), (36), (37), and (38), Theorem 8 is proven. ∎

REMARK 3 (1) Both Theorem 8 and Lemma 3 remain valid when one replaces $H^1(\mathbb{R}^n)$ and $L^2(\mathbb{R}^n)$ by $H_w^1(\mathbb{R}^n)$ and $L_w^2(\mathbb{R}^n)$, with w being in A_1 and A_2, respectively.

(2) By Theorem 8, Lemma 3, and interpolation, we obtain the corresponding L^p estimates, for $1 < p < \infty$. ∎

18.5   Non-convolution operators

We now consider operators that are not of convolution type. Let $P(x, y)$ be a polynomial in $\mathbb{R}^n \times \mathbb{R}^n$, $K(x, y)$ be a generalized Calderón-Zygmund kernel, T be defined by

$$Tf(x) = \text{P.V.} \int_{\mathbb{R}^n} e^{iP(x,y)} K(x, y) f(y) \, dy. \tag{39}$$

Since the $H^1 \to H^1$ boundedness does not hold in general, one seeks to prove the $H^1 \to L^1$ boundedness. The first result in this direction was obtained by Phong and Stein for operators with bilinear phase functions [PS].

Let $P(x, y) = (Bx, y)$ where B is an $n \times n$ real matrix. Phong and Stein introduced the space H_E^1, which is a variant of the standard H^1 and proved the $H_E^1 \to L^1$ boundedness of T.

DEFINITION 1   *A function $a(x)$ is called an H_E^1-atom if there is a cube $Q \subset \mathbb{R}^n$, such that*

  (i) a is supported in Q;

  (ii) $|a(x)| \leq \dfrac{1}{|Q|}$;

  (iii) $\displaystyle\int_Q e^{i(Bc_Q,x)} a(x) \, dx = 0$, where c_Q is the center of Q.

The space H_E^1 is given by

$$H_E^1 = \left\{ f \mid f = \sum_j \lambda_j a_j, \text{ with each } a_j \text{ an } H_E^1\text{-atom and } \sum_j |\lambda_j| < \infty \right\}.$$

THEOREM 9
(See [PS].) Let T be given as in (39) with $P(x, y) = (Bx, y)$. Then T is a bounded operator from H_E^1 to L^1. The bound of T is independent of B.

This theorem was extended to operators with polynomial phases in [P2]. Let $P(x, y)$ be a polynomial. We modify the definition of H_E^1 by replacing condition (iii) with

  (iii′) $\displaystyle\int_Q e^{iP(c_Q,y)} a(x) \, dx = 0$.

THEOREM 10

(See [P2].*) Let $P(x, y)$ be a real polynomial and T be given as in (39). Then T is bounded from H_E^1 to L^1 with a bound that can be taken to depend only on K and the degree of P, and is otherwise independent of P.*

Based on Theorem 3, it is interesting to note that when $P(x, y) = P(x - y)$, the space H_E^1 does not coincide with the standard H^1. More recent results can be found in [P4].

References

[CC]   S. Chanillo and M. Christ, *Weak (1,1) bounds for oscillatory singular integrals,* Duke Math. Jour. 55 (1987), 141–155.

[CKS]   S. Chanillo, D. Kurtz and G. Sampson, *Weighted weak (1,1) and weighted L^p estimates for oscillating kernels,* Trans. Amer. Math. Soc. 295 (1986), 127–145.

[Co]   R. Coifman, *A real variable characterization of H^p,* Studia Math. 51 (1974), 269–274.

[FS]   C. Fefferman and E. M. Stein, *H^p spaces of several variables,* Acta Math. 129 (1972), 137–193.

[H]   Y. Hu, *Oscillatory singular integrals on weighted Hardy spaces,* Studia Math. 102 (1992), 145–156.

[HP]   Y. Hu and Y. Pan, *Boundedness of oscillatory singular integrals on Hardy spaces,* Ark. Mat. 30 (1992), 311–320.

[NVWW]   A. Nagel, J. Vance, S. Wainger and D. Weinberg, *Hilbert transforms for convex curves,* Duke Math. Jour. 50 (1983), 735–744.

[NW]   A. Nagel and S. Wainger, *Hilbert transforms associated with plane curves,* Trans. Amer. Math. Soc. 223 (1976), 235–252.

[P1]   Y. Pan, *Uniform estimates for oscillatory integral operators,* Jour. Func. Anal. 100 (1991), 207–220.

[P2]   Y. Pan, *Hardy spaces and oscillatory singular integrals,* Rev. Mat. Iberoamericana 7 (1991), 55–64.

[P3]   Y. Pan, *Boundedness of oscillatory singular integrals on Hardy spaces: II,* Indiana U. Math. Jour. 41 (1992), 279–293.

[P4]   Y. Pan, preprint.

[PS]   D. H. Phong and E. M. Stein, *Hilbert integrals, singular integrals and Radon transforms I,* Acta Math. 157 (1986), 99–157.

[RS]   F. Ricci and E. M. Stein, *Harmonic analysis on nilpotent groups and singular integrals I,* Jour. Func. Anal. 73 (1987), 179–194.

[Sj]   P. Sjölin, *Convolution with oscillating kernels on H^p spaces,* J. London Math. Soc. 23 (1981), 442–454.

[St1]   E. M. Stein, *Oscillatory integrals in Fourier analysis,* in "Beijing Lectures in Harmonic Analysis", Princeton Univ. Press, Princeton, NJ, 1986.

[St2]   E. M. Stein, *Singular integrals and differentiability properties of functions,* Princeton Univ. Press, Princeton, NJ, 1970.

[SW]   E. M. Stein and G. Weiss, *Interpolation of operators with change of measure,* Trans. Amer. Math. Soc. 87 (1958), 159–172.

[ST]   J-O. Strömberg and A. Torchinsky, *Weighted Hardy Spaces,* Lecture Notes in Math. 1381, Springer-Verlag, 1989.

[Z]   A. Zygmund, *Trigonometric series,* Cambridge Univ. Press, Cambridge, 1959.

19

Three Types of Weighted Inequalities for Integral Operators

Luboš Pick

19.1 Three types of weighted inequalities

Let A be an integral operator and let ϱ be a *weight*, that is, measurable and a.e., positive and finite function in $\mathbb{R}^n$. In L_p space setting, the strong type inequality

$$\int |Af|^p \varrho \leq C \int |f|^p \varrho, \tag{1}$$

as well as the weak type inequality

$$\varrho(\{|Af| > \lambda\}) \leq C\lambda^{-p} \int |f|^p \varrho, \tag{2}$$

have been widely studied in connection with various operators (as usual, $\varrho(E)$ stands for $\int_E \varrho$ and ϱ_E stands for $\frac{\varrho(E)}{|E|}$, where $|E|$ is the Lebesgue measure of a set E). Replacing power function t^p by a general convex function $\Phi(t)$ (that is, passing from Lebesgue spaces to Orlicz spaces), a lot of obstacles occur because of the loss of homogeneity. For example, while (1) has its obvious analog

$$\int \Phi(|Af|)\varrho \leq \int \Phi(C|f|)\varrho, \tag{3}$$

there are at least two different possible analogies of (2), namely,

$$\Phi(\lambda)\varrho(\{|Af| > \lambda\}) \leq \int \Phi(C|f|)\varrho, \tag{4}$$

and

$$\varrho(\{|Af| > \lambda\}) \leq \int \Phi(C|f|\lambda^{-1})\varrho. \tag{5}$$

We say that (A, Φ, ϱ) *supports the strong type, weak type*, or *extra–weak type inequality*, if (3), (4), or (5) holds, respectively. If $\Phi(t) = t^p$, we write (A, p, ϱ). Our terminology is justified by the fact that (4) always implies (5) when A is positively homogeneous (take $\lambda = 1$), and this implication is not always reversible. The two types of inequalities (4) and (5) were apparently first compared together in [P2] although there have been partial results available earlier. We refer, for example, to [CCG], [G1], [K], and [P1]. While the usefulness of strong and weak type inequalities has been known for years, nice applications of extra-weak type inequalities to the interpolation of operators in Orlicz spaces were found recently by Bagby and Persons [BP] and [B].

We shall deal with the problem of characterizing ϱ and Φ, such that (3), (4), or (5) be valid. Weighted inequalities for integral operators have been studied by many authors and a lot of applications in Fourier analysis or PDE's is known. It is a part of analysis which has been rapidly developing, especially during the last two decades. Our standard general reference is to the comprehensive monographs [GR] and [KK].

19.2 Maximal operator—the basic result

The *Hardy–Littlewood maximal operator* is defined for $f \in L^1_{\text{loc}}$ by

$$Mf(x) = \sup_{Q \ni x} \frac{1}{|Q|} \int_Q |f(y)| \, dy,$$

where Q is a cube in $\mathbb{R}^n$ with sides parallel to coordinate axes. In 1972, Muckenhoupt [M] proved that if $p > 1$, then (M, p, ϱ) supports the strong type inequality (1), if and only if, $\varrho \in A_p$, that is,

$$\sup_Q \varrho_Q \left(\frac{1}{|Q|} \int_Q \varrho^{1-p'} \right)^{p-1} \leq C,$$

while if $p \geq 1$, then (M, p, ϱ) supports weak type inequality (2), if and only if, $\varrho \in A_p$, where $\varrho \in A_1$ means that $M\varrho \leq C\varrho$ a.e.

A substantial progress in the study of weighted inequalities for M in Orlicz spaces was done in the fundamental paper of Kerman and Torchinsky [KT], where the strong type inequality for M was characterized under the assumption that Φ, together with its *complementary function* $\tilde{\Phi}(t) = \sup_{s>0}(st - \Phi(s))$, satisfies the Δ_2 *condition* $\Phi(2t) \leq C\Phi(t)$. Under this assumption, (M, Φ, ϱ) supports strong type inequality, if and only if, $\varrho \in A_\Phi$, that is, there exists $C > 1$ such that

$$\sup_{\alpha>0} \sup_Q \left(\frac{1}{|Q|} \int_Q \alpha\varrho \right) \phi \left(\frac{1}{|Q|} \int_Q \phi^{-1} \left(\frac{1}{\alpha\varrho} \right) \right) = C < \infty, \qquad (6)$$

where $\phi = \Phi'$. Moreover, $\varrho \in A_\Phi$ is equivalent to $\varrho \in A_{i(\Phi)}$, where $i(\Phi)$ is the *lower index* of Φ. Let us still recall that the *lower* and the *upper index* $i(\Phi)$ and $I(\Phi)$ of Φ is defined as supremum over α and infimum over β, respectively, such that $\alpha \leq \beta$ and

$$\Phi(\lambda t) \leq C_{\alpha,\beta} \max(\lambda^\alpha, \lambda^\beta) \Phi(t), \qquad \lambda, t \geq 0.$$

In terms of indices, $\Phi \in \Delta_2$ means $I(\Phi) < \infty$ and $\tilde{\Phi} \in \Delta_2$ means $i(\Phi) > 1$.

In 1989, Gallardo [G1] proved that if $\Phi, \tilde{\Phi} \in \Delta_2$, then (M, Φ, ϱ) supports the weak type inequality, if and only if, $\varrho \in A_\Phi$.

While A_Φ is quite a natural analog of A_p, there is no obvious way to formulate an analog of A_p good for an extra-weak type inequality. However, results of [CCG], [K], and [P1] suggest that the reformulation of A_p,

$$\sup_Q \frac{1}{|Q|} \int\limits_Q \left(\frac{\varrho_Q}{\varrho}\right)^{p'-1} \leq C < \infty,$$

where $p' = \frac{p}{(p-1)}$, is correct. Our first goal is to characterize weak and extra-weak type inequalities for general functions Φ, in particular without any Δ_2 assumption.

We introduce auxiliary functions

$$R_\Phi(t) = \frac{\Phi(t)}{t}, \qquad S_\Phi(t) = \frac{\tilde{\Phi}(t)}{t}.$$

Apparently, the functions R_Φ and S_Φ are suitable substitutes for t^{p-1} and $t^{p'-1}$. We have an important estimate [P2]

$$\Phi(S_\Phi(t)) \leq \tilde{\Phi}(t). \tag{7}$$

Let Φ be a *Young function*, that is, convex increasing function on $[0, \infty)$, such that $\Phi(0) = 0$ and $\lim_{t \to 0+} R_\Phi(t) = 0$ and $\lim_{t \to \infty} R_\Phi(t) = \infty$. We state that $\varrho \in A_\Phi$ if there exists C, such that

$$\sup_{\alpha > 0} \sup_Q \alpha \varrho_Q R_\Phi \left(\frac{1}{C|Q|} \int\limits_Q S_\Phi\left(\frac{1}{\alpha\varrho}\right)\right) \leq C. \tag{8}$$

We state that $\varrho \in E_\Phi$ if there exists C, such that

$$\sup_Q \frac{1}{|Q|} \int\limits_Q S_\Phi\left(\frac{\varrho_Q}{C\varrho}\right) \leq C. \tag{9}$$

If $\Phi, \tilde{\Phi} \in \Delta_2$, then R_Φ is equivalent to ϕ and S_Φ is equivalent to ϕ^{-1}. Therefore, our definition of A_Φ is consistent with (6), the definition of Kerman and Torchinsky [KT].

Now let us state our basic result [P2].

THEOREM 1

Let Φ be a Young function. Then

(i)   *(M, Φ, ϱ) supports weak type inequality if and only if $\varrho \in A_\Phi$;*
(ii) *(M, Φ, ϱ) supports extra-weak type inequality if and only if $\varrho \in E_\Phi$.*

Taking $\alpha = \frac{1}{\varrho Q}$ in (8), we see that (8) always implies (9). This reflects the fact that (4) implies (5). Therefore, $A_\Phi \subset E_\Phi$ for every Φ.

While the proof of (i) is quite standard, it is the necessity of E_Φ in (ii), previously known only in case $\Phi(t) = t(1 + \log_+ t)^K$ ([CCG], [K], and [P1]), that causes the most trouble. In [P2] the proof uses some knowledge about Orlicz norm and saturation of the Hölder inequality in an Orlicz space. There is, however, a direct proof, a "pure" one, which uses only the Young inequality and (7). This proof is due to Gogatishvili and appears in [GP].

In [GP] we observed that in order for (M, Φ, ϱ) to support a weak type inequality, it is necessary that Φ be *quasiconvex*, that is, $C^{-1}\Phi(C^{-1}t) \leq \Phi_0(t) \leq C\Phi(Ct)$, where Φ_0 is convex. It is therefore reasonable to consider convex functions only. It is possible to extend Theorem 1 for general convex functions (not necessarily Young's). We state that Φ *is of bounded type near zero (near infinity)* and write $\Phi \in B_0$ ($\Phi \in B_\infty$) if $R_\Phi(t) \leq C$ ($R_\Phi(t) \geq C$). The extension of Theorem 1 is now obtained via the following two observations:

(1)   if $\varrho \in A_1$ and Φ is convex, then $\varrho \in A_\Phi$;
(2)   if $\Phi \in B_0$ or $\Phi \in B_\infty$ and (M, Φ, ϱ) supports a weak type inequality, then necessarily $\varrho \in A_1$.

Thus it only remains to modify the definition of A_Φ by stating that $\varrho \in A_\Phi$ if either Φ is Young's and (8) is true, or Φ is of bounded type and $\varrho \in A_1$. The second part of Theorem 1 remains unchanged, regardless of what Φ might be.

A similar result to (ii) was proven independently in [B].

Finally let us note that recently Bloom and Kerman [BKe2], and independently Gogatishvili [Go2], formulated the A_Φ condition in a different way,

$$\sup_{\alpha>0} \sup_Q \int_Q \tilde{\Phi}\left(\frac{R_\Phi(\lambda)\varrho q}{C\varrho x}\right) \varrho x \, dx \leq \Phi(\lambda)\varrho(Q) < \infty.$$

To illustrate that this condition is equivalent to (8), choose α (or λ) in such a way that $\alpha R_\Phi(\lambda)\varrho q = C$. Gogatishvili, moreover, proved a general theorem covering both weak and extra-weak case ([Go2], see also [KK] and Theorem 2.2.3).

19.3 The classes A_Φ and E_Φ

In the A_p scale, A_1 is the strongest condition. We have $A_1 \subset \bigcap\limits_{p>1} A_p$ and the inclusion is proper; this is seen, for example, when considering $\Phi_K(t) = t(1 + \log_+ t)^K$. Then $A_1 \subset E_{\Phi_K} \subset \bigcap\limits_{p>1} A_p$, and both the inclusions are proper (for details see [CCG]). An important corollary follows: in the result of Kerman and Torchinsky concerning the equivalence of A_Φ and $A_{i(\Phi)}$, it is essential that $i(\Phi) > 1$.

On the other hand, it follows from the result of Bloom and Kerman ([BKe2], Theorem 6) that $A_1 = \bigcap A_\Phi = \bigcap E_\Phi$.

At the other endpoint of the A_p scale we find a different circumstance. The weakest of A_p conditions is A_∞. We state that $\varrho \in A_\infty$, if for every $\varepsilon > 0$, there is $\delta > 0$ so that $\varrho(E) < \varepsilon\varrho(Q)$ whenever E is a measurable subset of a cube Q and $|E| < \delta|Q|$. Unlike the A_1 case, here we have $A_\infty = \bigcup A_p$. Many equivalent statements have been proven and remarkable applications of A_∞ have been discovered, see for example [BG], [C], [BK], or [CF]. The most important properties of A_∞ are its symmetry, its equivalence to the reverse Hölder inequality, and its equivalence to the exp-log-type conditions

$$\sup_Q \varrho_Q \exp\left(\frac{1}{|Q|}\int_Q \log\frac{1}{\varrho}\right) \le C, \tag{10}$$

(proven by Hruščev in [H] and independently by García–Cuerva and Rubio de Francía in [GR]), and

$$\sup_Q \int_Q \log_+\left(\frac{\varrho}{\varrho_Q}\right)\varrho \le C\varrho(Q), \tag{11}$$

(proven by a different method by Fujii in [F]).

Our approach can be applied to obtain new characterizations of A_∞. The idea is simple: First, for any Φ, we have $A_\Phi \subset E_\Phi \subset A_\infty$. On the other hand, if $\varrho \in A_\infty$, then there is a p such that $\varrho \in A_p$. Hence if we find a condition on Φ guaranteeing that Φ grows rapidly enough so that $A_p \subset A_\Phi$ for all p, then we obtain $A_\Phi = E_\Phi = A_\infty$ and both E_Φ and A_Φ are new characterizations of A_∞. The result reads as follows.

THEOREM 2
If Φ is a convex function such that $S_\Phi(t^\alpha)$ is quasiconcave for every $\alpha \ge \alpha_0$ with some $\alpha_0 \ge 1$, then $A_\Phi = E_\Phi = A_\infty$.

This theorem has a nice application: on taking $\tilde{\Phi}(t) = t(1 + \log_+ t)$ and using Theorem 2 we obtain both (10) and (11) as particular cases of Theorem 2.

Therefore, Theorem 2 provides a unifying approach (note that the equivalence of (10) and (11) is not obvious).

Furthermore, we can observe that if $\Phi \in B_\infty$, then $A_\Phi = E_\Phi = A_1$. Apart from the set of functions that satisfy the assumption of Theorem 2, B_∞ is thus another family of functions Φ, such that $A_\Phi = E_\Phi$. However, this identity is false in general. The simplest example is $\Phi(t) = t^p \chi_{[0.1]}(t) + t^q \chi_{(1.\infty)}(t)$ where $1 < p < q$. Then $i(\Phi) = p$ and hence $A_\Phi = A_p$ [KT]. However, $E_\Phi = A_q$, since (5) with $A = M$ is equivalent to

$$\varrho(\{Mf > \lambda\}) \le \int\limits_{\{f > \lambda/2\}} \Phi\left(C\frac{f}{\lambda}\right) \varrho,$$

and therefore the values of Φ near zero have no meaning for validity of (5). A more interesting example is due to Bagby [B]: For $\Phi(t) = t^p (1 + \log_+(t))^{-q}$, $p, q > 1$, the weight $\varrho x = x^{p-1}$ belongs to E_Φ but not to A_Φ.

For details concerning the results of this section see [GP].

REMARK 1   All the results of Sections 19.2 and 19.3 were "translated" to the context of one-sided maximal operator in [OP]. We have made substantial use of the "cutting-in-half" method of Martín–Reyes [MR] and employed the A_∞^+ condition introduced by Martín-Reyes, de la Torre, and the author in [MPT].   ∎

19.4   Hilbert transform and related operators

In this section all functions are defined on the real line. Also, we work with intervals I rather than cubes Q. Recall that Φ is convex.

Our next step will be to apply the approach of Section 19.2 to some more integral operators important in analysis.

The *Hilbert transform* is given for any function f, satisfying

$$\int\limits_{-\infty}^{\infty} |f(x)| \, (1 + |x|)^{-1} \, dx < \infty$$

by the Cauchy principal value integral

$$Hf(x) = \frac{1}{\pi} \lim_{\varepsilon \to 0+} \int\limits_{\mathbb{R}\setminus(x-\varepsilon.x+\varepsilon)} \frac{f(y)}{x - y} \, dy.$$

If f is an odd function, then Hf is even, and $Hf(x) = H_o f(|x|)$, where

$$H_o f(x) = \frac{2}{\pi} \int_0^\infty \frac{yf(y)}{x^2 - y^2} \, dy, \qquad x \in (0, \infty).$$

Now let us define the analog of A_Φ condition corresponding to the operator H_o. Let $dv(x) = x \, dx$.

We state that $\varrho \in A_\Phi^o$ if either Φ is a Young function and there exist positive C, ε such that

$$\sup_{\alpha, I} \left(\frac{\alpha}{v(I)} \int_I \frac{\varrho}{x} x \, dv \right) R_\Phi \left(\frac{\varepsilon}{v(I)} \int_I S_\Phi \left(\varepsilon \frac{x}{\alpha \varrho x} \right) dv \right) \leq C,$$

or $\Phi \in B_0 \cup B_\infty$ and $\varrho \in A_1^o$, that is,

$$\frac{\varrho(I)}{v(I)} \leq C \operatorname{ess\,inf}_I \frac{\varrho x}{x}.$$

We state that $\varrho \in E_\Phi^o$ if there exist positive C, ε such that

$$\sup_I \frac{1}{v(I)} \int_I S_\Phi \left(\varepsilon \frac{x}{\varrho} x \frac{\varrho(I)}{v(I)} \right) dx \leq C.$$

Similarly, as in the A_Φ case, both A_Φ^o and E_Φ^o coincide with the A_p^o when $\Phi(t) = t^p$ (A_p^o was introduced by Andersen in [A]).

For the above operators, we have the following result ([GP], [P3]).

THEOREM 3

(i) (H, Φ, ϱ) supports weak type inequality, if and only if, $\Phi \in \Delta_2$ and $\varrho \in A_\Phi$;
(ii) (H_o, Φ, ϱ) supports weak type inequality, if and only if, $\Phi \in \Delta_2$ and $\varrho \in A_\Phi^o$.

Let Φ satisfy the Δ_2^0 condition (that is, $\Phi(2t) \leq C\Phi(t)$ for $t \leq 1$). Then

(iii) (H, Φ, ϱ) supports extra-weak type inequality, if and only if, $\varrho \in E_\Phi$;
(iv) (H_o, Φ, ϱ) supports extra-weak type inequality, if and only if, $\varrho \in E_\Phi^o$.

Necessity of $\Phi \in \Delta_2$ for weak type inequalities involving Hilbert transform follows from the simple, but remarkable result of Gogatishvili [Go1]. Necessity of $\varrho \in A_\Phi$ follows from Theorem 1. For sufficiency in (i), we use Theorem 1, inclusion $A_\Phi \subset A_\infty$, and Coifman's good-λ inequality. As for (iii), the proof is rather long and not entirely analogous to the L_p-case. It is one of the main results of [GP]. Statements (ii) and (iv) are derived from (i) and (ii), respectively, in a manner similar to Andersen's approach [A].

Unfortunately, we have not been able to remove the Δ_2^0 assumption from (iii) and (iv). The other open problem is to treat even functions in a corresponding way to the odd ones (this works in L_p case, see [A]).

19.5   Strong type inequalities

Let us recall Kerman and Torchinsky's result again: If Φ, $\tilde{\Phi} \in \Delta_2$, then (M, Φ, ϱ) supports strong type inequality, if and only if, $\varrho \in A_\Phi$. The nonweighted theorem of Gallardo [G2], stating that M supports the strong type inequality (with $\varrho = 1$), if and only if, $\tilde{\Phi} \in \Delta_2$, suggests that the assumption $\tilde{\Phi} \in \Delta_2$ might be essential. On the other hand, Jensen's inequality shows that, roughly, the more rapid the growth of Φ, the more (3) with $A = M$ holds. Thus there is no point in restricting the growth of Φ from above. In other words, $\Phi \in \Delta_2$ is superfluous as far as the maximal operator is considered (recall that this is not true in the case of Hilbert transform, see Theorem 3, (i)).

Following the lines of [KT], we obtain Proposition 1.

PROPOSITION 1
If $\tilde{\Phi} \in \Delta_2$, and $\varrho \in A_\Phi$, then (M, Φ, ϱ) supports strong type inequality.

The next natural question is whether $\tilde{\Phi} \in \Delta_2$ is necessary for (3) with $A = M$. We do not know the full answer to this question, but we have a partial result.

PROPOSITION 2
If (M, Φ, ϱ) supports strong type inequality, then $\varrho \in A_\Phi$ and $\tilde{\Phi} \in \Delta_2^\infty$, that is, $\tilde{\Phi}(2t) \le C\tilde{\Phi}(t)$ for $t \ge 1$.

We thus have "almost" a characterization of strong type inequality for maximal operator, leaving us to question whether $\tilde{\Phi} \in \Delta_2^0$ is necessary. What is more interesting: Thanks to certain duality approach, unavailable for M, the corresponding problem for Hilbert transform (and in turn for H_o) allows full characterization.

THEOREM 4
(i)   *(H, Φ, ϱ) supports strong type inequality, if and only if, $\Phi \in \Delta_2$, $\tilde{\Phi} \in \Delta_2$, and $\varrho \in A_\Phi$.*

(ii) *(H_o, Φ, ϱ) supports strong type inequality, if and only if, $\Phi \in \Delta_2$, $\tilde{\Phi} \in \Delta_2$, and $\varrho \in A_\Phi^o$.*

The proof proceeds via Proposition 1 and Theorem 3; for details see [P3].

The Δ_2 condition appears as a statement of the above theorems rather than as an assumption. This might be of independent interest.

The proof of sufficiency in (i) moreover yields a corollary: *If* $\Phi \in \Delta_2$ *and* (M, Φ, ϱ) *supports strong type inequality, then* $\tilde{\Phi} \in \Delta_2$. This peculiar assertion leads us to the conjecture that the answer to the above question will be affirmative.

When this manuscript was written, we learned that an affirmative answer was finally given by Bloom and Kerman in [BKe2]. Therefore,

(M, Φ, ϱ) *supports strong type inequality, if and only if,* $\tilde{\Phi} \in \Delta_2$ *and* $\varrho \in A_\Phi$.

19.6 Hardy type operators

In this section we consider two weights, ϱ and σ, and all the functions are defined on $(0, \infty)$.

The example of the Hardy operator

$$Tf(x) = \int\limits_{0}^{x} f(t)\, dt$$

shows that, surprisingly, weak inequalities are in a certain sense closer to strong type ones rather than to extra-weak ones.

We write $(T, \Phi, \sigma, \varrho)$ instead of (T, Φ, ϱ) if ϱ at the right side of (3), (4), or (5) (with $A = T$) is replaced by σ.

THEOREM 5
The following statements are equivalent:

(i) $(T, \Phi, \sigma, \varrho)$ *supports strong type inequality;*
(ii) $(T, \Phi, \sigma, \varrho)$ *supports weak type inequality;*
(iii) *there exists* C *such that*

$$\sup_{\alpha, r > 0} \alpha \varrho(r, \infty) R_\Phi \left(C^{-1} \int\limits_{0}^{r} S_\Phi \left(\frac{1}{\alpha \sigma(x)} \right) dx \right) \le C$$

THEOREM 6
$(T, \Phi, \sigma, \varrho)$ *supports extra-weak type inequality, if and only if,*

$$\sup_{r > 0} \int\limits_{0}^{r} S_\Phi \left(\frac{\varrho(r, \infty)}{C \sigma(x)} \right) dx = C < \infty. \tag{12}$$

A suitable modification of our approach works except for the proof of (ii)$\Rightarrow$(i) in Theorem 5 and the proof of necessity of (12) in Theorem 6. To prove (ii) $\Rightarrow$ (i) in Theorem 5, we use the dyadic technique developed by Sawyer in [S]. As for necessity of (12) in Theorem 6, assume that B is the constant in the extra-weak type inequality, put $f = \chi_{(0,r)} S_\Phi \left(\frac{\varrho(r,\infty)}{(C\sigma)} \right)$, and assume that $Tf(r) > B$. Then, by (ii), (7), and convexity of Φ,

$$\varrho(r,\infty) \leq \int_0^r \Phi \left(\frac{B}{Tf(r)} f \right) \sigma$$

$$\leq \frac{B}{Tf(r)} \int_0^r \tilde{\Phi} \left(\frac{\varrho(r,\infty)}{C\sigma} \right) \sigma = BC^{-1} \varrho(r,\infty).$$

Choosing $C > B^{-1}$, we see a contradiction whence $Tf(r) \leq B$. This yields (12).

Similar results are obtained for weak and extra-weak type inequalities for the more general Volterra operators

$$Kf(x) = \int_0^x k(s,x) f(s) \, ds,$$

provided that the kernel k is nonnegative and nondecreasing in x. The equivalence of weak and strong inequality, however, will be lost. For details of proofs of the results in this section see [P4].

For more results on different types of modular inequalities involving Hardy-type operators see [BKe1].

References

[A]   K. F. Andersen, *Weighted norm inequalities for Hilbert transforms and conjugate functions of even and odd functions*, Proc. Amer. Math. Soc. 56, 4(1976), 99–107.

[B]   R. J. Bagby. *Weak bounds for the maximal function in weighted Orlicz spaces*, Studia Math. 95 (1990), 195–204.

[BK]   R. J. Bagby, D. S. Kurtz, *A rearranged good λ inequality*, Trans. Amer. Math. Soc. 293 (1986), 71–81.

[BP]   R. J. Bagby, J. D. Parsons, *Orlicz spaces and rearranged maximal functions*, Math. Nachr. 132 (1987), 15–27.

[BKe1] S. Bloom, R. Kerman, *Weighted L_Φ integral inequalities for operators of Hardy type*, preprint.

[BKe2] S. Bloom, R. Kerman, *Weighted Orlicz space integral inequalities for the Hardy–Littlewood maximal operator*, preprint.

[BG]   D. L. Burkholder, R. F. Gundy, *Extrapolation and interpolation of quasi-linear operators on martingales*, Acta Math. 124 (1970), 249–304.

[CCG]   A. Carbery, S. Y. Chang, J. Garnett, *Weights and L log L*, Pacific J. Math. 120-1 (1985), 33–45.

[C]   R. R. Coifman, *Distribution function inequalities for singular integrals*, Proc. Nat. Acad. Sci. USA 69(1972), 2838–2839.

[CF]   R. R. Coifman, C. Fefferman, *Weighted norm inequalities for maximal functions and singular integrals*, Studia Math. 51 (1974), 241–250.

[F]   N. Fujii, *Weighted bounded mean oscillation and singular integrals*, Math. Japonica 22-5 (1978), 529–534.

[GR]   J. García-Cuerva, J. L. Rubio de Francía, *Weighted norm inequalities and related topics*, North–Holland, 1985.

[G1]   D. Gallardo, *Weighted weak type inequalities for the Hardy–Littlewood maximal operator*, Israel J. Math. 67,1 (1989), 95–108.

[G2]   D. Gallardo, *Orlicz spaces for which the Hardy–Littlewood maximal operator is bounded*, Publ. Math. 32(1988), 261–266.

[Go1]   A. Gogatishvili, *Riesz transforms and maximal functions in $\Phi(L)$ classes*, Bull. Acad. Sci. Georgian SSR, 137, 3(1990), 489–492.

[Go2]   A. Gogatishvili, *General weak-type inequalities for the maximal operators and the singular integrals*, preprint.

[GP]   A. Gogatishvili, L. Pick, *Weighted inequalities of weak and extra-weak type for the maximal operator and the Hilbert transform*, to appear in Czechoslovak Math. J.

[H]   S. Hruščev, A description of weights satisfying the A_∞ condition of Muckenhoupt, Proc. Amer. Math. Soc. 90-2 (1984), 253–257.

[KT]   R. A. Kerman, A. Torchinsky, *Integral inequalities with weights for the Hardy maximal function*, Studia Math. 71(1982), 277–284.

[KK]   V. Kokilashvili, M. Krbec, *Weighted inequalities in Lorentz and Orlicz spaces*, World Scientific, Singapore 1991.

[K]   M. Krbec, *Two weights weak type inequalities for the maximal function in Zygmund class*, Function Spaces and Applications, Proc. Conf. Lund 1986, M. Cwikel et al. (Eds.), Lecture Notes in Math. 1302, Springer, Berlin, 1988, 317–320.

[MR]   F. J. Martín-Reyes, New proofs of weighted inequalities for the one-sided Hardy–Littlewood maximal functions, to appear in Proc. Amer. Math. Soc.

[MPT]   F. J. Martín-Reyes, L. Pick, A. de la Torre, A_∞ condition, to appear in Canad. J. Math.

[M]   B. Muckenhoupt, *Weighted norm inequalities for the Hardy maximal function*, Trans. Amer. Math. Soc. 165 (1972), 207–227.

[OP]   P. Ortega, L. Pick, *Two-weight weak and extra-weak type inequalities for the one-sided maximal operator*, to appear in Proc. Roy. Soc. Edinburgh.

[P1]   L. Pick, *Two weights weak type inequality for the maximal function in* $L(1 + log^+L)^K$, Constructive Theory of Functions, Proc. Conf. Varna 1987, B. Sendov et al. (Eds.), Publ. House Bulg. Acad. Sci., Sofia 1988, 377–381.

[P2]   L. Pick, *Two weight weak type maximal inequalities in Orlicz classes*, Studia Math. 100-1 (1991), 207–218.

[P3]   L. Pick, *Weighted estimates for the Hilbert transform of odd functions*, Preprint.

[P4]   L. Pick, *Weighted modular estimates for the Hardy-type operators*, Preprint.

[S]   E. Sawyer, *Weighted Lebesgue and Lorentz norm inequalities for the Hardy operator*. Trans. Amer. Math. Soc. 281, 1 (1984), 329–337.

20

Boundary Value Problems for Higher Order Operators in Lipschitz and C^1 Domains

Jill Pipher[1]

20.1 Introduction

In this chapter we discuss some recent progress in the theory of higher order homogeneous elliptic operators. These operators have the general form $L = \sum a_\alpha D^\alpha$, where α is a multiindex. Ellipticity, or strong ellipticity, for L is the requirement that there exists a constant C such that

$$\forall \xi = (\xi_1, \ldots, \xi_n) \text{ in } \mathbb{R}^n, \quad C^{-1}|\xi|^{2m} \geq \sum_{|\alpha|=2m} a_\alpha \xi^\alpha \geq C|\xi|^{2m},$$

where α is a multiindex, and m is an integer. The coefficients a_α are assumed to be real and this, together with the ellipticity condition, forces the order $(2m)$ of the operator to be even. If $m \geq 2$ the operator is said to be of higher order.

The behavior of solutions to higher order operators is vastly different from that of solutions to second-order operators, even in smooth domains. Solutions need not satisfy a Harnack inequality or a maximum principle; the Green's function need not be of one sign and the fundamental solution may even change sign, all unlike the second-order situation. Indeed, the property of unique continuation for such an operator may fail. In 1961, Plis [Pl] constructed an example of a fourth-order homogeneous elliptic operator with smooth coefficients (and constant coefficients outside the unit ball), which has a nontrivial solution supported in the unit ball. Thus, since we are interested in the unique solvability of the problem $Lu = 0$ in a domain Ω with Dirichlet conditions on the boundary of Ω, we shall henceforth assume that the coefficients a_α of L are constant. This guarantees that unique continuation holds, but yet gives rise to a theory which is much different from the second-order one, exhibiting still all the aforementioned pathology of solutions and of Green's functions.

Such operators arise naturally in physical problems, for instance in the theory of elastostatics. One well known problem involving the biharmonic operator is

[1] Supported in part by an A. P. Sloan Foundation Fellowship and the NSF.

the clamped plate problem: to solve $\Delta^2 u = f$ in Ω with zero Dirichlet conditions on $\partial\Omega$. (We shall be more specific about the boundary conditions later on.) The function f represents the force acting on a clamped plate, and the solution u is the displacement of that plate. Hadamard conjectured that positive f should give rise to positive u. Physically this means that the displacement should take place in one direction if the force acts in one direction. And it would mean that the Green's function for Δ^2 in Ω would be of one sign. This is true if Ω is a ball. But Duffin [Du] showed that the Green's function for Δ^2 will change sign in an infinite strip and, later, Garabedian [G] showed that a sign change occurs if the domain Ω is a sufficiently eccentric ellipse. Near the vertex of some infinite cones, the Green's function may even change sign infinitely often [O].

We turn now to a discussion of boundary value problems associated with solving $Lu = 0$, where L has constant coefficients. In the late 1950s and early 1960s a rather complete theory was developed by Agmon, Douglis, and Nirenberg and by Browder in [ADN1], [ADN2], and [B] for the upper half space and for domains with smooth boundary, and for very general boundary conditions. On the the the upper half space, in [ADN1], explicit Poisson kernels are constructed, L^p estimates up to the boundary and extensions of the maximum principle are proven, and interior estimates and Schauder estimates are obtained. The techniques and results of the work cited above (see also [A]) lead to solvability of the Dirichlet problem, in the sense of nontangential estimates, when the domain is sufficiently smooth (and the smoothness depends on the order of the operator).

When the domain fails to be smooth, these boundary value problems have been less well understood. Recently, Verchota and I have shown [PV5] that, in Lipschitz domains in $\mathbb{R}^n$, the Dirichlet and regularity problems with data in L^p, for p near 2, are uniquely solvable with appropriate nontangential estimates, for all higher order operators that are constant coefficient homogeneous and elliptic (CCHE). The main goals of the remainder of this article are to explain the formulation of this problem in nonsmooth domains, give the background and the precursors of this result, describe the difficulty that arises in the higher order case, and sketch the argument that overcomes this difficulty. Briefly, the problem consists of finding the appropriate substitute for the Rellich identity (see [JK1]) which, in the second-order case, allows one to control all derivatives of a solution on the boundary by a conormal derivative.

Acknowledgements The work described here is joint work with Verchota, and I am grateful for this fruitful collaboration over the past several years. This article is a synopsis of a talk given at the Fourier Analysis and PDE conference held at Miraflores de la Sierra, June 1992. I am grateful to the organizers for having had the opportunity to participate in this conference and to be a part of these proceedings.

20.2 The Dirichlet problem on nonsmooth domains

If $\varphi : \mathbb{R}^{n-1} \to \mathbb{R}$ is a Lipschitz function, then $D = \{(x, y) \in \mathbb{R}^{n-1} \times \mathbb{R} : y > \varphi(x)\}$ is an infinite Lipschitz domain in $\mathbb{R}^n$. If φ is C^1, then D is called a C^1 domain. A bounded domain $\Omega \subseteq \mathbb{R}^n$ is Lipschitz if the boundary of D is given, locally and uniformly, by the graph of a Lipschitz function. (For a more precise definition see [JK2].) Alternatively, such a domain satisfies a uniform interior and exterior cone condition. Thus there exists a family of truncated cones $\{\Gamma(Q) : Q \in \partial\Omega\}$ such that $\Gamma(Q)$ is compactly contained in Ω, and these truncated cones are the appropriate nontangential approach regions to a point on the boundary of the domain. For a function v defined in Ω, the nontangential maximal function of v is $v^*(Q) = \sup\{v(X) : X \in \Gamma(Q)\}$. The normal vector $N(Q)$ to $Q \in \partial\Omega$ exists almost everywhere. A function v belongs to L_1^p if it has tangential derivatives in $L^p(\partial\Omega, d\sigma)$. Above a graph, $\{y > \varphi(x)\}$, this simply means that $\nabla_x v(x, \varphi(x))$ belongs to $L^p(dx, \mathbb{R}^{n-1})$, and there is a natural localization of this definition to bounded domains. (See [DK], for example.)

The Dirichlet problem in L^p for Laplace's equation in a Lipschitz domain is the problem of solving $\Delta u = 0$ in Ω, $u|_{\partial\Omega} = f \in L^p(d\sigma)$, with the estimate $\|u^*\|_{L^p(d\sigma)} \leq C\|f\|_{L^p(d\sigma)}$. Dahlberg ([D]) showed that this problem was uniquely solvable if $p > 2 - \epsilon$, for $\epsilon = \epsilon(\Omega)$. Moreover, for any $p < 2$, there exists a domain, depending on this p, on which this fails to be uniquely solvable. (Note that the range of solvability $2 < p < \infty$ follows from the $p = 2$ case and the maximum principle by interpolation.) The regularity problem for Laplace's equation is that of solving $\Delta u = 0$ in Ω, $u|_{\partial\Omega} = f \in L_1^p(d\sigma)$ with the estimate $\|(\nabla u)^*\|_{L^p(d\sigma)} \leq C\|f\|_{L_1^p(d\sigma)}$. Jerison and Kenig ([JK1]) solved this problem for $p = 2$, then Verchota [V1] solved this problem for $1 < p < 2$, by the method of layer potentials, and again this range of p, $1 < p < 2 + \epsilon$, is sharp.

The formulation of the Dirichlet problem with data in L^p (D_p) for a fourth-order CCHE operator is straightforward. We need to specify two pieces of boundary data. The problem is to solve

$$Lu = \begin{cases} 0 & \text{in } \Omega, \\ u = f \in L_1^p(d\sigma), & \text{on } \partial\Omega, \\ \frac{\partial u}{\partial N} = g \in L^p(d\sigma) & \text{on } \partial\Omega, \end{cases}$$

with the estimate

$$\|(\nabla u)^*\|_{L^p(\partial\Omega)} \leq C\{\|f\|_{L_1^p} + \|g\|_{L^p}\}.$$

The constant C should depend only on the Lipschitz character of Ω, and the normal derivative $\frac{\partial u}{\partial N}$ is understood in the sense of nontangential limits, viz. $\nabla u(X) \cdot N(Q) \to g(Q)$ as $X \to Q$, $X \in \Gamma(Q)$ for a.e., Q. To formulate the L^p regularity problem (R_p), which involves a condition on two derivatives on the boundary, more care is required since the boundary of our domain is only differentiable once. To specify the boundary conditions for this problem, one can stipulate the existence of a $C_0^\infty(\mathbb{R}^n)$ function F such that

$$Lu = \begin{cases} 0 & \text{in } \Omega, \\ u = F & \text{on } \partial\Omega, \\ \frac{\partial u}{\partial N}(Q) = \sum_j N^j(Q) D_j F & \text{on } \partial\Omega, \end{cases}$$

with the *a priori* estimates

$$\|(\nabla\nabla u)^*\|_{L^p(\partial\Omega, d\sigma)} \leq C \sum \|D_j F\|_{L_1^p(\partial\Omega, d\sigma)},$$

and where N^j denotes the jth component of the normal vector. The problem also has an intrinsic formulation involving arrays of functions defined on the boundary of the domain satisfying certain compatibility conditions. See [CG1] and [V2]. In terms of these arrays, or by solving the BV problem associated with the restriction of such an F and its derivatives to $\partial\Omega$, the problems (D_p) and (R_p) for any $2m$-order operator may be formulated on nonsmooth domains in order to give meaning to the data $u, \ldots, \frac{\partial^{m-1} u}{\partial N^{m-1}}$, when restricted to the boundary of Ω.

In 1982, Cohen and Gosselin [CG] solved (D_p) and (R_p), $1 < p < \infty$, for the biharmonic operator on C^1 domains in the plane. In 1984, via a special representation for solutions to Δ^2, Dahlberg, Kenig, and Verchota [DKV] solved the problem (D_2) for the biharmonic equation on Lipschitz domains in $\mathbb{R}^n$. Subsequently, Verchota ([V2] and [V3]) was able to generalize this representation to solve (D_p) on C^1 domains in $\mathbb{R}^n$ for any $1 < p < \infty$ and to solve (D_2) and (R_2) for the polyharmonic operators Δ^m in Lipschitz domains. As in the case of the Laplacian, the L^p Dirichlet problems on Lipschitz domains are not uniquely solvable if $p < 2$ [DKV]. Unlike the second-order case, there is no maximum principle (and therefore no automatic solution to the L^p Dirichlet problem for $p = \infty$) and therefore from solvability of (D_2), one cannot conclude solvability of (D_p) for $p > 2$.

In [PV1], Verchota and I established that (D_p) was solvable in Lipschitz domains in $\mathbb{R}^3$ if $2 < p < \infty$, but may fail for some $p > 2$ if the dimension is larger than 3. In [PV3], we showed that this positive result is in fact a consequence of a weak maximum principle (the $p = \infty$) case of (D_p), which holds for Δ^2 in dimension 3, but fails in higher dimensions. The positive and negative results were also extended to include the polyharmonic operators Δ^m, $m \geq 4$, in [PV4]. In certain dimensions, depending on the order of the operator, these counterexamples can be obtained from a construction in [MNP]. Indeed, parallel to the development and progress on general Lipschitz and C^1 domains described above, is a series of remarkable papers by Maz'ya et al., analyzing the behavior of solutions to Δ^2 (and more general higher order operators and elliptic systems) on conical domains and polyhedra. See, for example, [KoM1], [KoM2], [KoM3], [KrM], [MN], [MNP], [MP], [MNP1], and [MR]. For related work, and additional sources, the following papers are a small, but representative, sample of the available literature: [Da], [Gr], [KO], [Ko1], [Ko2], and [S].

We now wish to describe one means of solving the problem (D_2) for Laplace's equation. It will then be apparent how readily it extends to all constant coefficient second-order elliptic operators. The heart of this proof, or of any other proof, is a Rellich identity, or boundary Garding inequality. And this is precisely where the difficulty lies in solving (D_2) for higher order operators. For simplicity and convenience, we work above a graph, and we will also ignore the required limiting arguments needed to make this proof rigorous.

To solve $\Delta u = 0$ in Ω with $u|_{\partial\Omega} = f \in L^2(d\sigma)$, we assume f continuous, obtain a solution, and need only derive the *a priori* estimate $\|u^*\|_{L^2(\partial\Omega)} \leq C\|f\|_{L^2(\partial\Omega)}$. We assume that $\partial\Omega = \{(x, y) : y > \varphi(x)\}$. By Green's identity, with $\Gamma(X, Y) = c_n|X - Y|^{2-n}$ the fundamental solution of Δ,

$$u(X) = \iint \Delta_Y \Gamma(X, Y)u(Y)\,dY$$

$$= \int_{\partial\Omega} \frac{\partial\Gamma}{\partial N}(X, Q)f(Q)\,d\sigma - \int_{\partial\Omega} \Gamma(X, Q)\frac{\partial u}{\partial N}(Q)\,d\sigma(Q)$$

$$= A + B.$$

Term **A** has the desired nontangential estimate in virtue of the theorem of Coifman, McIntosh, and Meyer on the Cauchy integral on Lipschitz curves, [CMM]. That is, $\|A^*\|_{L^2(d\sigma)} \leq C\|f\|_{L^2(d\sigma)}$. But term **B** involves an extra derivative on u (and not enough derivatives on Γ). Define a harmonic function v by $u = D_n v$, where $D_n = \frac{\partial}{\partial y}$. Then, if N^j denotes the jth component of the normal vector, on the boundary we have

$$\frac{\partial u}{\partial N} = \sum_j N^j D_j u$$

$$= \sum_j N^j D_j D_n v$$

$$= \sum_j (N^j D_n - N^n D_j)D_j v$$

where we have made use of the fact that $\sum_j D_j D_j v = 0$. But $N^j D_n - N^n D_j$ is a tangential derivative, which we now denote as T_j. That is, $\frac{\partial u}{\partial N} = \sum_j \frac{\partial}{\partial T_j} D_j v$ and so term **B** becomes, after an integration by parts,

$$\mathbf{B}(X) = \int_{\partial\Omega} \frac{\partial\Gamma}{\partial T_j}(X, Q)D_j v(Q)\,d\sigma(Q),$$

and again by [CMM],

$$\|(\mathbf{B})^*\|_{L^2(d\sigma)} \leq C\|D_j v\|_{L^2(d\sigma)}.$$

To complete the proof, one needs the Riesz transform inequality,

$$\sum_j \|D_j v\|_{L^2(\partial\Omega.d\sigma)} \le C\|D_n v\|_{L^2(\partial\Omega.d\sigma)} = C\|u\|_{L^2(\partial\Omega.d\sigma)},$$

and this is the Rellich identity alluded to earlier. The proof is as follows. First, since Ω is Lipschitz, N^n is bounded from below. Thus, dropping the summation, we obtain

$$\int_{\partial\Omega} |D_j v|^2 \, d\sigma \le c_0 \int_{\partial\Omega} |D_j v|^2 N^n \, d\sigma$$

$$= c_0 \int\int_{\Omega} D_n (D_j v)^2 \, dX$$

$$= 2c_0 \int\int_{\Omega} D_n D_j v D_j v \, dX$$

$$= 2c_0 \int\int_{\Omega} D_j (D_n v D_j v) \, dX.$$

The last inequality uses the equation for v. Another integration by parts gives

$$\int\int_{\Omega} D_j (D_n v D_j v) \, dX = \int_{\partial\Omega} N^i D_n v D_j v \, d\sigma$$

$$\le \left(\int_{\partial\Omega} |D_n v|^2 \, d\sigma \right)^{1/2} \left(\int_{\partial\Omega} |D_j v|^2 \, d\sigma \right)^{1/2},$$

by Cauchy–Schwarz. Thus we have the inequality $\int |D_j v|^2 \, d\sigma \le \int |D_n v|^2 \, d\sigma$.

The method works just as well for a general constant coefficient second-order operator, which we may write as $L = div A \nabla$, where $A = (a_{ij})$ is elliptic, i.e., $A\xi \cdot \xi > C|\xi|^2$. That is, to solve (D_2) for L, one begins by expressing the solution u in terms of a potential involving the fundamental solution of L. An integration by parts yields two boundary integrals, one of which contains the data $u|_{\partial\Omega}$, and is thus readily estimated. Finally, it is only the Rellich identity that is needed to finish the argument. The essential element needed for the Rellich identity is to introduce a form on the boundary that enables one to make use of the equation satisfied by v, where $u = D_n v$. In the second-order situation, ellipticity is a very strong condition, for we may apply it to the vector ∇v. Hence

$$\int_{\partial\Omega} |D_j v|^2 N^n \, d\sigma \le C' \int_{\partial\Omega} A\nabla v \cdot \nabla v N^n \, d\sigma \tag{1}$$

holds because $A\nabla v \cdot \nabla v \ge C|\nabla v|^2$ pointwise. The rest of the argument goes through just as is the case of the Laplacian, for in the solid integral $\int\int D_n (A\nabla v \cdot \nabla v) \, dX$ one will be able to use the equation $Lv = 0$ as before. Now, it is exactly this pointwise estimate, $A\nabla v \cdot \nabla v \ge C|\nabla v|^2$, that has no analogue in the case of higher order elliptic equations, and the desired version of inequality (1) need not be true. There is a substitute, however, that makes this method work, and, in what follows, we shall describe the method used in [PV5] to obtain these Riesz transform type inequalities and to solve the Dirichlet and regularity problems for any CCHE operator in such domains.

Briefly, the setup is as follows. I shall describe only the fourth-order case, although the necessary boundary Garding identity is valid in all dimensions. Let $L = \sum_{|\alpha|=4} a_\alpha D^\alpha$ be a constant coefficient and let $\Gamma(X, Y)$ denote the fundamental solution, which has size $|X - Y|^{4-n}$ in dimensions $n = 3$ and $n \geq 5$. The solution $u(X)$ is given by

$$u(X) = \int\int_\Omega L_Y \Gamma(X, Y) u(Y) \, dY,$$

and the Dirichlet conditions on the boundary mean that $|\nabla u| \in L^2(\partial\Omega, d\sigma)$. The solid integral gives rise to four boundary integrals, one of which has the form

$$\mathbf{A} = \int_{\partial\Omega} D^2 \Gamma(X, Q) Du(Q) \, d\sigma,$$

and D denotes some derivative in Q that is explicit from the integration by parts. We recall now that the desired estimate involves the nontangential maximal function of the gradient of the solution in the fourth-order situation. Again, by the theory of singular integrals and the theorem of Coifman, McIntosh, and Meyer [CMM], we have the estimate $||(\nabla \mathbf{A})^*||_{L^2(\partial\Omega)} \leq C||\nabla u||_{L^2(d\sigma)}$. (Note that it is three derivatives of Γ that satisfies the estimates for which the theory of [CMM] applies.) There are three other boundary integrals involving too few or too many derivatives on u. To handle such terms, we introduce v by setting $u = D_n D_n v$, so that $Lv = 0$. (The number of D_n's introduced here is connected with the order of the operator.) The claim is that the following boundary inequality, the analogue of the Riesz transform inequality for solutions of second-order operators, is the key element in the proof of the L^2 estimate:

$$||\nabla\nabla\nabla v||_{L^2(\partial\Omega, d\sigma)} \leq C||\nabla D_n D_n v||_{L^2(\partial\Omega, d\sigma)}. \tag{2}$$

The expression $|\nabla\nabla\nabla v|^2$ abbreviates the sum over all $j, k, l \leq n$ of $|D_j D_k D_l v|^2$.

Let $w = D_j v$, and consider one of the terms arising in (2). We first want to obtain the inequality

$$||\nabla\nabla w||_{L^2(d\sigma)} \leq C||\nabla D_n w||_{L^2(d\sigma)}.$$

Iteration of this step yields (2). The problem here is the introduction of a bilinear form on the boundary that permits one to make use of the equation satisfied by w (or v) in the solid integral. The substitute is the following boundary Garding inequality [PV5] stated here in the fourth-order case only, but valid as well, with appropriate modifications, for operators of any order.

$$\int_{\partial\Omega} |\nabla\nabla w|^2 \, d\sigma$$

$$\leq C\left(\int_{\partial\Omega} |\nabla D_n w|^2 \, d\sigma + \sum_{|\alpha|=|\beta|=2,\alpha_n=0=\beta_n} \int_{\partial\Omega} D^\alpha w a_{\alpha\beta} D^\beta w N^n \, d\sigma\right),$$

where $Lw = 0$ and $L = \sum_{|\alpha|=|\beta|=2} a_{\alpha\beta} D^\alpha D^\beta$.

Before sketching a proof of this inequality, we describe the new algebraic identities that underlie this in the general situation. The idea is to make use of the Fourier transform by passing from an integral on the boundary of our Lipschitz domain to an integral over $\mathbb{R}^{n-1}$. Toward this end, we define a quadratic form

$$Q(m, \xi, \eta) = \frac{1}{2} \sum_{i,j=1}^{r} \sum_{|\alpha|=m-2} |\xi_i \eta(\alpha + e_j) - \xi_j \eta(\alpha + e_i)|^2,$$

where η is complex valued. In applications, r is the dimension $(n - 1)$ and $2m$ is the order of the operator. Given a positive definite form on $\mathbb{R}^r$, let us write the constants as $a_{\alpha\beta}^{ij}$; that is, we are assuming the existence of a constant C, such that

$$C^{-1}|\xi|^{2r} \geq \sum_{|\alpha|=|\beta|=m-1} \sum_{i,j=1}^{r} \xi_i \xi^\alpha a_{\alpha\beta}^{ij} \xi_j \xi^\beta \geq C|\xi|^{2r}.$$

We then claim

(1) $Q(m, \xi, \eta) = 0$ iff there exists a constant $c \in \mathbf{C}$ such that $\eta(\beta) = c\xi^\beta$.
(2)   There are constants E and E' such that

$$EQ(m, \xi, \eta) + \text{Re}\left(\sum_{|\alpha|=|\beta|=m-1} \sum_{i,j=1}^{r} \xi_i a_{\alpha\beta}^{ij} \xi_j \eta(\alpha)\overline{\eta(\beta)}\right) \geq E'|\xi|^2 |\eta|_{m-1}^2,$$

where we define

$$|\eta|_{m-1}^2 = \sum_{|\alpha|=m-1} |\eta(\alpha)|^2.$$

Statement (1) is proven by induction and reduces to knowing when equality holds in the Cauchy–Schwarz inequality. Statement (2) is a quantitative version of (1) combined with the ellipticity condition. See [PV5] for details.

Consider now the boundary Garding inequality in the fourth-order case. Take $r = n - 1$ above. Then, if a_{ijkl} are the coefficients of the positive definite bilinear form associated to L, we have

$$\sum_{l,k=1}^{n} \sum_{i,j=1}^{n-1} \int_{\partial\Omega} D_i D_k w a_{ijkl} D_j D_l w N^n \, d\sigma$$

$$= \int_{\partial\Omega} \left(\frac{\partial}{\partial x_i} D_k w - \frac{\partial\varphi}{\partial x_i} D_n D_k w\right) a_{ijkl} \left(\frac{\partial}{\partial x_j} D_l w - \frac{\partial\varphi}{\partial x_j} D_n D_l w\right) d\sigma,$$

and all terms with a DD_n component are good terms for the purposes of our inequality. It therefore suffices to estimate the integral

$$\int_{\partial\Omega} \frac{\partial}{\partial x_i} D_k w a_{ijkl} \frac{\partial}{\partial x_j} D_l w \, d\sigma,$$

which, by Plancherel, equals

$$\int_{\xi\in\mathbb{R}^{n-1}} \xi_i \eta(k) a_{ijkl} \xi_j \overline{\eta(l)} \, d\xi$$

with $\eta(k) = \widehat{D_k w}$. To this integral we add and subtract the quantity

$$\int_{\mathbb{R}^{n-1}} Q \, d\xi = \int_{\mathbb{R}^{n-1}} \left| \frac{\partial}{\partial x_i} D_k w - \frac{\partial}{\partial x_k} D_i w \right|^2 dx$$

$$= \int_{\mathbb{R}^{n-1}} \left| D_n D_k w \frac{\partial\varphi}{\partial x_i} - \frac{\partial\varphi}{\partial x_k} D_n D_i w \right|^2 dx,$$

which is again a good term for the purposes of our inequality, since it contains terms involving $D_n w$. Hence an application of (2) and Parseval's theorem yields the inequality. We conclude then, with a precise statememt of the main results of [PV5]. (See [V2] for a precise definition of the boundary array space $WA^p_{m-1}(\partial\Omega)$.)

THEOREM 1

Let $\Omega \subset \mathbb{R}^n$ be a bounded Lipschitz domain with nontangential approach regions $\Gamma^\alpha(Q)$ for all $Q \in \partial\Omega$ for α large enough, depending on the Lipschitz character of Ω. Let L be a homogeneous real constant coefficient elliptic partial differential operator of order $2m$ in $\mathbb{R}^n$ with ellipticity constant E. Then there is an $\epsilon > 0$ depending on n, on the Lipschitz character of Ω, and on E so that, if $2 - \epsilon < p < 2 + \epsilon$, $g \in L^p(\partial\Omega)$ and $\dot{f} \in WA^p_{m-1}(\partial\Omega)$, there is a unique real analytic solution u to $Lu = 0$ in Ω so that

(i) $(\nabla^{m-1} u)^* \in L^p(\partial\Omega)$

(ii) $\lim \frac{\partial^{m-1} u(X)}{\partial N_Q^{m-1}} = g(Q)$ *a.e. as* $X \to Q$, $X \in \Gamma^\alpha(Q)$

(iii) $\lim D^\gamma u(X) = f_\gamma(Q)$ *a.e. as* $X \to Q$, $X \in \Gamma^\alpha(Q)$ *for* $0 \le |\gamma| \le m - 2$.

In addition

(iv) $\|(\nabla^{m-1} u)^*\|_{L^p(\partial\Omega)} \le \|g\|_{L^p(\partial\Omega)} + C \sum_{|\gamma|=m-2} \|\nabla_T f_\gamma\|_{L^p(\partial\Omega)}.$

and

(v)   *the nontangential limit of $D^\gamma u(X)$ exists a.e. for $|\gamma| = m - 1$, so that $\nabla_T D^\gamma u(X) \to \nabla_T f_\gamma(Q)$ a.e., as $X \to Q$, $X \in \Gamma^\alpha(Q)$, for $|\gamma| = m - 2$ where C depends only on n, m, E, p, and the Lipschitz character of Ω.*

THEOREM 2

With the same hypotheses as Theorem 1, there is an $\epsilon > 0$ depending on n, E, and the Lipschitz character of Ω so that, if $2 - \epsilon < p < 2 + \epsilon$ and $\dot{f} \in WA_m^p(\partial\Omega)$, then there is a unique real analytic solution u to $Lu = 0$ in Ω so that

(i) $(\nabla^m u)^* \in L^p(\partial\Omega)$,

and

(ii) $\lim D^\gamma u(X) = f_\gamma(Q)$ as $X \to Q$, $X \in \Gamma^\alpha(Q)$ a.e., for $0 \leq |\gamma| \leq m - 1$.

In addition

(iii) $\|(\nabla^m u)^*\| \leq C \sum_{|\gamma|=m-1} \|\nabla_T f_\gamma\|_{L^p(\partial\Omega)}$,

(iv) $\|(\nabla^{m-1} u)^*\|_{L^p(\partial\Omega)} \leq C \sum_{|\gamma|=m-1} \|f_\gamma\|_{L^p(\partial\Omega)}$,

and

(v) *the nontangential limit of $D^\gamma u(X)$ exists a.e. for $|\gamma| = m$ so that $\lim \nabla_T D^\gamma u(X) = \nabla_T f_\gamma(Q)$ as $X \to Q$, $X \in \Gamma^\alpha(Q)$ for $|\gamma| = m - 1$ where C depends only on n, m, E, p, and the Lipschitz character of Ω.*

References

[A]   S. Agmon, *Lectures on elliptic boundary value problems,* Van Nostrand, 1965.

[ADN1] S. Agmon, A. Douglis, and L. Nirenberg, *Estimates near the boundary for solutions of elliptic partial differential equations satisfying general boundary conditions I,* Comm. Pure Appl. Math. 12 (1959), 623–727.

[ADN2] S. Agmon, A. Douglis, and L. Nirenberg, *ibid II,* Comm. Pure Appl. Math. 22 (1964), 35–92.

[B]   F. Browder, *On the regularity of properties of solutions of elliptic boundary value problems,* Comm. Pure Appl. Math. 9 (1956), 351–361.

[C]   J. Cohen, *BMO estimates for biharmonic multiple layer potentials* Studia Math. XCI (1988), 109–123.

[CG1]   J. Cohen and J. Gosselin, *The Dirichlet problem for the biharmonic equation in a C^1 domain in the plane,* Ind. U. Math. J. 32 (1983), 635–685.

[CG2]   J. Cohen and J. Gosselin, *Stress potentials on C^1 domains,* Jour. Math. Anal. Appl 125 no. 1 (1987), 22–46.

[CMM]   R. Coifman, A. McIntRosh, and Y. Meyer *L'intégrale de Cauchy définit un operateur borné sur L^2 pour les courbes lipschitziennes,* Ann. of Math. 116 (1982), 361–387.

[Da]   M. Dauge, *Elliptic boundary value problems on corner domains, Lecture Notes in Math. 1341,* Springer–Verlag, 1988.

[D]   B. E. J. Dahlberg, *On estimates for harmonic measure,* Arch. Rat. Mech. Anal. 65 (1977), 272–288.

[DK]   B. E. J. Dahlberg and C. Kenig, *Hardy spaces and the L^p-Neumann problem for Laplace's equation in a Lipschitz domain,* Annals. of Math. 125 (1987), 437–465.

[DKV]   B. Dahlberg, C. Kenig, and G. Verchota *The Dirichlet problem for the biharmonic equation in a Lipschitz domain,* Ann. de l'Inst. Fourier 36 (1986), 109–134.

[G]   P. Garabedian, *A partial differential equation arising in conformal mapping.,* Pac. J. Math. 1 (1951), 485–524.

[Gr]   P. Grisvard, *Elliptic problems in nonsmooth domains,* Pitman Publishing, 1985.

[JK1]   D. Jerison and C. Kenig, *The Dirichlet problem in non-smooth domains,* Ann. of Math. 113 (1981), 367–382.

[JK2]   D. Jerison and C. Kenig, *Boundary value problems on Lipschitz domains,* MAA Studies in Math. 23 (1982).

[KO]   V. Kondratiev and O. Oleinik, *Estimates near the boundary for 2nd order derivatives of solutions of the Dirichlet problem for the biharmonic equation,* Att. Accad. Naz. Lincei Rend. Cl. Sci. Fis. Mat. Natur. 80 (8) (1986), 525–529.

[Ko1]   V. Kozlov, *The strong zero theorem for an elliptic boundary value problem in a corner,* Dokl. Akad. nauk. SSSR 309 (6) (1989), 1299–1301.

[Ko2]   V. Kozlov, *The Dirichlet problem for elliptic equations in domains with conical points,* Diff. Eq. 26 (1990), 739–747.

[KoM1] V. Kozlov and V. Maz'ya, *Estimates of the L^p means and asymptotic behavior of solutions of elliptic boundary value problems in a cone, I,* Seminar analysis, Akad. Wiss. DDR, Berlin (1986), 55–91.

[KoM2] V. Kozlov and V. Maz'ya, *ibid, II,* Math. Nachr. 137 (1988), 113–139.

[KoM3] V. Kozlov and V. Maz'ya, *Spectral properties of operator pencils generated by elliptic boundary value problems in a cone,* Funct. Anal. Appl. 22 (1988), 114–121.

[KrM]   G. Kresin and V. Maz'ya, *A sharp constant in a Miranda-Agmon type inequality for solutions to elliptic equations,* Soviet Math. 32 (1988), 49–59.

[MN]   V. Maz'ya and S. Nazarov, *The apex of a cone can be irregular in Wiener's sense for a fourth-order elliptic equation,* Mat. Zametki 39 (1986), 24–28.

[MNP]   V. Maz'ya, S. Nazarov, and B. Plamenevskiĭ, *On the singularities of solutions of the Dirichlet problem in the exterior of a slender cone,* Math. Sb. 50 (1980), 415–437.

[MP]   V. G. Maz'ya and B. A. Plamenevskiĭ, *Estimates in L_p and in Hölder classes and the Miranda–Agmon maximum principle for solutions of elliptic boundary value problems in domains with singular points on the boundary,* Amer. Math. Soc. Transl. (2) 123 (1984), 1–56 Transl. from Math. Nachr. 81 (1978), 25–82.

[MNP1]   V. Maz'ya, S. Nazarov, and B. Plamenevskiĭ, *Asymptotische Theorie Elliptischer Randwertaufgaben in singular gestorten Gebieten I,* Mathematische Lehrbrucher und Monographien, Akademie–Verlag, Berlin 1991.

[MR]   V. Maz'ya and J. Rossman, *On the Miranda–Agmon maximum principle for solutions of elliptic equations in polyhedral and polygonal domains,* preprint.

[O]   S. Osher, *On Green's function for the biharmonic equation in a right angle wedge,* J. Math. Anal. Appl. 43 (1973), 705–716.

[PV1]   J. Pipher and G. Verchota, *The Dirichlet problem in L^p for the biharmonic operator on Lipschitz domains,* Amer. J. Math. 114 (1992), 923–972.

[PV2]   J. Pipher and G. Verchota, *Area integral estimates for the biharmonic operator in Lipschitz domains,* Trans. Amer. Math. Soc. 327 (2) (1991), 903–917.

[PV3]   J. Pipher and G. Verchota, *A maximum principle for biharmonic functions in non-smooth domains,* Comm. Math. Helv., 68 (1993), 385–414.

[PV4]   J. Pipher and G. Verchota, *Maximum principles for the polyharmonic equation on Lipschitz domains,* to appear, J. of Pot. Analysis.

[PV5]   J. Pipher and G. Verchota, *Dilation invariant estimates and the boundary Garding inequality for higher order elliptic operators,* to appear, Annals of Math.

[S]   V. Slobodin, *Homogeneous boundary value problems for a polyharmonic operator with boundary conditions on thin sets,* Izv. Vyssh. Vchebn. Zaved. Mat. 10 (1989), 57–63.

[V1]   G. Verchota, *Layer potentials and regularity for the Dirichlet problem for Laplace's equation,* J. of Funct. Anal. 59 (1984), 572–611.

[V2]   G. Verchota, *the Dirichlet problem for the biharmonic equation in C^1 domains,* Ind. U. Math. J. 36 (1987), 867–895.

[V3]   G. Verchota, *The Dirichlet problem for the polyharmonic equation in Lipschitz domains,* Ind. U. Math. J. 39 (1990), 671–702.

21

A_p and Approach Regions

A. Sánchez-Colomer
J. Soria[1]

ABSTRACT   *We study the relationship between weighted inequalities of certain maximal operators associated to approach regions and the geometry of these sets.*

21.1   Introduction

In [NS] Nagel and Stein studied the weak-$(1, 1)$ boundedness of a maximal operator M_Ω associated to a general approach region Ω in $\mathbb{R}_+^{n+1}$. Sueiro [Su] gave the analogous result for spaces of homogeneous type. Pan [P] studied the weighted estimates for M_Ω following the ideas of [Su]. In [SS] we obtained these weights in the case of $\mathbb{R}^n$ as an application of another kind of general maximal operator, introduced by Ruiz and Torrea in [RT]. We also mentioned that whenever the cross section of the approach region was (in some sense) comparable to a ball, then the classes of weights A_p^Ω (i.e., those w for which $M_\Omega : L^p(w) \to L^{p,\infty}(w)$ is bounded) were exactly the classical A_p (see [GR]).

In this note we give a converse result; that is, if A_p^Ω equals A_p, then Ω is essentially a cone, therefore these classes of weights determine the shape of the approach region.

21.2   Definitions and previous results

Let $\{\Omega_x\}_{x \in \mathbb{R}^n}$ be a family of measurable sets in $\mathbb{R}_+^{n+1} = \{(x, r) : x \in \mathbb{R}^n, r > 0\}$ satisfying that if $(y, t) \in \Omega_x$, then $(y, s) \in \Omega_x$ for all $s > t$. We define the maximal

---

[1] Partially supported by Grant DGICYT PB91-0259.

operator

$$M_\Omega f(x) = \sup_{(y,t)\in\Omega_x} \frac{1}{|B(y,t)|} \int_{B(y,t)} |f(\xi)| \, d\xi$$

where $B(y, t)$ is the ball of center y and radius r in $\mathbb{R}^n$ and $|\cdot|$ is the Lebesgue measure in $\mathbb{R}^n$. We state that $w \in L^1_{\text{loc}}(\mathbb{R}^n)$, $w \geq 0$ is in A_p^Ω if the operator $M_\Omega : L^p(w) \to L^{p,\infty}(w)$ is bounded. Associated to Ω_x, we consider the following sets (see [Su]):

$$\Omega_x(r) = \{y \in \mathbb{R}^n : (y, r) \in \Omega_x\},$$

$$S(x, r) = \{y \in \mathbb{R}^n : B(x, r) \cap \Omega_y(r) \neq \emptyset\}.$$

In [SS] we obtained the following characterization $\big(w(E)$ stands for $\int_E w(\xi) \, d\xi\big)$.

THEOREM 1

(a) $w \in A_1^\Omega$, if and only if, there exists $C > 0$ such that

$$\frac{w(S(x, r))}{|B(x, r)|} \leq C \operatorname{ess\,inf}_{B(x,r)} w, \quad \text{for all } (x, r) \in \mathbb{R}_+^{n+1}.$$

(b)   If $p > 1$, then $w \in A_p^\Omega$, if and only if,

$$\sup_{(x,r)\in\mathbb{R}_+^{n+1}} \frac{w(S(x, r))}{|B(x, r)|} \left(\frac{w^{-1/(p-1)}(S(x, r))}{|B(x, r)|} \right)^{p-1} < \infty.$$

If $(x, 0) \in \overline{\Omega_x}$ for all $x \in \mathbb{R}^n$, we say that $\{\Omega_x\}_{x\in\mathbb{R}^n}$ is a family of approach regions. In this case we may assume, with no loss of generality, the inclusion $\{(x, r) : r > 0\} \subset \Omega_x$. Using the above characterization and the fact that $Mf(x) \leq M_\Omega f(x)$, where M is the centered Hardy–Littlewood maximal operator, one obtains Lemma 1.

LEMMA 1

If $(x, 0) \in \overline{\Omega_x}$ for all $x \in \mathbb{R}^n$, then w is in A_p^Ω, $(p \geq 1)$, if and only if, w is in A_p and there exists $C > 0$ such that

$$w(S(x, r)) \leq Cw(B(x, r)) \quad \text{for all} \quad (x, r) \in \mathbb{R}_+^{n+1}.$$

Finally, we say that $S(x, r)$ is comparable to a ball if there exists a constant $C > 0$, such that for every (x, r) we can find a ball B containing $S(x, r)$ and $|B| \leq C|S(x, r)|$. In [SS] we have the following proposition.

PROPOSITION 1

If $(x, 0) \in \overline{\Omega_x}$ and $S(x, r)$ is comparable to a ball, then the following are equivalent:

(i) $w \equiv 1$ *is in* A_1^Ω;

(ii) $A_1^\Omega \neq \emptyset$;

(iii)   *There exists* $q \geq 1$ *such that* $A_q^\Omega \neq \emptyset$;

(iv) $A_p^\Omega = A_p$ *for all* $p \geq 1$.

PROOF   It is sufficient to see that (iii) $\Longrightarrow$ (i) $\Longrightarrow$ (iv).

First, suppose $w \in A_q^\Omega$ for some $q \geq 1$. Then, by Lemma 1, $w \in A_\infty$, so there exists C_1, C_2, δ_1, and δ_2 such that for every ball, $B \subset \mathbb{R}^n$, and for every measurable set, $E \subset B$ (see [GR])

$$C_1 \left(\frac{|E|}{|B|} \right)^{\delta_1} \leq \frac{w(E)}{w(B)} \leq C_2 \left(\frac{|E|}{|B|} \right)^{\delta_2}.$$

It is easy to see that if $\{(x, r); r > 0\} \subset \Omega_x$, then $B(x, r) \subset S(x, r)$. Using these two conditions and the fact that $S(x, r)$ is comparable to a ball B, we obtain

$$\frac{|B(x, r)|}{|S(x, r)|} \geq \frac{|B(x, r)|}{|B|} \geq \left(\frac{1}{C_2} \frac{w(B(x, r))}{w(B)} \right)^{1/\delta_2}$$

$$\geq \left(\frac{1}{C_2} \frac{1}{C} \frac{w(S(x, r))}{w(B)} \right)^{1/\delta_2}$$

$$\geq \left(\frac{1}{C_2} \frac{1}{C} C_1 \left(\frac{|S(x, r)|}{|B|} \right)^{\delta_1} \right)^{1/\delta_2} \geq C',$$

and this shows that $1 \in A_1^\Omega$.

Secondly, suppose that $|S(x, r)| \leq C |B(x, r)|$; this implies $A_1 = A_1^\Omega$ and by the extrapolation theorem of Rubio de Francia, we have $A_p^\Omega = A_p$ for all $p \geq 1$. In fact, if $w \in A_1$, then

$$w(S(x, r)) \leq w(B) \leq C(\operatorname{ess\,inf}_B w) |B|$$

$$\leq C(\operatorname{ess\,inf}_{S(x,r)} w) |S(x, r)|$$

$$\leq C(\operatorname{ess\,inf}_{B(x,r)} w) |B(x, r)|$$

$$\leq Cw(B(x, r)),$$

and, by Lemma 1, we obtain that $w \in A_1^\Omega$. ∎

REMARK 1   In fact, (i) and $S(x, r)$ comparable to a ball is equivalent to the condition $\Omega_x \subset \Gamma_x^\alpha = \{(y, t) \in \mathbb{R}_+^{n+1} : |x - y| < \alpha t\}$, for some $\alpha > 0$. ∎

The question now is whether we can find a region Ω, such that $A_p^\Omega \neq \emptyset$ and $A_p^\Omega \neq A_p$.

21.3   Translations of a region

Let $\Omega \subset \mathbb{R}^{n+1}_+$ be an approach region at $0 \in \mathbb{R}^n$; i.e., Ω contains the line $\{(0, r) \in \mathbb{R}^{n+1}_+ : r > 0\}$. Define $\Omega_x = (x, 0) + \Omega$. With this choice, we can now provide a complete answer to the question raised above, by improving Proposition 1.

PROPOSITION 2

If $\Omega_x = (x, 0) + \Omega$ is as before, then $A_p = A_p^\Omega$ for some $p \geq 1$, if and only if, Ω is contained in a cone with vertex at 0.

REMARK 2   We can equivalently say that $A_p = A_p^\Omega$ for all $p \geq 1$ via the extrapolation theorem of Rubio de Francia [GR].   ∎

As a corollary, we determine that there exists a region Ω (see [Su]) for which A_p^Ω is not A_p, but $w \equiv 1 \in A_p^\Omega$ for $p \geq 1$; i.e., $\emptyset \neq A_p^\Omega \neq A_p$.

PROOF   Clearly if $\Omega \subset \Gamma$ (a cone with vertex at 0), then $A_p \subset A_p^\Omega$, since $M_\Omega f(x) \leq C\, Mf(x)$. Conversely, we can assume $p > 1$. Thus there exists $\beta > 0$ such that $w(\xi) = |\xi|^\beta \in A_p^\Omega$. Observe that if Ω_x are the translates of Ω, then $S(x, r) = B(x, r) - \Omega(r)$ (see [Su]); i.e.,

$$S(x, r) = \bigcup_{y \in \Omega(r)} B(x - y, r). \tag{1}$$

Now, if Ω is not contained in a cone, for every $m > 1$, there exists $(x_m, r_m) \in \Omega$ such that $|x_m| \geq m r_m$. Take $w(\xi) = |\xi|^\beta$, then by (1)

$$\frac{w(S(0, r_m))}{w(B(0, r_m))} \geq \frac{\int_{B(-x_m, r_m)} |\xi|^\beta \, d\xi}{\int_{B(0, r_m)} |\xi|^\beta \, d\xi}$$

$$\geq \frac{\int_{B(-x_m, r_m)} (|x_m| - r_m)^\beta \, d\xi}{\int_{B(0, r_m)} |\xi|^\beta \, d\xi}$$

$$= C_n \frac{(|x_m| - r_m)^\beta r_m{}^n}{(r_m)^{\beta+n}} \geq C_n (m - 1)^\beta.$$

This is a contradiction, since by Lemma 1, $w \in A_p^\Omega$ implies $w(S(x, r)) \leq Cw(B(x, r))$ for all $(x, r) \in \mathbb{R}^{n+1}_+$.   ∎

References

[GR] J. García-Cuerva and J. L. Rubio de Francia, *Weighted Norm Inequalities and Related Topics*, Mathematical Studies, vol.116, North–Holland, 1985.

[NS] A. Nagel and E. Stein, *On certain maximal functions and approach regions*, Adv. in Math. 54 (1984), 83–106.

[P] Pan Wenjie, *Weighted norm inequalities for certain maximal operators with approach regions*, Lecture Notes in Mathematics, vol. 1494, (1992), 169–175.

[RT] F. J. Ruiz and J. L. Torrea, *Weighted norm inequalities for a general maximal operator*, Ark. Mat. 26 (1988), 327–340.

[SS] A. Sánchez-Colomer and J. Soria, *Weighted norm inequalities for general maximal operators and approach regions*, to appear in *Math. Nachr.*

[Su] J. Sueiro, *On maximal functions and Poisson-Szegö integrals*, Trans. Am. Math. Soc. 298 (1986), 653–669.

22

Maximal Operators Associated to Hypersurfaces with One Nonvanishing Principal Curvature

Christopher D. Sogge

The purpose of this paper is to answer a question raised by Stein regarding averages over hypersurfaces with one nonvanishing principal curvature. Specifically, we shall prove the following result using the local smoothing estimates for Fourier integral operators in Mockenhaupt, Seeger, and Sogge [MSS].

THEOREM 1
Let $S \subset \mathbb{R}^n$, $n \geq 2$, be a C^∞ hypersurface having the property that, at each $x \in S$, at least one principal curvature is nonzero. Then if $d\sigma$ denotes Lebesgue measure on S and if $\rho \in C_0^\infty(\mathbb{R}^n)$

$$\left\| \sup_{t>0} \left| \int_S f(x + t^{-1}y)\, \rho(y)\, d\sigma(y) \right| \right\|_{L^p(\mathbb{R}^n)} \leq C_p \|f\|_{L^p(\mathbb{R}^n)}, \text{ if } p > 2. \quad (1)$$

When $n = 2$, this result is due to Bourgain [B], so we shall assume that $n \geq 3$ in what follows. Our result also generalizes an estimate of Stein [St] and Greenleaf [G] that says (1) holds for $p > \frac{k+1}{k}$, if at least $k \geq 2$, principal curvatures are assumed to be nonzero. The counterexample in Stein [St] shows that (1) is sharp if, for instance, S is $S^1 \times \mathbb{R}^{n-2}$.

To prove (1), after perhaps contracting $\sup \rho$ using a partition of unity, we see that we may assume that on $\sup \rho$

$$S = \{x : \Phi(x) = 0\},$$

where Φ is a C^∞ function satisfying $\nabla \Phi \neq 0$. Since (1) holds when 2 or more principal curvatures are nonzero, we may also assume that ρ is supported near a point $a \in S$ at which *exactly* one principal curvature is nonzero. This will simplify the calculations later on.

If we then abuse the notation a bit by replacing ρ by $|\nabla \Phi|\rho$, and let $Af(x, t)$ denote the averaging operator in (1), then, using the defining function Φ, we can write

$$Af(x,t) = \int_{\mathbb{R}^n} t^n \delta_0\big(\Phi(t(x+y))\big)\rho(t(x+y))f(y)\,dy$$

$$= (2\pi)^{-1} \int_{\mathbb{R}^n} \int_{-\infty}^{\infty} t^n e^{i\tau \Phi(t(x+y))} \rho(t(x+y))f(y)\,d\tau\,dy. \qquad (2)$$

Here δ_0 of course denotes the one-dimensional Dirac delta function.

Using this Fourier integral representation, we shall break up the operators dyadically. For this purpose, let us fix $\beta \in C_0^\infty(\mathbb{R}\backslash 0)$ satisfying $\sum_{-\infty}^{\infty} \beta(2^{-j}s) = 1$, $s \neq 0$. We then define the dyadic operator A_j by

$$A_j f(x,t) = (2\pi)^{-1} \int_{\mathbb{R}^n} \int_{-\infty}^{\infty} t^n e^{i\tau \Phi(t(x+y))} \beta(2^{-j}\tau)\rho(t(x+y))f(y)\,d\tau\,dy.$$

Then since $\sup_t |\sum_{j\le 0} A_j f(x,t)|$ is dominated by the Hardy–Littlewood maximal function, f^*, (1) would follow from showing that when $2 < p < \infty$, there is an $\varepsilon_p > 0$ such that

$$\big\| \sup_{t>0} |A_j f(x,t)| \big\|_{L^p(\mathbb{R}^n)} \le C2^{-j\varepsilon_p} \|f\|_{L^p(\mathbb{R}^n)}. \qquad (1')$$

Next we claim that this in turn would follow from the seemingly weaker estimates

$$\big\| \sup_{t\in[1,2]} |A_j f(x,t)| \big\|_{L^p(\mathbb{R}^n)} \le C2^{-j\varepsilon_p} \|f\|_{L^p(\mathbb{R}^n)}. \qquad (1'')$$

To show that $(1')$ implies $(1'')$ we need to use Littlewood–Paley operators L_k, which are defined via the $\mathbb{R}^n$-Fourier transform by $(L_k f)^\wedge(\xi) = \beta(2^{-k}|\xi|)\hat{f}(\xi)$. Then if we use the fact that we are assuming $\nabla\Phi \neq 0$ on $\sup\rho$, it is not hard to argue that there is an absolute constant C_0 such that, when $t \in [1,2]$,

$$A_j f(x,t) = A_j \left(\sum_{|j-k|\le C_0} L_k f \right)(x,t) + R_j f(x,t),$$

where, for any N, there is a uniform constant C_N such that $|R_j f(x,t)| \le C_N 2^{-jN} f^*(x)$, $1 \le t \le 2$. Thus, if $(1'')$ held, a dilation argument would yield

$$\int \sup_{t>0} |A_j f(x,t)|^p \, dx$$

$$\le \sum_{\ell=-\infty}^{\infty} \int \sup_{t\in[2^\ell,2^{\ell+1}]} \left| A_j \left(\sum_{|k+\ell-j|\le C_0} L_k f \right)(x,t) \right|^p dx + C_N^p 2^{-jNp} \int |f^*|^p\,dx$$

$$\le C^p C_0 2^{-j\varepsilon_p p} \int \sum_{k=-\infty}^{\infty} |L_k f(x)|^p dx + C_N^p 2^{-jNp} \int |f^*|^p\,dx$$

$$\le C^p C_0 2^{-j\varepsilon_p p} \int \left(\sum_{-\infty}^{\infty} |L_k f|^2 \right)^{p/2} dx + C_N^p 2^{-jNp} \int |f^*|^p\,dx.$$

In the last step we have used the fact that $p > 2$. If we now use the L^p boundedness of Littlewood–Paley square functions and the Hardy–Littlewood maximal theorem, we finish our proof of the claim.

To prove (1″) we shall use the fact that *if $F \in C^1([1, 2])$, $p > 1$ and $p' = \frac{p}{p-1}$, then*

$$\sup_{t \in [1,2]} |F(t)|^p \le |F(1)|^p + p \left(\int_1^2 |F(t)|^p \, dt \right)^{1/p'} \left(\int_1^2 |F'(t)|^p \, dt \right)^{1/p}.$$

One easily verifies this using the fundamental theorem of calculus and Hölder's inequality. Applying this, we find that

$$\int_{\mathbb{R}^n} \sup_{t \in [1,2]} |A_j f(x, t)|^p \, dx$$

$$\le p \int_{\mathbb{R}^n} \left[\left(\int_1^2 |A_j f(x, t)|^p \, dt \right)^{1/p'} \cdot \left(\int_1^2 \left| \frac{d}{dt} A_j f(x, t) \right|^p \, dt \right)^{1/p} \right] dx$$

$$+ \int_{\mathbb{R}^n} |A_j f(x, 1)|^p \, dx$$

$$\le p \left(\int_1^2 \int_{\mathbb{R}^n} |A_j f(x, t)|^p \, dx \, dt \right)^{1/p'} \cdot \left(\int_1^2 \int_{\mathbb{R}^n} \left| \frac{d}{dt} A_j f(x, t) \right|^p \, dx \, dt \right)^{1/p}$$

$$+ \int_{\mathbb{R}^n} |A_j f(x, 1)|^p \, dx.$$

But $\|A_j f(x, 1)\|_p \le C 2^{-j/p} \|f\|_p$, $p \ge 2$, since this inequality trivially holds for $p = \infty$ and it holds for $p = 2$, as well due to our assumption of at least one nonvanishing principal curvature. Taking this fixed time estimate into account, we conclude that (1″) would follow from

$$\left(\int_1^2 \int_{\mathbb{R}^n} \left| \left(\frac{d}{dt} \right)^\alpha A_j f(x, t) \right|^p \, dx \, dt \right)^{1/p}$$

$$\le C 2^{-j[(1/p)+\varepsilon_p-\alpha]} \|f\|_{L^p(\mathbb{R}^n)}, \qquad \alpha = 0, 1. \tag{3}$$

Since $\left(\frac{d}{dt} \right) A_j f(x, t)$ behaves like $2^j A_j f(x, t)$, we shall only prove the estimate for $\alpha = 0$.

To prove (3) we recall that we are assuming that ρ is supported in a small neighborhood of $a \in S$ at which exactly one principal curvature is nonzero. Thus we may assume that, if $x' = (x_2, \ldots, x_n)$, then

$$\Phi(x) = x_1 - h(x') \text{ and } \nabla h(a') = 0, \tag{4}$$

and that

$$\frac{\partial^2 h(a')}{\partial x_2^2} \ne 0, \text{ but } \frac{\partial^2 h(a')}{\partial x_j \partial x_k} = 0 \text{ if either } j \text{ or } k \ne 2. \tag{5}$$

Since we shall need to make use of cinematic curvature and since this condition involves third derivatives of h, we notice that, after perhaps applying a rotation in the last $n - 2$ variables, we may assume that either

$$\frac{\partial^3 h(a')}{\partial x_2^2 \partial x_j} = 0 \quad \forall j \geq 3, \tag{6}$$

or that

$$\frac{\partial^3 h(a')}{\partial x_2^2 \partial x_3} \neq 0, \text{ but } \frac{\partial^3 h(a')}{\partial x_2^2 \partial x_j} = 0 \quad \forall j > 3. \tag{7}$$

If, say $n = 3$, then in the first case S would look like a cylinder near $(h(a'), a')$, while in the second case S would locally look like a cone.

In proving (3) we shall only use the first two y variables and the first two x variables together with either t or x_3, depending on whether we are assuming (6) or (7). With this in mind, we set

$$\Phi(x, t; y) = t(x_1 + y_1) - h(t(x' + (y_2, 0))),$$

$$\rho(x, t; y) = \rho(t(x + (y_1, y_2, 0))).$$

Since we may assume that f in (3) has fixed compact support, we conclude that if we define

$$B_j g(x, t) = \int_{\mathbb{R}^2} \int_{-\infty}^{\infty} e^{i\tau \Phi(x, t; y)} \beta(2^{-j}\tau)\rho(x, t; y)g(y)\, d\tau\, dy,$$

then it suffices to show that these "frozen operators," sending functions of two variables to functions of $n + 1$ variables, satisfy

$$\left(\int_1^2 \int_{\mathbb{R}^n} |B_j g(x, t)|^p \, dx\, dt\right)^{1/p} \leq C 2^{-j[(1/p) + \varepsilon_p]} \|g\|_{L^p(\mathbb{R}^2)}. \tag{3'}$$

Let us first prove this under the assumption that (6) holds. We then claim that, if $\varepsilon_p > 0$ is small enough and if $\sup \rho$ is also small, there must be a uniform constant C so that we have the stronger inequality

$$\left(\int_1^2 \int\int |B_j g(x, t)|^p \, dx_1\, dx_2\, dt\right)^{1/p}$$
$$\leq C 2^{-j[(1/p) + \varepsilon_p]} \|g\|_{L^p(\mathbb{R}^2)} \quad \forall x'' = (x_3, \ldots, x_n). \tag{3''}$$

Notice that, for fixed t and x'', $(x_1, x_2) \to B_j g(x, t)$ is a dyadic Fourier integral operator of order $-\frac{1}{2}$, which is locally a canonical graph by (5). Hence we have the bounds $O(2^{-j/p})$ using results in [SSS]. Moreover, if

$$\varepsilon_p < \begin{cases} \dfrac{1}{2p}, & 4 \leq p < \infty, \\[2mm] \dfrac{1}{2}\left(\dfrac{1}{2} - \dfrac{1}{p}\right), & 2 < p < 4, \end{cases}$$

then the local smoothing estimates in [MSS] yield $(3'')$ if, for each fixed x'', the operators $(x_1, x_2, t) \to B_j g(x, t)$ satisfy the cinematic curvature condition defined in [S1], [MSS].

To verify this, we must show that, for fixed x'' near a'', the canonical relations of the operators $(x_1, x_2, t) \to B_j g(x, t)$; i.e.,

$$C_{x''} = \left\{ \left(x_1, x_2, t, \tau \frac{\partial \Phi}{\partial x_1}, \tau \frac{\partial \Phi}{\partial x_2}, \tau \frac{\partial \Phi}{\partial t}, y_1, y_2, -\tau \frac{\partial \Phi}{\partial y_1}, -\tau \frac{\partial \Phi}{\partial y_2} \right) : \tau \in \mathbb{R} \backslash 0, \right.$$

$$\left. \Phi(x, t; y) = 0 \quad \text{and} \quad \rho(x, t; y) \neq 0 \right\},$$

satisfy the nondegeneracy condition and the cone condition defined in [MSS]. If we call $Z = \{(x_1, x_2, t)\}$ and $Y = \{(y_1, y_2)\}$, then the nondegeneracy condition is that the two projections

$$C_{x''} \to Z \quad \text{and} \quad C_{x''} \to T^* Y \backslash 0$$

be submersions. If we let $\Pi_{z_0} : T^* Z \backslash 0 \times T^* Y \backslash 0 \to T^*_{z_0} Z \backslash 0$ denote the natural projection, then the nondegeneracy condition guarantees that, if z_0 is in the image of the projection from $C_{x''}$ to Z, then $\Gamma_{z_0} = \Pi_{z_0}(C_{x''}) \subset T^*_{z_0} Z \backslash 0$ must be a conic hypersurface. The cone condition is that Γ_{z_0} have one nonvanishing principal curvature at every point.

Since the projection from $C_{x''}$ to (x_1, x_2, t) is a submersion the nondegeneracy condition amounts to the condition that the projection from the relation to $T^* Y \backslash 0$ be a submersion. But this would clearly be the case if

$$\det \begin{pmatrix} 0 & \frac{\partial \Phi}{\partial y} \\ \frac{\partial \Phi}{\partial (x_1, x_2)} & \frac{\partial^2 \Phi}{\partial (x_1, x_2) \partial y} \end{pmatrix} \neq 0 \tag{8}$$

on $\sup \rho(x, t; y)$. And since (5) implies that this holds if ρ is supported in a small neighborhood of $(h(a'), a')$, we conclude that the nondegeneracy condition is satisfied.

To verify the cone condition, we must show that, for fixed (x, t), the two dimensional cones given by

$$\Gamma_{x.t} = \left\{ \tau \left(\frac{\partial \Phi}{\partial x_1}, \frac{\partial \Phi}{\partial x_2}, \frac{\partial \Phi}{\partial t} \right) : \Phi(x, t; y) = 0, \rho(x, t; y) \neq 0 \right\}$$

$$= \left\{ \tau \left(t, -t \frac{\partial h(t(x' + (y_2, 0)))}{\partial x_2}, h(t(x' + (y_2, 0))) - (x' + (y_2, 0)) \right. \right.$$

$$\left. \left. \cdot \nabla h(t(x' + (y_2, 0))) \right) : t(x' + (y_2, 0)) \text{ near } a' \right\}$$

have one nonvanishing principal curvature. In checking this we may take $t = 1$. Then it suffices to show that, for x' near a' and y_2 near 0, the curves

$y_2 \to \gamma_{x'}(y_2)$

$$= \left(-\frac{\partial h}{\partial x_2}(x' + (y_2, 0)), h(x' + (y_2, 0)) - (x' + (y_2, 0)) \cdot \nabla h(x' + (y_2, 0))\right)$$

have nonvanishing curvature. Since (4) and (5) imply that $(a_2, 1)$ is normal to $\gamma_{a'}$ at $\gamma_{a'}(0)$, the curvature at this point is comparable to

$$a_2 \frac{\partial^3 h}{\partial y_2^3}(a') - \frac{\partial^2 h}{\partial y_2^2} - \sum_{j=2}^{n} a_j \frac{\partial^3 h}{\partial y_2^2 \partial y_j}(a').$$

But our assumption (6) implies that this equals $-\frac{\partial^2 h}{\partial y_2^2}(a') \neq 0$, and hence the cone condition must also be satisfied if sup ρ is small. This finishes the proof of (3″).

To prove (3′) under assumption (7), we shall argue that in this case we have the stronger estimate

$$\left(\int\int\int |B_j g(x, t)|^p \, dx_1 \, dx_2 \, dx_3\right)^{1/p}$$

$$\leq C2^{-j[(1/p)+\varepsilon p]} \|g\|_{L^p(\mathbb{R}^2)} \quad \forall (x_4, \ldots, x_n, t)$$

if sup ρ is small. Repeating the above arguments, this amounts to showing that the canonical relations

$$C_{x_4, \ldots, x_n, t} = \left\{\left(x_1, x_2, x_3, \tau \frac{\partial \Phi}{\partial x_1}, \tau \frac{\partial \Phi}{\partial x_2}, \tau \frac{\partial \Phi}{\partial x_3}, y_1, y_2, -\tau \frac{\partial \Phi}{\partial y_1}, -\tau \frac{\partial \Phi}{\partial y_2}\right) : \tau \in \mathbb{R}\backslash 0, \right.$$

$$\left. \Phi(x, t; y) = 0 \quad \text{and} \quad \rho(x, t; y) \neq 0\right\}$$

satisfy the nondegeneracy and cone conditions. Since (8) still holds, we obtain the first requirement and therefore need only verify the cone condition. Here, this amounts to showing that, for fixed x near $(h(a'), a')$, the two-dimensional cones

$$\Gamma_{x.1} = \left\{\tau \nabla_{x_1.x_2.x_3} \Phi : \Phi(x, 1; y) = 0, \rho(x, 1; y) \neq 0\right\}$$

have one nonvanishing principal curvature. If we argue as before, we conclude that this would follow from showing that, for x' near a' and y_2 near 0, the curves

$$y_2 \to \gamma_{x'}(y_2) = \left(\frac{\partial h}{\partial x_2}(x' + (y_2, 0)), \frac{\partial h}{\partial x_3}(x' + (y_2, 0))\right)$$

have nonvanishing curvature. This time (5) implies that $(0, 1)$ is normal to $\gamma_{a'}$ at $\gamma_{a'}(0)$. Hence the curvature at this point must be $\frac{\partial^3 h}{\partial x_2^2 \partial x_3}(a') \neq 0$, giving us the cone condition, and finishing the proof of the estimates for the remaining case.

Acknowledgements: We are grateful to E. M. Stein for suggesting this problem and for a helpful conversation.

References

[B]   J. Bourgain, *Averages in the plane over convex curves and maximal operators,* J. Analyse Math. 47 (1986), 69–85.

[G]   A. Greenleaf, *Principal curvature and harmonic analysis,* Indiana Math. J. 30 (1982), 519–536.

[MSS] G. Mockenhaupt, A. Seeger, and C. D. Sogge, *Local smoothing of Fourier integral operators and Carleson–Sjölin estimates,* J. Amer. Math. Soc. 6 (1993), 65–130.

[SSS] A. Seeger, C. D. Sogge, and E. M. Stein, *Regularity properties of Fourier integral operators,* Annals of Math. 134 (1991), 231–251.

[S1]   C. D. Sogge, *Propagation of singularities and maximal functions in the plane,* Inv. Math. 104 (1991), 349–376.

[S2]   _____, *Fourier integrals in classical analysis,* Cambridge Univ. Press, Cambridge, 1993.

[St]   E. M. Stein, *Maximal functions: spherical means,* Proc. Nat. Acad. Sci. 73 (1976), 2174–2175.

[SW]   E. M. Stein and S. Wainger, *Problems in harmonic analysis related to curvature,* Bull. Amer. Math. Soc. 84 (1978), 1239–1295.

70

Date Due